D1544257

# BIOLOGICAL MONITORING OF
# AQUATIC SYSTEMS

# BIOLOGICAL MONITORING OF
# AQUATIC SYSTEMS

### Edited by
## Stanford L. Loeb
## Anne Spacie

**LEWIS PUBLISHERS**
Boca Raton    Ann Arbor    London    Tokyo

**Library of Congress Cataloging-in-Publication Data**

Biological monitoring of aquatic systems / edited by Stanford L. Loeb and Anne Spacie
    p.  cm.
    Papers presented at a symposium held Nov. 29–Dec. 1, 1990 at Purdue University
    Includes bibliographical references and index.
    ISBN 0-87371-910-7
    1. Water quality bioassays—Congresses.   2. Freshwater ecology—Congresses.
I. Loeb, Stanford L.   II. Spacie, Anne.
QH96.8.B5B547     1993
574.5′2632′0287—dc20                            93-34962
                                                                         CIP

© 1994 by CRC Press, Inc.
Lewis Publishers is an imprint of CRC Press

No claim to original U.S. Government works
International Standard Book Number 0-87371-910-7
Library of Congress Card Number 93-34962
Printed in the United States of America    2 3 4 5 6 7 8 9 0
Printed on acid-free paper

# Preface

Freshwater ecosystems are now undergoing dramatic changes in response to human population growth and economic development. Regional and global processes such as climate change, acid deposition, urbanization, deforestation, stream flow alteration, and the transport of exotic species affect both the structure and functioning of lake and stream systems and jeopardize their resource values. The large spatial and temporal scales of such processes make it difficult to evaluate or predict the ecological effects of even a single disturbance, much less the combined effects of multiple stressors. Most regulatory water quality programs are designed either to protect human health or to protect aquatic organisms from identifiable toxic chemicals. Yet many of the observed changes in biological communities relate to resource exploitation, nonpoint pollutant interactions, and habitat alteration — factors that are missed by routine chemical sampling. Consequently, there is a growing recognition among water resource planners, policy makers, and researchers that new ecologically-based approaches are urgently needed to monitor the health of aquatic ecosystems.

Because of the great importance of this issue, many resource agencies and ecological societies are currently working to develop the scientific basis of ecological monitoring. The International Society of Limnology (Societas Internationalis Limnologiae), Biological Monitoring Working Group organized a symposium on November 29 – December 1, 1990, at Purdue University to promote an interchange of ideas among leading researchers on freshwater monitoring theory and application. This book, in part, represents a synthesis of formal papers and round-table discussions which took place at that meeting. Ecological monitoring involves a wide variety of approaches and there is no single taxon, community index, or statistical design, that is "best" for all purposes. Nevertheless, discussions among the participants showed broad areas of agreement on philosophy and approach. The editors believe this book portrays freshwater research at an exciting stage in its development — as governmental agencies begin to incorporate modern ecosystem theory into their planning, assessment and regulatory programs.

This book is divided into six sections: I. Introduction; II. Background; III. Experimental Design; IV. Community Response; V. Program Considerations; and VI. Conclusion. The material is meant to provide the reader with an understanding of how biological monitoring can be used to assess the health of water resources. As efforts to maintain and restore the world's water resources intensify, the need to develop accurate methods to assess the health of these resources becomes critical. Biological monitoring will be an important component of any water quality assessment program.

**Stanford L. Loeb**
**Anne Spacie**

# Acknowledgments

The editors would like to thank the chapter authors for their generous contributions to this text and for their cooperation in the editorial process. The activities leading to the publication of this text were supported by an Educational Service Agreement (#OBO795NAEX) with the United States Environmental Protection Agency. We especially appreciate the encouragement and support of this effort provided by Dr. Dixon H. Landers, U.S. EPA Research Laboratory, Corvallis, Oregon. Susan T. Umberger served as Conference Coordinator at Purdue University. Janet E. Elder assisted with the editing of the manuscripts. Judy A. Wiglesworth was of great assistance with the word processing associated with this text. Additional support for the symposium and manuscript preparation was provided by the University of Kansas and Purdue University. The views expressed in this book are those of the authors and/or editors and do not necessarily represent the opinions, policies or recommendations of the U.S. Environmental Protection Agency.

 **Stanford L. Loeb** is Adjunct Professor with the Department of Systematics and Ecology at the University of Kansas. He received a B.A. degree (1969) and an M.A. degree (1972) in biology from the University of California, Santa Barbara, and a Ph.D. in ecology from the University of California, Davis (1980). From 1980 to 1987 he served as a research ecologist with the Institute of Ecology, University of California Davis Lake Tahoe Research Program. He has published over 30 journal articles, book chapters, and technical reports on limnology and related water quality topics. He is a principal reviewer for the National Park Service Global Change Program and is chairperson of Science Review Panels for the Crater Lake National Park Limnological Program and the Northern Cascades National Park Service Complex Aquatic Research Program. Also, he is chairperson of the Extra-European Section of the Biological Monitoring Working Group for the Societas Internationalis Limnologiae (International Society of Limnology).

 **Anne Spacie** is Professor of Fisheries and Aquatic Science in the Department of Forestry and Natural Resources at Purdue University. She holds an A.B. degree in chemistry from Mount Holyoke College, an M.S. in marine chemistry from the University of California, San Diego, and a Ph.D. in fisheries and limnology from Purdue University.

Her research interests include aquatic biology, limnology, and water pollution processes with specialization in the fate and accumulation of xenobiotics in aquatic food chains. Her publications include papers on uptake of contaminants by aquatic organisms, sediment-water interactions, long-term lake monitoring, and bioaccumulation theory and test methods. Dr. Spacie coordinates the aquatic studies for a 200 km$^2$ experimental watershed instrumented for landscape and climatological research. She has also conducted research as a visiting scientist at the University of Bayreuth in Germany, the Savannah River Ecology Laboratory in South Carolina, and the U.S. EPA Environmental Research Laboratory in Duluth, Minnesota. As an educator, she advises graduate students and teaches courses in limnology, fisheries, aquatic toxicology, field methods, and global contaminant processes.

Dr. Spacie has held offices in several professional societies including two terms on the Board of Directors of the Society of Environmental Toxicology and Chemistry. She has also served on numerous academic and governmental panels including the Great Lakes Science Advisory Board, National Research Council review panels, and U.S. Environmental Protection Agency, Army Corps of Engineers, U.S. Forest Service and Fish and Wildlife Service advisory groups.

# Contributors

*Michael T. Barbour*, Tetra Tech, Owens Mills, MD

*Barbara J. Benson*, Center for Limnology, University of Wisconsin, Madison, WI

*Dr. John Cairns, Jr.*, University Center for Environmental and Hazardous Materials Studies, Virginia Polytechnic Institute and State University, Blacksburg, VA

*Stephen R. Carpenter*, Center for Limnology, University of Wisconsin, Madison, WI

*Donald F. Charles*, Patrick Center for Environmental Research, The Academy of Natural Sciences of Philadelphia, Philadelphia, PA

*Loveday L. Conquest*, Center for Quantitative Science in Forestry, Fisheries and Wildlife, University of Washington, Seattle, WA

*Kenneth W. Cummins*, South Florida Water Management District, West Palm Beach, FL

*Daniel R. Engstrom*, Limnological Research Center, University of Minnesota, Minneapolis, MN

*Thomas M. Frost*, Center for Limnology, University of Wisconsin, Madison, WI

*Dr. Pier Francesco Ghetti*, Dipartimento di Scienze Ambientali, Universitá degli Studi di Venezia, Venezia, ITALY

*Martin E. Gurtz*, U.S. Geological Survey, Water Resources Division, Raleigh, NC

*Steven A. Heiskary*, Nonpoint Source Section, Minnesota Pollution Control Agency, St. Paul, MN

*Robert M. Hughes*, ManTech Environmental Technology, U.S. EPA Environmental Research Laboratory, Corvallis, OR

*H. B. Noel Hynes*, Professor Emeritus, Department of Biology, Waterloo, Ontario, CANADA

*James R. Karr*, Institute for Environmental Studies, University of Washington, Seattle, WA

*Timothy K. Kratz*, Center for Limnology, University of Wisconsin, Madison, WI

*David R. Lenat*, North Carolina Division of Environmental Management, Water Quality Section, Raleigh, NC

*Rick A. Linthurst*, U.S. EPA, Research Triangle Park, NC

*James M. Loar*, Environmental Sciences Division, Oak Ridge National Laboratory, Oak Ridge, TN

*John J. Magnuson*, Center for Limnology, University of Wisconsin, Madison, WI

*William J. Matthews*, Biological Station, University of Oklahoma, Kingston, OK

*Peter B. Moyle*, Department of Wildlife and Fisheries Biology, University of California, Davis, CA

*Robert J. Naiman*, Center for Streamside Studies, University of Washington, Seattle, WA

*Ruth Patrick*, The Academy of Natural Sciences of Philadelphia, Philadelphia, PA

*Steven G. Paulsen*, Environmental Research Center, University of Nevada, c/o U.S. EPA, Corvallis, OR

*Stephen C. Ralph*, Center for Streamside Studies, University of Washington, Seattle, WA

*Dr. Oscar Ravera*, Dipartimento di Scienze Ambientali, Universitá degli Studi di Venezia, Venezia, ITALY

*Eric P. Smith*, Department of Statistics, Virginia Polytechnic Institute and State University, Blacksburg, VA

*John P. Smol*, Paleoecological Environmental Assessment and Research Laboratory, Department of Biology, Queen's University, Kingston, Ontario, CANADA

*Arthur J. Stewart*, Environmental Sciences Division, Oak Ridge National Laboratory, Oak Ridge, TN

*Chris O. Yoder*, Ecological Assessment Section, Ohio Environmental Protection Agency, Columbus, OH

# Contents

## IV. Community Response

## V. Program Considerations

## VI. Conclusion

Dedicated to the accomplishments and in memory of

G. Evelyn Hutchinson   1903–1991

# SECTION I

# Introduction

# CHAPTER 1

# An Ecological Context for Biological Monitoring

**Stanford L. Loeb,** Department of Systematics and Ecology, University of Kansas, Lawrence, Kansas.

The availability of freshwater resources is essential to the survival of human populations throughout the world; however, human populations exert an inordinate stress on these resources resulting in their continuing degradation. Although awareness of the diminishing availability of unpolluted freshwater is widespread, methods for evaluating the health of aquatic ecosystems have not been fully developed to accurately identify their quality or protect and restore their health.

The health of an aquatic ecosystem is degraded when the ecosystem's assimilative capacity to absorb a stress has been exceeded. A healthy ecosystem is composed of biotic communities and abiotic characteristics, which form a self-regulating and self-sustaining unit. Although changes within an ecosystem can result from naturally occurring events, anthropogenic activities often impose stresses on these systems. When organisms of an ecosystem are exposed to stress, their resistance to displacement from that ecosystem may be exceeded. Depending upon the magnitude and temporal nature of the stress, the organisms may not be sufficiently resilient to reestablish their pre-stress community structure.

The community structure of an aquatic ecosystem is sensitive to, and determined by, the conditions and resources available within a habitat. Conditions include abiotic environmental factors, which vary with time and space (e.g., temperature, salinity, and flow) (Begon et al. 1990).

Resources are all things consumed by an organism (e.g., food, light, and space) (Tilman 1982). Organisms that come to make up an aquatic community are those that can endure, tolerate, compete, reproduce, and persist within a given habitat. If a habitat is characterized by conditions that are within acceptable limits and provides all necessary resources for a given species, that species could potentially occur there (Begon et al. 1990).

In essence, the above account defines the niche space of an organism, that is, an n-dimensional hypervolume of all ecological factors relative to a species' ability to survive and multiply (Hutchinson 1958). Hutchinson considered two general types of niche axes or categories of variables: scenopoetic and bionomic. Scenopoetic variables have no direct relationship to competition, rather they are conditions that an organism must be able to tolerate. Bionomic variables involve resources for which there may be competition. If an aquatic organism is exposed to a stress that changes the conditions or resources of a habitat, the niche space within which an organism survives is changed and a new hypervolume is created. The niche space can be described as a hypervolume with a response structure within which there is a point of optimal survival surrounded by areas of less than optimal survival (Maguire 1973). A stress that leads to the alteration of any environmental characteristic of an aquatic ecosystem would affect the survivorship of organisms living within that ecosystem. The result would be a restructuring of the biotic community that was originally present.

A stress on an aquatic ecosystem can be categorized into one of three types: (1) physical; (2) chemical; or (3) biological alterations. Physical alterations include changes in water temperature, water flow, substrate/habitat type, and light availability. Chemical alterations include changes in the loading rates of biostimulatory nutrients, oxygen consuming materials, and toxins. Biological alterations include the introduction of exotic species. Activities that result in a change in any of these environmental characteristics can lead to the deformation of an organism's niche space, possibly leading to its extinction.

A decision must be made concerning which attributes (i.e., variables) to measure when designing an assessment program to evaluate the health of an aquatic ecosystem. An attribute needs to be quantifiable such that the variance or uncertainty surrounding the measurement of that attribute can be determined. An attribute should be diagnostic in its ability to assist in the determination of whether an aquatic ecosystem is healthy or not. It also needs to be responsive to any change in resource availability and/or habitat conditions. A variety of attributes may be required to adequately identify the transitional stages that a healthy aquatic ecosystem goes through as it suffers impairment and as a system recovers from an impaired state.

Chemical variables have been the most common attributes selected when designing environmental assessment programs of aquatic ecosystems. Chemical analyses are quantifiable and the variance surrounding any specific sampling episode is usually small. An understanding of the water chemistry can assist in an evaluation of the type of stress that may exist (e.g., biostimulatory chemicals or toxins); however, diagnosis of the impact of these chemical substances on the aquatic ecosystem cannot be determined based solely on water chemistry. When used alone, the utility of water chemical analyses is limited. A static concentration of a chemical represents the residual of that element or compound after biological organisms have ingested, absorbed, stored or transformed it. Furthermore, abiotic materials (i.e., sediments) also modify the chemistry of waters through adsorption. Chemical analyses fail to account for the rates of transformation of a chemical by biological organisms and the possible relocation of a chemical within the aquatic ecosystem due to abiotic adsorption and sedimentation. Chemical attributes lack the responsiveness necessary to evaluate the health of an aquatic ecosystem. The dynamic interactions (physical, chemical, and biological) that characterize an ecosystem are not adequately assessed through the sole use of chemical attributes.

Biological monitoring is an essential element needed to assess the environmental health of aquatic ecosystems. Biological organisms are diagnostic in determining the health of aquatic ecosystems and they can be measured quantitatively. Ecologically, the concept of niche space provides the theoretical framework for understanding the importance of biological monitoring to any evaluation of environmental health. The organisms that inhabit aquatic ecosystems are the fundamental sensors that respond to any stress affecting that system. The health of an aquatic ecosystem is reflected in the health of the organisms that inhabit it. Any stress imposed on an aquatic ecosystem manifests its impact on the biological organisms living within that ecosystem.

The purpose of this text is to provide an understanding of the efficacy of biological monitoring assessing the environmental health of aquatic ecosystems. The authors of the following chapters discuss the importance of including biological monitoring in assessments of water quality. In Section II, Background, a historical perspective concerning the use of biological monitoring (Hynes, Chapter 2), requirements for an effective biological monitor (Patrick, Chapter 3), and a European perspective on the use of biological monitoring (Ghetti and Ravera, Chapter 4) are presented. In Section III, Experimental Design, analytical issues concerning biological monitoring are addressed including their statistical validity (Cairns and Smith, Chapter 5), sampling design and data analysis (Conquest et al., Chapter 6), spatial and temporal variation (Stewart and Loar, Chapter 7), and ecoregional application of biological monitoring data (Hughes et al., Chapter 8). Section IV, Community Response, presents examples of biotic community that have been used to assess the health of aquatic ecosystems. Biological assessment of lotic systems using an analysis of functional organization (Cummins, Chapter 9), fish community structure (Moyle, Chapter 10), and benthic macroinvertebrates (Lenat and Barbour, Chapter 11) are discussed. Biological assessment of lentic systems is discussed in relation to landscape position, scaling, spatial temporal variability (Kratz et al., Chapter 12), and paleore-construction of biological communities (Charles et al., Chapter 13). In Section V, Program Considerations, two major environmental assessment programs involving aquatic ecosystems are described. These programs are the United States Environmental Protection Agency's Environmental Monitoring and Assessment Program (EMAP) (Paulsen and Linthurst, Chapter 14) and the United States Geological Survey's National Water-Quality Assessment (NAWQA) Program (Gurtz, Chapter 15). The Conclusion (Section VI) discusses the challenges associated with the goal to assess the health of aquatic ecosystems and how biological monitoring can be used to achieve that goal (Karr, Chapter 16).

With the dwindling amount of healthy aquatic ecosystems, all efforts to preserve our remaining water resources and restore those that are impaired must receive a high priority. An ecological approach using biological monitoring can facilitate that effort. The following chapters address the underlying principles concerning biological monitoring and clarify many of the questions about its successful implementation.

# REFERENCES

Begon, M., J. L. Harper, and C. R. Townsend. 1990. *Ecology.* 2nd Ed. Blackwell Scientific Publication, Oxford, England. 945 p.

Hutchinson, G. E. 1958. Concluding remarks. *Cold Springs Harbor Symp. Quant. Biol.* 22: 415-27.

Maguire, B., Jr. 1973. Niche response structure and the analytical potentials of its relationship to the habitat. *American Naturalist* 107: 213-246.

Tilman, D. 1982. *Resource Competition and Community Structure.* Princeton University Press, Princeton, New Jersey. 296 p.

Fitzpatrick, R. (??). Books complaints in second floor library gathering dust in the shelves. . . . . . . . . some floor administration 312-315.

Jacob, H. (1993). Survey to ??? . . . . How to use trade fields at University, Nate. Some state.

# SECTION II

# Background and Perspective

# CHAPTER 2

# Historical Perspective and Future Direction of Biological Monitoring of Aquatic Systems

**H. B. N. Hynes,** Department of Biology, University of Waterloo, Ontario, Canada

Our hunter-gatherer ancestors must have been excellent ecologists because their very survival depended upon their knowing just where to look for the various species of animals and plants they needed. Today's naturalists and biologists also usually know exactly where to look for species they seek. That knowledge of habitat is the original idea behind the use of the biota to determine the conditions at a particular locality.

We have been abusing our surroundings for many millennia, and even as far back as in the Old Testament, Jeremiah (2,7) bewails the transformation of God's "heritage" into "an abomination". I have also previously pointed out that Aristotle was aware of the effects of organic pollution, and that Spenser's *Faerie Queen*, written in the sixteenth century, mentions pollution by tin mines (Hynes 1960). Therefore the application of general ecological principles to awareness of the conditions of habitats has a very long history.

However, it was less than a century ago that Kolkwitz and Marsson (1908, 1909) codified the study of microbiota into a system that could be used to gauge the severity of organic pollution. This "Saprobiensystem" was further expanded and explained by Kolkwitz in 1950, and then numerous Europeans, changed it into quite a complex edifice in which individual species of animals and plants were allotted points on a scale of tolerance to organic pollution. These points were then added up to give a score to the habitat being examined. The literature on this activity is enormous, but notable contributions are those of Liebmann (1951, 1961), von Tümpling (1960, 1974), and Sládecek (1973, 1985). These were all efforts to reduce biological estimations of the conditions of pollution to numbers or grades that could be used to compare different places and that could also used by regulatory agencies and engineers.

A little later than the early German studies, Richardson (1928) was looking at the macroinvertebrates of the Illinois River in relation to organic pollution by the cattle industry in Chicago, and he produced similar ideas of zones of degradation and recovery that could be characterized by the biota of the benthos. This concept was elaborated by Whipple, Fair, and Whipple (1927) in their book on drinking water; however, as was so often true in that era, North Americans and Europeans seemed unaware of each other's work.

The British were similarly insular, although they were aware of the American studies. Carpenter had done much work on heavy-metal mine pollution in Wales, which is summarized in her book of 1928, and she brought first-hand knowledge of American work to Britain. Thus, by 1960 it was possible for me to publish a book on the biology of pollution, which was, however, concerned mainly with the British Isles. Workers in Britain then went on to develop numerical indices rather similar to those of the Saprobiensystem, although based much more on macroinvertebrates than on microbiota. The reader is referred to Woodiwiss (1964) and to discussions by Hellawell (1978, 1986) for details.

These early attempts at designing codified systems that could produce numerical results were primarily floristic and faunistic, although Patrick (1949, 1951) had already introduced the idea that it is the structure of the community that is relevant, not the mere list of species. Nevertheless, the idea that a list of the numbers of species is informative persists, although it is often combined with measures of abundance of each, and many biologists have used their special knowledge of particular groups to assess pollution. Recent examples are Kann (1986), Watanabe et al. (1986), Somashakar (1988), and Sumita (1988) on algae; Bazzanti and Bambacigno (1987) on Chironomidae; Lang et al. (1989) on benthic invertebrates; and Schmedtje and Kohlmann (1987) on macrophytes. Rather similarly, limnologists make great use of planktonic algae in their assessment of eutrophication resulting from the nutrient enrichment caused by pollution. The search for indices of various types continues on. That is, I suspect, because the engineers and regulators who are responsible for controlling pollution believe that science is numbers (God has been described as a mathematician!) and it is easy to discuss them, impress the general public with them, and design purification plants and regulations to them, no matter how shaky their basis is. This probably arises from the fact that the early history of pollution studies was dominated by chemists, who could analyze to absolute numbers—parts per million, or even trillion with modern tools—of this and that. However, very soon it was also appreciated that even simple sewage pollution was a complex phenomenon (God nowadays seems to be thought of as more untidy-minded, like an ecologist, as Gaia replaces Yahweh in the public conscience). So chemists, under the auspices of the British Royal Commission of 1910, devised the Biochemical Oxygen Demand test (BOD), which is an indirect, and hence, inherently unreliable, measure of polluting potential. I stress potential rather than actuality, because the effect varies greatly between water bodies for a variety of reasons, which biologists understand fairly well. We have a similar example in the *Escherischia coli* count, which is used to regulate the uses of water, quite frequently for the wrong reasons; only in this instance it is medicine that thinks it has all the answers.

The point is that numbers are regarded as being secure and real, and they can be used in designing treatment plants, whereas opinions are just opinions. However, it is easy to show that BOD does not always tell one the truth—a high oxygen demand coupled with a bactericide gives a low reading, and many industrial effluents disrupt it. Similarly, *E. coli* may be abundant or scarce for a variety of reasons that are not directly connected with pollution by man or domestic animals.

Nevertheless, the search for a perfect index continues. The Saprobiensystem has been much modified, as stated above, and the whole topic has been reviewed by several Europeans, and in English by Bick (1963). The so-called Trent Index from Britain has been modified for other areas (Andersen et al. 1984), as is inevitable for indices based upon the biota, which varies across the map because of biogeographical and geological changes. The North Americans have produced their own local indices (e.g., Cairns et al. 1968, Hilsenhoff 1987) and the concept has spread all over the planet, for example, to New Zealand (Hirsch 1958), South Africa (Chutter 1972), India (Tiwarti and Ali 1988, Bick et al. 1967), and Japan (Komatsu 1964). There are even specialized indices for such things as acidification (Engblom and Lingdell 1984, Wade et al. 1989) and pulp mill effluents (Hendricks et al. 1974). Here again there is now a huge literature.

There have also been moves into indirect measures of community metabolism as parameters of pollution. For example, Frutiger (1985) advocates the use of production quotients, Fontvieille and Fevotte (1981) consider the DNA content of sediments, and there are other examples of similar ideas.

All of these measures are attempts to quantify stress on the biotic community, thus modern ecological theory has inevitably been brought to bear upon the problem. The current dogma is that stress reduces biological diversity, which is just another way of expressing Thienemann's second ecological principle first published in 1920 (Thienemann 1954). Biologists are rather better at reinventing wheels than most scientists! We publish more and longer papers, so older seminal ideas, like fossils in geological strata, tend to become quickly buried out of sight. This is especially true nowadays, when nobody seems prepared to read a paper that is not either in English or in their native tongue.

Thus we now have an additional large literature on the effects of pollution on various types of diversity index, and such studies have invoked fairly extensive and complex statistical analysis, and also, increasingly, another type of numerical expression in the form of cluster analysis. This approach has been reviewed by Wilhm (1972) and Washington (1984), and it forms part of the general discussion of indices by Abel (1989) and Hellawell (1986). It has been used in North America (e.g., Chance and Deutsch 1980, Hughes 1978, Narf et al. 1984, Stoneburner et al. 1976), the former U.S.S.R (e.g., Kozhova et al. 1979, Morokov 1986), Australia (e.g., Marchant et al. 1984), and doubtless in many places elsewhere on the planet.

However, not only do many things affect diversity indices and similarities or differences between biotic communities, but the acquisition of a number by such a study does not really tell one very much about the pollution itself, nor of the actual biological conditions that prevail. Such numbers really indicate only differences between stations over distance or time. They enable one to compare upstream with downstream or one place in different years or seasons. To that extent they enable one to determine change, but if there is any relationship between that change and some sort of pollution, that relationship lies only in the minds of the investigator and fellow professionals. The number remains just a number, and I feel that it is a waste of resources to continue to search for some universal index, the reality of which is at least questionable.

The point I am making is that biotic data, no matter how they are massaged, give rise only to diagnoses, not to anything that is really absolute. They are very different from the chemical analyses of the concentration of a toxin or of dissolved oxygen. They are even more subject to informed interpretation than is, for example, BOD. The biologist investigating or monitoring pollution is similar in many respects to a physician; a few attributes can be measured numerically (fever, pulse rate, weight), but these are only aids to diagnosis. That point I find is also made by Karr (1991). We do not expect the medical profession to diagnose by numbers, and we all know that it would be ridiculous to say that someone was 25% unwell. We do not write out formulae for diagnosis that can be handed to a technician who can make the measurements, use those formulae, and thus arrive at a valid diagnosis. Were we able to do this, medicine would be a job for computers, which do that sort of thing much better than people. But, unfortunately, the chemical aspects of pollution monitoring do lend themselves, to a large extent, to that kind of approach and manufacturers, municipal officials, and regulatory agencies have become used to the idea that all that is needed is a good system to be operated by reliable technical personnel.

We are indeed able to arrange systems that permit such an approach for one particular type of effluent in one small geographical area and that can be used for monitoring whether things are getting better or worse in the aquatic system, but by themselves they tell one nothing about reasons. Was it a change in the effluent, the temperature, or the rainfall, or even something else that we did not include in the protocol? In some respects the designing of such systems is a bit regrettable, as it leads to false impressions among the uninformed public. We need an alert mind applied to the problem at all times. No kind of formula or numerical analysis is likely to function as a substitute. We accept that fact about medicine but seem unwilling to do so about biology, which is likely to be far more complex because it involves so many more levels of biological organization.

Looking ahead, my feeling is that we should be concentrating on acquiring a fundamental understanding of the ecologies of inland waters—I use the plural because I think that lakes, streams, and wetlands have some fundamental differences. We should not allow ourselves to be diverted by the search for an elusive, and probably non-existent, basic index or system. I have the impression that we are not really too far from understanding the process in the fairly closed systems of lakes, but running water is far behind in that respect, and really serious work on wetlands has hardly started. In due course then, we will have well-researched theories or models to assist our overall understanding of inland waters.

Thus, we shall, for a long time to come, need well-trained and continuously interested practitioners to do our monitoring; and there, I think, lies an important difficulty. Medical practitioners are in little danger of becoming bored with their work; they always have fresh patients, new problems, and above all, the human reward of the personal contact with fellow men and women. Chemical analysis is fairly easily automated, and thus can be left in the hands of reliable personnel. Moreover, chemical results are the same no matter where they are applied. Biological data have to be considered against the local background of flora and fauna, as well as such things as regional water chemistry. This was recognized by Karr (1991) and Hughes et al. (1986, 1987) in their efforts to find a baseline for setting up an index based upon fishes, and the British have realized that baseline data are needed for stream monitoring (Moss et al. 1988) and that even they are not particularly stable (Weatherley and Ormerod 1989). However, while such data are not too hard to acquire for a small discrete area like Britain, they are much more difficult to deal with on a continental scale, where judgment over regional boundaries must be exercised. Also, the background information has simply not been amassed for most continents, and at the present rate of acquisition it will not be available for a long while.

So my thesis is that biological monitoring will probably always, or at least for the foreseeable future, have a research element to it, and that it will continue to require the employment of rather high-grade personnel. One cannot leave a technician to do it because biological monitoring needs constant evaluation, just as medical practice does. How then can one prevent the biological monitor from becoming bored with a job for which, most of the time, he/she is overqualified? I believe this to be an important problem for which we have no ready solution. Research scientists tend to regard monitoring as not worthy of their time, and the folk who will undertake it often do not incorporate sufficient imagination and motivation into the task. Here I believe our science needs imaginative restructuring.

If we are able to solve that problem then we can go ahead and learn much more about the responses of aquatic systems to stress, and there I believe we have the potential for the development of a magnificent tool for environmental control of the landscape. Already in well-worked areas we are beginning to understand something of river ecology. We know about the effects of many types of pollution, deforestation, agriculture, poor land-use, impoundment, etc., almost to the point of being able to predict what is going on uphill (upstream) of a site. We have the potential to develop a universal and cheap pulse on the landscape—a sort of annual check-up on the drainage basins, which could be applied cheaply at very few points within the catchment.

However, even here we have problems. We have in Europe and North America, and probably in Africa and Asia also, virtually no large rivers that are pristine, and hence no certainty of being able to determine what pristine biota were. Perhaps we should already write off most of the really large rivers, as we cannot restore them to conditions that we do not know.

We also, as a species, seem very ambivalent about just what we expect of the world. At the present time much is made of the loss of diversity, the felling of rainforests, and the potential of a greenhouse effect; but we continue to subject enormous areas to dreadful devastation by agriculture, which is certainly our most damaging ecological activity, and we continue to increase in numbers at an unsustainable rate. Should we be looking much more closely at population control and at technical alternatives to agriculture?

I also believe that the ordinary person has only a sentimental and poorly understood desire to live with really wild nature. He or she is prepared to spend time in well-tended national parks and nature reserves, but really likes a pleasant simple landscape with trees and fields, and a bit of fishing or hunting that can be reached along easy paths with no spiders or snakes! Even that much alteration of an original landscape would, and indeed does, show up in the biota of the local stream. The fields cause nutrient enrichment, soil erosion and loss of diversity, the paths interfere with forest development and become runoff channels, the streams become regulated to produce recreational lakes, and the riparian vegetation is disturbed to make for good fishing.

However, it does seem that this sort of "Garden of Eden" is what most people would like to live in, and a lot of the fuss about wild nature is, I believe, just talk and TV images. Perhaps the garden landscape is the sort of baseline we should be working towards, and not towards the preservation of Lewis Carroll's tulgey woods infested with dangerous Jubjub birds and Jabberwocks, not to mention Bandersnatches, but where the water bodies remain in their pristine state.

# REFERENCES

Abel, P. D. 1989. *Water Pollution Biology*. Wiley, Chichester. 231 p.

Andersen, M. M., F. F. Reget, and H. Sparholt. 1984. A modification of the Trent Index for use in Denmark. *Water Res.* 18:145-151.

Bazzanti, M. and F. Bambacigno. 1987. Chironomids as water quality indicators in the River Mignone (Central Italy). *Hydrobiol. Bull.* 21:213-222.

Bick, H. 1963. A review of Central European methods for the biological estimation of water pollution levels. *Bull. Wld. Hlth. Org.* 29:401-413.

Bick, H., J. S. S. Lakshmin Arayana, and K. P. Krishnamoorthi. 1967. Preliminary findings concerning the potentialities of the European saprobity system for monitoring water quality under tropical conditions in India. WHO/EP/67.6.

Carpenter, K. E. 1928. *Life in Inland Waters*. Sidgwick-Jackson, London. 267 p.

Cairns, J., D. W. Albaugh, F. Busey, and M. D. Chanay. 1968. The sequential comparison index. A simplified method for non-biologists to estimate relative differences in biological diversity in stream pollution studies. *J. Wat. Poll. Control Fed.* 40:1607-1613.

Chance, J. M. and W. G. Deutsch. 1980. A comparison of four similarity indexes in the cluster analysis of Susquehanna River, U.S.A., macro-benthic samples. *Proc. Penn. Acad. Sci.* 54:169-173.

Chutter, F. M. 1972. An empirical biotic index of the quality of water in South African streams and rivers. *Water Res.* 6:19-30.

Engblom, E. and P.-E. Lingdell. 1984. The mapping of short-term acidification with the help of biological indicators. *Rep. Inst. Freshwat. Res. Drottning-holm* 61:60-68.

Fontvieille, D. and G. Fevotte. 1981. DNA content of the sediment in relation to self-purification in streams polluted by organic wastes. *Verh. int. Verein. theor. angew. Limnol.* 2:221-226.

Frutiger, A. 1985. The production quotient PQ: A new approach for quality determination of slightly to moderately polluted running waters. *Arch. Hydrobiol.* 104:513-526.

Hellawell, J. M. 1978. *Biological Surveillance of Rivers. A Biological Monitoring Handbook*. Water Research Centre, Stevenage. 332 p.

Hellawell, J. M. 1986. *Biological Indicators of Freshwater Pollution and Environmental Management*. Elsevier, London. 546 p.

Hendricks, A. et al. 1974. Utilization of diversity indices in evaluating the effect of a paper mill effluent on bottom fauna. *Hydrobiologia*. 44:463-474.

Hilsenhoff, W. L. 1987. An improved biotic index of organic stream pollution. *Gt. Lakes Ent.* 20:31-39.

Hirsch, A. 1958. Biological evaluation of organic pollution of New Zealand streams. *N.Z. J. Sci. Technol.* 1:500-553.

Hughes, B. D. 1978. The influence of factors other than pollution on the value of Shannon's Diversity Index for benthic macro-invertebrates in streams. *Water Res.* 12:359-364.

Hughes, R. M., D. P. Larsen, and J. M. Omernik. 1986. Regional reference sites: a method for assessing stream pollution. *Environ. Mgmt.* 10:629-635.

Hughes, R. M., E. Rexstad, and C. E. Bond. 1987. The relationship of aquatic ecoregions, river basins, and physiographic provinces to the ichthyographic regions of Oregon. *Copeia* 1987:423-432.

Hynes, H. B. N. 1960. *The Biology of Polluted Waters*. Liverpool University Press, Liverpool. 202 p.

Kann, E. 1986. Können benthische Algen zur Wassergütebestimmung herangezogen werden? *Arch. Hydrobiol. Suppl.* 73:405-423.

Karr, J. R. 1991. Biological integrity: A long-neglected aspect of water resource management. *Ecol. Applications* 1:66-85.

Kolkwitz, R. 1950. Oekologie der Saprobien. Über die Beziehungen der Wasserorganismen zur Umwelt. *Schr. Ver. Wasserhyg* 4:64.

Kolkwitz, R. and M. Marsson. 1908. Ökologie der pflanzlichen Saprobien. *Ber. Dt. Bot. Ges.* 26:505-519.

Kolkwitz, R. and M. Marsson. 1909. Ökologie der tierische Saprobien. Beitrage zur Lehre von der biologische Gewässerbeurteilung. *Int. Revue Hydrobiol. Hydrogr.* 2:126-152.

Komatsu, T. 1964. Aquatic insect communities and the biotic index of the rivers which flow into the southeastern part of the Lake Biwa. *Jap. J. Ecol.* 14:182-189. (Japanese, English summary.)

Kozhova, O. M., N. A. Shastina, and N. G. Melnik. 1979. Statistical methods of assessing water ecosystem condition. *Gidrobiol. Zh. Kiev* 15:3-13. (Russian, English summary.)

Lang, C., G. l'Eplattenier, and O. Reymond. 1989. Water quality in rivers of western Switzerland: application of an adaptable index based on benthic invertebrates. *Aquat. Sci.* 51:224-234.

Liebmann, H. 1951. *Handbuch der Frischwasser und Abwasserbiologie*. Oldenburg, Munich.

Liebmann, H. 1961. *Handbuch der Frischwasser und Abwasserbiologie*. Band II. Oldenburg, Munich.

Marchant, R., P. Mitchell, and R. Norris. 1984. Distribution of benthic invertebrates along a disturbed section of the La Trobe River, Victoria: An analysis based on numerical classification. *Aust. J. Mar. Freshwat. Res.* 35:355-374.

Morokov, V. V. 1986. Experience in the development of an integrated index of the human impact on rivers. *Vod. Resur.* 1986:92-101. (Russian.)

Moss, D., M. T. Furze, J. E. Wright, and P. D. Armitage. 1988. The prediction of the macro-invertebrate fauna of unpolluted running-water sites in Great Britain using environmental data. *Freshwat. Biol.* 17:41-52.

Narf, R. P., E. L. Lange, and R. C. Wildman. 1984. Statistical procedures for applying Hilsenhoff's biotic index. *J. Freshwat. Ecol.* 2:441-448.

Patrick, R. 1949. A proposed biological measure of stream conditions, based on a survey of the Conestoga Basin, Lancaster County, Pennsylvania. *Proc. Acad. Nat. Sci. Philad.* 101:277-341.

Patrick, R. 1951. A proposed biological measure of stream conditions. *Verh. Int. Verein. Theor. Angew. Limnol.* 11:299-307.

Richardson, R. E. 1928. The bottom fauna of the Middle Illinois River, 1913-1925. Its distribution, abundance, valuation, and index value in the study of stream pollution. *Bull. Ill. St. Nat. Hist. Surv.* 17:387-475.

Schmedtje, U. and F. Kohlmann. 1987. Bioindication by macrophytes — can macrophytes indicate sparobity? *Arch. Hydrobiol.* 109:455-469. (German, English summary.)

Sládecek, V. 1985. Scale of saprobity. *Verh. int. Verein. theor. angew. Limnol.* 22:2337-2341.

Sládecek, V. 1973. System of water quality from the biological point of view. *Ergebn. Limnol.* 7. 218 p.

Somashakar, R. K. 1988. On the possible utilization of diatoms as indicators of water quality — a study of River Chaudry. pp. 375-382. *In:* R. K. Trivedy, ed., *Ecology and Pollution of Indian Rivers.* Ashish Publishing House, New Delhi. 304 p.

Stoneburner, D. L., L. A. Smock, and H. C. Eichhorn. 1976. A comparison of two diversity indexes used in wastewater impact assessments. *J. Wat. Pollut. Control Fed.* 48:736-741.

Sumita, M. 1988. A numerical water assessment of rivers in Hokuriku District using epilithic diatom assemblage on river bed as a biological indicator 1. The value of diatom assemblage index to organic water pollution (DIA po). *Jap J. Limnol.* 49:299-308. (Japanese, English summary).

Thienemann, A. 1954. Ein drittes biozönotisches Grundprinzip. *Arch. Hydrobiol.* 49:421-422.

Tiwarti, T. N. and M. Ali. 1988. Water quality index for Indian rivers. pp. 271-286 *In:* R. K. Trivedy, ed., *Ecology and Pollution of Indian Rivers.* Ashish Publishing House, New Delhi.

Von Tümpling, W. 1960. Probleme, Methoden und Ergebnisse Biologische Güteuntersuchungen an Vorflutern, dargestellt am Beispiel der Werra. *Int. Revue ges. Hydrobiol. Hydrogr.* 45:513-534.

Von Tümpling, W. 1974. Zur ökologischen Charakteristik der Wasserbeschaffenheit von Fliessgewässern. *Acta Hydrochim. Hydrobiol.* 6:543-549.

Wade, K. R., S. J. Ormerod, and A. S. Gee. 1989. Classification and ordination of macroinvertebrate assemblages to predict stream acidity in upland Wales. *Hydrobiologia* 171:59-78.

Washington, H. G. 1984. Diversity, biotic and similarity indices: a review with special reference to aquatic ecosystems. *Water Res.* 18:653-694.

Watanabe, T., Kizuka, T. and S. Tanaka. 1986. Water quality chart of the river Yamato-gawa using diatom assemblage index to organic water pollution (DIApo) based on attached diatom assemblage in river. *Diatom* 2:125-131. (Japanese, English summary.)

Weatherley, N. S. and S. J. Ormerod. 1989. Modelling ecological impacts of the acidication of Welsh streams: temporal changes in the occurrence of macroflora and invertebrates. *Hydrobiologia* 185:163-174.

Whipple, G. C., G. M. Fair, and M.C. Whipple. 1927. *The Microscopy of Drinking Water*. John Wiley. 586 p.

Wilhm, J. 1972. Graphic and mathematical analyses of biotic communities in polluted streams. *A. Rev. Ent.* 17:223-252.

Woodiwiss, F. S. 1964. The biological system of stream classification used by the Trent River Board. *Chem. Indust.* 11:443-447.

# CHAPTER 3

## What are the Requirements for an Effective Biomonitor?

**Ruth Patrick**, Ph.D., Francis Boyer Chair of Limnology, Academy of Natural Sciences, Philadelphia, PA.

Biological organisms have been used by many researchers to determine change over geological time as well as present day changes. The need for current biological monitoring of the air, soil, and water is of increasing interest. This is largely because of the vast number of chemicals that are entering our air, water, and land, whose effects upon the environment we do not understand. Also, we do not know their effects upon human life.

To date, most of the monitoring of effluents has been at the end of the pipe or smokestack. The interaction with the media into which they are discharged is not fully realized. For example, SOx is discharged at the end of the smokestack. It is only when it interacts with the air and moisture that it forms $H_2SO_4$ and drops out as acid rain. Likewise, there are many chemicals discharged into water that may dissolve or precipitate, depending upon the characteristics of the receiving stream or groundwater. Furthermore, chemicals interact with organisms in water, often changing their chemical characteristics, and they may become less toxic or, in some cases, more toxic.

To date, the monitoring that has been done has been largely for chemical parameters, and the response of organisms to the chemicals is not determined. Knowledge of the latter is important because it is the communities of organisms that convert wastes into usable products and cycle them as nutrients through the biosphere.

The organisms in water, or on or in the land, differ greatly in their physiological sensitivity to various chemicals. Also, they differ in the assortment of chemicals necessary for growth. The most effective method is to study all organisms in the food web that live in a given area. The more lines of evidence from which one can draw conclusions, the more reliable the conclusions. Shifts in the relative number of species belonging to different groups indicate changes before they become severe (Patrick 1949). If one cannot monitor all organisms within the ecosystem, be they on land or in the water, then it is important to select organisms that are important in the transfer of nutrients and energy through the food web. These organisms, if eliminated or greatly reduced, will have grave effects upon the cycling of nutrients.

For biological monitoring it is important to select organisms that will continuously accumulate over time. It is important that one knows the time period in which such accumulation occurs. Therefore, one must be able to place a monitoring apparatus within the ecosystem being monitored. Some organisms, such as clams used in the mussel watch started by EPA, are only effective if one knows the period of time during which accumulation takes place before depuration occurs. Not knowing this, one might draw false conclusions about the concentrations of chemicals in the receiving body of water.

Some species, be they in water or part of the landscape, are only active during certain periods of the year and therefore they would not be effective for continually monitoring the condition of the environment in which the ecosystem lives.

If possible, the group of organisms selected as biomonitors should be widely distributed so that one can compare the findings from one body of water to another. Of course, each species has specific physiological requirements, but they are likely to be more similar between species belonging to the same major group of organisms than between species in different major groups. For example, diatom species are usually more similar in their reactions to given substances than are species belonging to different major groups of organisms. It would not be wise to use oligochaetes as biological monitors in one body of water, mayflies in another body of water, and diatoms in a third, if one wants to compare the amount of pollution existing in different water habitats.

The design of the biological monitoring system is very important. First the collector should collect species living in the area in which the monitor is placed. It should be an effective monitor throughout the year and under different environmental conditions. This applies to organisms that monitor chemicals that are in the air and fall out on the vegetation as well as monitoring organisms in streams. Second, the monitor must collect species important in the transfer of nutrients in the food web (Patrick 1988). Third, the organisms must be active throughout the year.

In determining the appropriate biological monitor, one should first decide the questions that they want to answer. For example, if one wants to monitor the concentration of radionuclides in the fallout on the vegetation of the landscape or present in the water, one needs to select organisms that will accumulate over the time period of interest and in all types of weather conditions. Examples are lichens for monitoring aerial fallout and diatoms for monitoring radionuclides in water.

The analyses of the data from the biological monitor also depend on the questions to be answered. For example, if one simply wants to determine the amount of a chemical that passes a given point in the water during a period of time, one is not interested in the species but rather in the amount of chemicals that the species accumulate in relation to the concentrations in the receiving stream. This type of monitoring is effective in determining the chemicals in an effluent or from fallout.

Often, ambient concentrations are too low to determine by using available instruments. However, organisms that bioconcentrate will bring the concentrations up to higher levels and make them much easier to identify. This type of monitoring is qualitative monitoring. Quantitative monitoring means developing a K factor for bioconcentration of the organism in relation to concentrations in the water.

Often chemicals interact with the water or are changed by the metabolism of organisms in the stream, such as the reduction of CBOD by bacteria, or the changing of the form of metals from simple inorganic to methylated forms by bacteria. Therefore, by determining changes in the monitoring organisms, one is able to estimate the effects of these introduced substances on the aquatic ecosystem. One must first determine the relative tolerance of the monitored organisms to the tolerance of other species in the ecosystem, e.g., fish and mayflies. In other words, determine a K factor. Obviously, different organisms will not have similar sensitivities to different chemicals. This type of research must be done in order to equate the effect from a single group of organisms to the ecosystem as a whole.

When one is studying the effects on the association of monitored species, it is important that statistical models be developed in order to relate changes to the whole ecosystems. It is important that a model, diversity index, or the cluster analysis not be used by itself. Although these are valuable ways of showing differences, it is only by studying the auto-ecology of the species that one can determine whether the changes observed are caused by the pollutant or a causal factor that should be monitored.

Diatoms fit many of the requirements for an excellent monitoring system of aquatic ecosystems (Patrick 1986). First, they are very important in the transfer of nutrients and energy in the food web. Second, in water they are probably the most important group in the food web of insects and fish larvae, and even young adult fish. Third, they are the main food of organisms such as *Daphnia* and other zooplankton, which feed to a large extent upon diatoms.

Furthermore, one can construct an instrument for collecting diatoms that collects 75–85 % of the species in the area where the collector is located. This has been verified in many streams where hand collections were made at the same time that the diatometers were collecting diatoms (Patrick and Holm 1956, Patrick and Strawbridge 1963).

The sensitivity of diatoms to many pollutants is quite similar to that of fish and invertebrates such as snails. As shown in toxicity tests, diatoms are often slightly more sensitive (Doudoroff et al. 1951, Patrick et al. 1968).

Several different ways have been developed to treat biological monitoring results mathematically in order to objectively show change. The first methodology developed was a truncated curve model (Patrick et al. 1954). This research showed that a natural diatom community corresponded to a truncated normal curve in structure and that pollution of various types affected the structure of the community in various ways. The first effect of moderate nutrient pollution is to lengthen the tail of the curve. This means that certain species become much more common than others. Toxic chemicals often result in only a few species being able to live, and their populations are very small. When the community is subjected to severe pollution the truncated curve becomes a less accurate model.

The same sort of data, which is the numbers of species and the sizes of their populations, have been expressed as Whittaker's Importance Curves (Whittaker 1970). This is simply another way of plotting the same data as that plotted in the truncated normal curve.

Another method of comparing communities is to plot "sigma square" versus the number of species for several communities that are natural or not adversely affected by pollution. Using the ellipses formed by these plots, one can determine the similarity of other communities with 95 % or 99 % confidence. Thus, any other community can be examined by testing its sigma square and numbers of species against these ellipses to see whether or not they are natural or polluted. The further away the point is from the structure of the ellipse, the greater the pollution.

Diversity indices, such as the Shannon Weaver Index, are easier to construct than a truncated normal curve. However, unless one identifies and records the populations of a large number of species, the chance of error increases. Several thousand specimens need to be counted in order to have a reasonable degree of confidence that the community structure being developed is truly representative of the community.

There really is no shortcut to accuracy. However, if one does not need a high degree of accuracy, shorter methods such as diversity indices from counting a few hundred specimens, can be used. If one only counts specimens until the asymptote of a curve is developed—that is, counting species and specimens until only one or two species per 100 specimens are added to the curve, therefore, counting only about one-third of the species present—the result may be very erroneous concerning the true structure of the diatom community.

To compare various communities, other types of mathematical analyses, such as a Jaccard Cluster Analysis, can be used. This methodology will determine whether or not two communities are different as to species composition. The way the clusters fall out determines how different they are from each other. However, as stated above, the difference may not be the answer to the question asked.

All of these mathematical analyses show differences, but the differences portrayed by the analyses may not be caused by the factors set forth by one's hypothesis. For example, in a study that we made in the Flint River, it is very evident that Station 1-right side and Station 1-left side were very different according to cluster analyses. The hypothesis is that this difference is due to the presence of pollution. However, when one carries out a detailed study of the autoecology of the species, one finds that although the communities differ in kinds of species, this difference has nothing to do with the tolerance of species toward pollution. In other words, other environmental factors, (e.g., light, water flow, and substratum) are causing different species to form the communities, not the presence of pollution. These types of false conclusions are very common in all types of mathematical analyses unless one examines the autoecology of the species forming the communities.

Another use of diatoms is for the bioconcentration of specific types of pollutants. In this type of study one collects the diatoms and analyzes them for the chemical in question. Diatoms are excellent organisms for this type of study. They are covered by a colloidal gelatin that rapidly accumulates metals and also some organic compounds. Since they typically divide every day, a new surface of colloid material on half of the frustule is presented for absorption. Thus, over time the community accumulates large amounts of various pollutants.

The chemicals accumulated are often so dilute in the ambient medium that they are below the limits of detection. However, when bioconcentrated by diatoms they are easily determined. An example of these types of studies is as follows. Recently, we were concerned about the presence of metals in the Flint River. We identified the presence of mercury. Mercury was not being used by the plant and the great concern was how the mercury got into the water. It was not present in the river above the plant. A detailed analyses of all the processes showed that the chemical compound being used for neutralization had been treated by a mercury process in creating the chemical. Minute amounts of this contaminant had carried over into the final product and, when released into the river, were picked up by the diatoms through bioconcentration.

We have also used diatoms extensively for detecting the presence of radionuclides. In 1967, R. S. Harvey and Patrick (unpublished data) studied the accumulating factors of diatoms for various radionuclides in the test materials. The amount present in ambient effluent was known, therefore bioconcentration factors could be determined. Diatoms were compared with other types of algae in this study. It was shown that diatoms bioaccumulate metals that are radioactive in greater concentrations than other algae. They are believed to be capable of bioaccumulation in larger amounts than other organisms in the ecosystem.

It was interesting to note that when the Chernobyl plumes passed over the Susquehanna River in Pennsylvania, the diatoms were more effective in recording various types of radionuclides associated with those plumes than were many of the instruments used in our most sophisticated laboratories.

# CONCLUSION

From these studies it is very evident that one must be specific in the questions being asked about a given ecosystem before designing the monitoring system. Although diatoms to date have been the most thoroughly studied organism in this regard, no doubt other groups of organisms in the future will be carefully developed to be reliable biological monitors.

# REFERENCES

Doudoroff, P., B. G. Anderson, G. E. Burdick, P. S. Galtsoff, W. B. Hart, R. Patrick, E. R. Strong, E. W. Surber, and W. M. Van Horn. 1951. Bioassay methods for the evaluation of acute toxicity of industrial wastes to fish. *Sew. & Indus. Wastes* 23(11):1380-1397.

Patrick, R. 1949. A proposed biological measure of stream conditions based on a survey of Conestoga Basin, Lancaster Co., PA. *Proc. Acad. Nat. Sci. Philadelphia.* 101:277-341.

Patrick, R., M. H. Hohn, and J. H. Wallace. 1954. A new method for determining the pattern of the diatom flora. *Not. Nat. Acad. Nat. Sci. Philadelphia* 259:1-12.

Patrick, R. and M. H. Hohn. 1956. The diatometer—a method for indicating the conditions of aquatic life. *Proc. Amer. Petroleum Inst.* 36(3):332-339.

Patrick, R. and D. Strawbridge. 1963. Variation in the structure of natural diatom populations. *Amer. Nat.* 97(892):51-57.

Patrick, R., John Cairns, Jr., and A. Scheier. 1968. The relative sensitivity of diatoms, snails, and fish to twenty common constituents of industrial wastes. *Prog. Fish-Cult.* 30(3):137-140.

Patrick, R. 1986. Diatoms as indicators of changes. pp. 759-799. *In*: M. Ricard, ed., *Water Quality Proc. 8th Intern. Diatom Symposium.*

Patrick, R. 1988. Importance of diversity in the functioning and structure of riverine communities. *Limnol. Oceanogr.* 33:1304-1307.

Whittaker, R. H. 1970. *Communities and Ecosystems*. MacMillan Co. 158 p.

# CHAPTER 4

# European Perspective on Biological Monitoring

**Pier Francesco Ghetti** and **Oscar Ravera**, Dipartimento di Scienze Ambientali, Università degli Studi di Venezia, Venezia, Italy

## EARLY DEVELOPMENT OF BIOLOGICAL INDICATORS

Pre-technological societies were closely attuned to the plants and animals of their surroundings, since the condition of the environment largely determined the patterns of everyday life and, ultimately, their survival.

". . . Biological indication has been applied for a very long time. Within living memory hunters, shepherds and farmers have recognized the quality of land by the growth of plants and by behavior of animals, when they choose the place of their settlements, their hunting grounds, their fields and pastures." (Best and Haeck 1982).

To insure the safety of the public water supply of ancient Rome, Sesto Giulio Frontino, *curator aquae* in 30-103 A.D., closely monitored the health of the local residents near the sources for the city's aqueduct system (Farrington 1953). By the 19th century, according to Cohn (1875), water supply managers recognized that ". . . microscopic analysis of the water for potable use, if correctly carried out, integrates chemical analysis and, unlike chemist's reagents, can give answers to certain questions." Air quality was also associated with the condition of biological indicators. Thus, in the introduction to a study on the lichens of the Garden of Luxenbourg in Paris, Nylander (1866) stated that ". . . lichens provide in their own way, a measure of the wholesomeness of the air." To monitor the pollution of German rivers, Kolkwitz and Marsson (1902) set up a system for defining the quality of surface waters based mainly on the presence of indicator species belonging to the plankton and periphyton communities. This system, called "Saprobiensystem", is still the basis for a series of biological indices (De Pauw, et al. 1991).

In recent decades the ecological approach has shown that the biotic and abiotic components of an ecosystem are closely related and functionally integrated. Although their responses are not teleological, biological communities do tend toward some kind of steady state, either through acclimation of individuals or adaptations of populations. When disturbed, therefore, these systems tend to reestablish an optimum pattern of energy use for the new conditions. Nevertheless, a deteriorating environment may trigger a positive feedback, causing the system's new equilibrium to be substantially altered both in structure and function. Loss of species diversity and a deterioration of the quality of the resource for human use may ensue.

The importance of conserving biological resources and detecting community change must be brought to the attention of environmental policymakers. One possible way to do this is to develop ecological quality criteria that may lead to environmental policies more objectively related to the actual degradation processes involved.

Biological indicators (e.g., different species) and indices (e.g., Saprobic Index, Extended Biotic Index, Diatom Index) represent an essential tool in programs of ecological monitoring, particularly if they can detect or even anticipate the negative effects of pollution. Such indices provide a measure to evaluate environmental health and can be used as the basis for the protection of species and trophic complexes characteristic of the high degree of functional quality of a healthy ecosystem (Ravera 1979, Ghetti 1980).

# BIOLOGICAL INDICATORS AS
# TOOLS IN "ENVIRONMENTAL SURVEILLANCE"

In the past, environmental indicators focused mainly on the quality of natural resources (i.e., water, air, and soil) for human use. Broader aspects of ecosystem structure and regulation were largely ignored. Chemical, physical, and human hygiene indicators were chosen on the basis of their value to resource management. These non-biological criteria provide inadequate information for successful management: a) of quantity and quality of the environmental stress causes; b) because understanding of ecosystem responses to stress is very limited; and c) because many ecosystems have already been degraded so that their original structure is lost and resource quality is reduced.

Environmental monitoring should, therefore, give answers to the following questions:

1) What is the current functional state of the ecosystem from which the resource will be taken?
2) For what purposes is it possible to use the resource and how much of it can be drawn without modifying the functionality of the ecosystem?

Hence, there is a need to develop techniques to assess "normal" conditions of functionality of each investigated ecosystem and the structure of populations living in it. The intrinsic biological quality of a given environment is, by definition, independent of any anthropocentric evaluation and it can be drawn from the properties of the ecosystem under examination.

It is important to determine the deviation from the normal condition and the capability of the ecosystem to absorb external stresses through internal regulatory processes without modifying the overall character of the environment. Many problems are met in this field owing to:

a) poor consolidation of the approach to ecological problems;
b) the lack of environments having "natural" conditions;
c) the difficulty in determining the carrying capacity of the various environments relative to human manipulations; and

    d) the difficulty in translating diagnoses, typically based on qualitative evaluations, into a screening system that should be quantitative and universally understandable.

Despite such problems, there is a great demand for information on the quality of the environment, often involving large, complex areas, that can be obtained with limited time and instruments.

Modern field investigations commonly generate large amounts of data, but distillation of that data to a final evaluation and recommendation is extremely difficult. If it is true that simple answers to complex problems cannot be given, it is also true that the opposite risk—having no answers—also exists. Thus, we believe that the use of biological indices in environmental assessment may be important because they can provide a practical way to evaluate the "state of functionality" of a given environment, as well as the actual level of "risk" of ecosystem damage.

A biological indicator is meant to give a useful biological measure, that if sensitive enough, can be used for diagnoses, control, prevention, and reclamation. Therefore, a selection has to be made from among all biological responses in order to have a restricted range of indicators. Biological indicators were used by humans in the past from necessity (e.g., on agricultural, fishing, hunting activities); at present they can become a powerful tool that may alert humans to the impacts resulting from their manipulations of the environment. This can be achieved by enhancement of monitoring techniques, the acceptance of these techniques worldwide, and by improved education and information availability. A common criticism of the use of biological indicators is that they are often based on qualitative methods and that their terminology is not comprehensible to non-specialists (e.g., taxonomic lists). Thus, an effort must be made to have quantitative methods and to present the data in non-specialist language (e.g., statistics, mathematical models). Information drawn from life is the result of an intelligent, sophisticated process that has been carried out at different levels of complexities. The real problem is to correctly interpret these signals so to evaluate their importance and to communicate this information to other experts.

# BIOLOGICAL INDICATORS OF
# FRESHWATER QUALITY

Blandin (1986) recently analyzed 690 titles of scientific papers published in one year on the use of biological indicators in investigations on freshwater, marine, and terrestrial environments. Of these works, 43% dealt with inland waters, running waters in particular. Two prevalent types of methods were used: laboratory and field toxicity tests (including bioaccumulation tests), and field studies on natural populations (mainly bacteria, benthic invertebrates and, to a less extent, algae and fishes). In addition, some biological indices are currently used in routine analyses by public agencies for environmental enforcement and control. Such results are rarely published (Provincia Autonoma di Trento 1988). In Europe, several categories of biological monitoring are used for freshwater:

a) analysis of natural communities to evaluate current conditions (particularly for lotic) (Ghetti 1985, De Pauw et al. 1991);
b) toxicity tests (laboratory and field) to measure the environmental hazard of discharges;
c) alarm biological assays (Garric and Larbaigt 1986). The methodology of alarm control is mainly characterized by its utility in providing rapid feedback information for effluent control processes and facilitating prevention of accidental spills.
d) bioaccumulation tests (laboratory and field) (Solbe 1986); and
e) use of biological indicators in Environmental Impact Assessments (Institute for System Engineering and Informatics 1990).

Today biological monitoring has become an integral part of water quality monitoring. Two major categories are of importance in biological monitoring of aquatic environments. The first category comprises the bioassays (early warning or alarm systems, ecotoxicological tests, bioaccumulation tests, biodegradation tests, eutrophication tests). The second category, bioassessments, covers the methodologies related to analysis of the biological communities. Bioassays are experimental, while bioassessments are observational in approach.

Bioassays and bioassessments, in turn, are linked to the choices of environmental policymakers concerning the defining and updating of "water quality criteria" and "effluent standards". The evaluation of environmental impacts requires a precise characterization of the biological environment before the setting up of a proposed activity, and of any possible side-effects deriving from collaterial activities.

## SYSTEMS FOR BIOLOGICAL ASSESSMENT OF WATER QUALITY

Biological assessment can be considered from a taxonomical or a functional/structural point of view (De Pauw 1988). Bioindicators may consist of several groups of micro-organisms (bacteria, fungi, microalgae, protozoans) as well as macro-organisms (macrophytes, insects, molluscs, worms, fishes). Since no assessment is capable of representing the entire biota, any one system usually analyzes a single segment of the community. Those typically considered in water quality assessment are:

> plankton,
> periphyton,
> microbenthos,
> macrobenthos, and
> nekton

Advantages and disadvantages of various groups of water quality indicators are summarized by Hellawell (1978). More than 90 different methods based on classification of lotic communities for the assessment of running water quality are used (Table 4.1). This has persuaded the Commission of the European Communities (C.E.C.) to organize three tests for the intercalibration (Tittizer 1975, Woodiwiss 1976, Ghetti 1978).

**Table 4.1.** Indices for assessment of running waters based on analysis of biota in their natural environment (from De Pauw et al. 1991).

| Indices | Communities* | References** |
|---|---|---|
| *Saprobic indices* | | |
| Biol. Effect. Org. Load. (BEOL) | PAMFV | Knöpp 1954 |
| Coupling Analysis | M | Buck 1974 |
| Relative Purity | PAM | Knöpp 1954 |
| Saprobic Index (S) | PAMFV | Pantle & Buck 1955; DIN 38-410 |
| Saprobic Index (S) | PAMFV | Zelinka & Marvan 1991 |
| Saprobic Index (S) | D | Sladecek 1984, 1986 |
| Saprobic Quotient (SQ) | P | Dresscher & Van der Mark 1975 |

**Table 4.1.** Continued

| Indices | Communities* | References** |
|---|---|---|
| *Biotic indices* | | |
| Average Score Per Taxo (ASPT) | M | Armitage et al. 1983 |
| Belgian Biotic Index (BBI) | M | De Pauw & Vanhooren 1983; NBN T92-402 |
| Biol. Index of Pollut. (BIP) | M | Graham 1965 |
| Biotic Index (IB) | M | Tuffery & Verneaux 1968 |
| Biotic Index (IB) | M | Tuffery & Davaine 1970 |
| Biotic Index (BI) | M | Chutter 1971 |
| Biotic Index (BI) | M | Hawmiller & Scott 1977 |
| Biotic Index (BI) | M | Winget & Mangun 1977 |
| Biotic Index (BI) | M | Hilsenhoff 1982 |
| Biotic Index for Duero Basin | M | Gonzalez del Tanago & Garcia Jalon 1984 |
| Biotic Index modif. Rio Segre | M | Palau & Palomes 1985 |
| Biotic Score (BS) | M | Chandler 1970 |
| Biotic Score modif. La Mancha | M | Gonzalez del Tanago et al. 1979 |
| Biotic Score modif. Rio Jarama | M | Gonzalez del Tanago & Garcia Jalon 1980 |
| BMWP-Score (BMWP) | M | Chesters 1980; Armitage et al. 1983 |
| BMWP Spanish modif. (BMWP) | M | Alba-Tercedor & Sanchez-Ortega 1988 |
| Cemagref Diatom Index (IDC) | PAD | Cemagref 1984 |
| Chironomid Index (Ch.I.) | M | Bazerque et al. 1989 |
| Ch.I. based on pupal exuviae | M | Wilson & McGill 1977 |
| Damage Rating | V | Haslam & Wolseley 1981 |
| Dept. of Environm. Class. | MF | DOE UK 1970 |
| Diatom Index (IDD) | AD | Descy 1979 |
| Diatom Index (ILB) | AD | Lange-Bertelot 1979 |
| Diatom Index (IPS) | AD | Cemagref 1982-1084 |
| Diatom Index (IFL) | AD | Fabri & Leclerq 1984-1986 |
| Diatom Index (ILM) | AD | Leclercq & Maquet 1987 |
| Diatom Index (CEC) | AD | Descy & Coste 1991 |
| Extended Biotic Index (EBI) | M | Woodiwiss 1978 |
| EBI Italian modif. (EBI) | M | Ghetti 1986 |
| EBI Spanish modif. (BILL) | M | Prat et al. 1983, 1986 |
| Index of Biotic Integrity (IBI) | F | Karr et al. 1986 |
| Family Biotic Index (FBI) | M | Hilsenhoff 1987, 1988 |
| Generic Diatom Index (IDG) | AD | Rumeaux & Coste 1988 |
| Global Biotic Index (IBG) | M | Verneaux et al. 1984; AFNOR T 90-350 |
| Glob. Biot. Qual. Index (IQBG) | M | Verneaux et al. 1976 |
| Ichthyological Index | F | Badino et al. 1991 |
| Lincoln Quality Index (LQI) | M | Extance et al. 1987 |
| Macroindex | M | Perret 1977 |
| Median Diatomic Index (MI) | AD | Bazerque et al. 1989 |
| Pollution index (I) | M | Beck 1955 |
| Quality Index (K135, K12345) | M | Tolkamp & Gardeniers 1977 |
| Quality Rating System (Q-value) | M | Flanagan & Toner 1972 |
| Simplified Biotic Index (SBI) | MF | Jordana et al. 1989 |
| Trent Biotic Index (TBI) | M | Woodiwiss 1964 |

**Table 4.1.** Continued

| Indices | Communities* | References** |
|---|---|---|
| *Diversity indices* | | |
| | | |
| Diversity Index (d) | | Simpson 1949, Pielou 1969 |
| Diversity Index (H') | | Shannon & Weaver 1949, 1963 |
| Diversity Index (d) | | Margalef 1951 |
| Diversity Index (d) | | Menhinick 1964 |
| Diversity Index (d) | | McIntosh 1967 |
| Diversity Index (d) | | Wilhm & Dorris 1968 |
| Diversity Index (d) | | Lloyd et al. 1968 |
| Equitability (e) | DM | Lloyd & Ghelardi 1964 |
| Ephem., Plec., Trich. (EPT) index | M | Plafkin et al. 1989 |
| Index of Well Being (IWB) Score | F | Gammon 1980; Platkin et al. 1989 |
| Lognormal Distribution | D | Preston 1948; Patrick 1973 |
| Number of individuals per taxon | PAMFV | Helawell 1986; Plafkin et al. 1989 |
| Percent dominant family | M | Shakleford 1988 |
| Ratio tubificids/other macroinv. | M | Goodnight & Whitley 1960 |
| Ratio insects/tubificids | M | King & Ball 1964 |
| Ratio L. hoffmeisteri/other tub. | M | Brinkhurst 1966 |
| Ratio Gammarus/Asellus | M | Hawkes & Davies 1971 |
| Ratio Scrapers/Filt. Collectors | M | Merritt & Cummins 1984 |
| Ratio Shredders/Total macroinv. | M | Swift et al. 1988ab |
| Ratio EPT/Chironomidae | M | Ferrington 1987 |
| Sequential Compar. Index (SCI) | AM | Cairns et al. 1968 |
| Taxa Richness (S) | PAMFV | Helawell 1986; Plafkin et al. 1989 |
| Total Number of Individuals (N) | PAMFV | Helawell 1986; Plafkin et al. 1989 |
| Williams' Alpha Index | | Fischer et al. 1943 |
| | | |
| *Comparative indices* | | |
| | | |
| Biological Condition Score | M | Plafkin et al. 1989 |
| Biotic Condition Index | M | Winget & Mangum 1979 |
| Cluster Analysis | M | Plafkin et al. 1989 |
| Coefficient of Association | | Looman & Campbell 1960 |
| Coefficient of Similarity | M | Jaccard 1912; Boesch 1977 |
| Coefficient of Similarity | M | Kulezynski 1948 |
| Community Loss Index | M | Courtemanch & Davies 1987 |
| Community Similarity Index | M | Bray & Curtis 1957 |
| Community Similarity Index | M | Morisita 1959 |
| Community Similarity Index | M | Pinkham & Pearson 1976 |
| Comparative Measure | | Czekanowski 1913 |
| Comparative Measure | | Raabe 1952 |
| Distance Measure | | Sokal 1961 |
| Environment. Quality Index (EQI) | M | Sweeting et al. 1992 |
| Fluctuation Index (D) | D | Dubois 1973 |
| Indicatgor Assemblage Index (IAI) | M | Shackleford 1988 |
| Index of Similarity (IS) | | Mountford 1962 |
| Quotient of Similarity (QS) | | Sorensen 1948 |
| Species Deficit (SP) | M | Kothé 1962 |

**Table 4.1.** Continued

| Indices | Communities* | References** |
|---|---|---|
| Rank Correlation Coefficient | M | Spearman 1913 |
| Rank Correlation Coefficient Tau | M | Kendall 1962 |

*Communities considered:
  P = Plankton; A = Periphyton (Aufwuchs); M = Macroinvertebrates; F = Fish;
  V = Aquatic Vegetation; D = Diatoms

**References: see DePauw, Ghetti, Rauzini, Spaggiari (1991).

The results of this comparison can be summarized as follows:

1) A biological classification system would be suitable for mapping the state of European rivers; methods of biological assessment must become an integral part of water monitoring procedures.
2) Chemical and biological data each have a distinct role to play in the measurement of water quality, and both must be taken into account to obtain a complete water quality assessment.
3) The three Technical Seminars on "Biological Water Assessment Methods" (Institute for Systems Engineering and Informatics 1990) revealed a growing emphasis on the use of assessment methods based upon the study of macroinvertebrate populations.
4) Some of the assessment methods used were significantly correlated, even when applied to environments that were different from those for which they were originally devised.
5) Relationship studies between pairs of assessment methods demonstrate that intercalibration is possible despite the differences of scaling and sensitivity.
6) Two methods, BEOL (Biologically Effective Organic Load [Knopp 1954]) (adapted) and EBI (Extended Biotic Index [Woodiwiss 1978]), agree very closely and can be used to define convenient intervals of water quality.

Based on this experience, biological methods for running water quality assessment have been adopted by European nations (Table 4.2). Though these methods are usually up to date, little attention has been paid to a possible intercalibration among them. Still, the greatly increased use

**Table 4.2.** Application of major index methods for assessment of running waters in E.C. countries based on macroinvertebrates (from De Pauw et al. 1991).

| Country | Index method[1] | Sampline[2] | Analysis[2] | Identification[3] | Standard[4] | Range |
|---------|------------|-----------|-----------|-----------------|-----------|-------|
| Belgium | BBI | Qual | Qual | O F G | N | 0–10 |
| Denmark | S | Qual | Qual | S | N | 1–4 |
| France | IBG | Quant | Qual | F | N | 0–20 |
| Germany | BEOL/S | Qual | Quant | S | N | 0–100/1–4 |
| Greece | – | – | – | – | – | – |
| Ireland | Q-rating | Qual | Qual | F G S | N | 0–5 |
| Italy | EBI | Qual | Qual | O F G | R | 0–14 |
| Luxemburg | IB | Qual | Qual | O F | N | 0–10 |
| Netherlands | K135 | Qual | Qual | F G S | R | 100–500 |
| Portugal | – | Qual | Qual | O F G | – | 0–10 |
| Spain | BMWP | Qual | Qual | F | – | 0– >150 |
| UK | BMWP/ASPT | Qual | Qual | F | N | 0– >150/0–10 |

[1]See References in Table 4.1.
[2]Qual = Qualitatively; Quant = Quantitatively.
[3]Required identification level: O = Order; F = Family; G = Genus; S = Species.
[4]N = National; R = Regional.

of these indicators and their routine applications in the policy of water quality control of many nations is an encouraging trend (Provincia Autonoma di Trento 1988).

# ASSESSMENT OF INLAND WATER QUALITY BY BIOLOGICAL ASSAYS

The aim of the bioassays is to certify the quality of water to guarantee its safe use by humans and for the protection of natural environments. In order to reach and maintain good levels of quality, it is important to know:

1) the concentrations of toxic substances in the effluents and the mechanism of their action, and
2) the persistence and extension of toxic actions of the effluents on their natural recipient environment.

Wastes are increasingly often mixtures of chemicals in varying proportions. Chemical monitoring is usually carried out on the basis of some parameters (e.g., $NH_3$, Cd, Cu, and Pb) aimed at detecting small quantities of products and subproducts in the wastes. Although some of these substances are present at trace concentrations, they may exhibit high toxicity. Therefore, it would be very complex and expensive to carry out direct chemical analyses on all possible toxicants in waste water. It would also be difficult to predict overall wastewater toxicity on the basis of the toxicities of individual components in simple solution because of chemical and physical interactions in the complex mixture (Pantani et al. 1989).

To overcome such difficulties, toxicity tests on whole effluent can be used to complement chemical analyses and to regulate the discharge of toxic wastes. The application of biological tests can be particularly successful in the following cases:

1) the assessment of chronic and acute effects of effluents on communities in the receiving water, and
2) the regulation of toxic effluents and the design of wastewater control technology to reduce wastewater toxicity.

Such bioassays are obviously unable to meet all requirements of control, particularly in the following cases:

a) The assessment of direct effects on human health (e.g., carcino-genetic and teratogenetic effects)
b) Bioaccumulation tests
c) Eutrophication by nutrient addition
d) Coliform and other microbial indicators of human pathogens (OECD, CEE 1986, Department of Civil Engineering 1978)

Some European nations are supplementing chemical monitoring with toxicity tests using *Daphnia magna* to detect accidental spills in wastewater discharges. In 1985 the C.E.M.A.G.R.E.F. (France) carried out 132 wastewater evaluations with both chemical and biological analyses (Garric and Larbaigt 1986). No toxicity was detected in 10% of the cases by chemical analyses, while toxicological tests showed significant environmental damage.

Short term tests, commonly used for waste monitoring can be listed as follows:

1) Tests with *Daphnia* (which is the most commonly used)
2) Microtox (T.M.) assays, based on the inhibition of the biolumines-cence of bacteria (photobacterium *Phosphoreum*)
3) Tests with fish, particularly on *Brachydanio rerio* and rainbow trout (*Onchorhynchus mykiss*)
4) Toxicity tests with algae ("*Selenastrum* bottle test")

Ideally, short term assays should be integrated with chronic toxicity tests, although chronic testing requires considerable additional cost and effort.

Furthermore, a correlation should be drawn between the information obtained in toxicity tests and studies on the biological quality of the natural environments. Thus, a close link is needed between laboratory assays and observed or predicted effects on receiving water communities. Biological indices are particularly useful for comparing the quality of upstream and downstream sites in order to assess the effects caused by toxicants in different environmental conditions (e.g., temperature, water hardness, and flow variations) (Garric and Larbaigt 1986).

Biological monitoring in England is conducted primarily by the water authorities and some chemical industries, which use a variety of test methods (Solbe 1986). The main applications of biological assays are:

a) Continuous monitoring of water that is to be abstracted for potable supply. The Water Research Center has developed an intake-protection system based on physiological measurements of captive rainbow trout (*Salmo gairdneri*). The apparatus, called Mark III, is commercially produced and installed upstream of water intakes around the United Kingdom.

b) Continuous and intermittent monitoring of industrial discharges. Discharges are controlled (limits based on flow, physico-chemical and biological parameters of the effluent). The regulatory authority has access to a sampling facility and data are published in registers open to the public. Traditionally, biological monitoring by regulatory authorities and some industries has been by ecological survey, usually measuring benthic fauna diversity (some data bases extend back seventeen years). Routine monitoring of chemicals (e.g., metals in animals and plants) is undertaken. Toxicity testing by regulatory authorities has been on an *ad hoc* basis. Some industrial companies have a history (in one instance over 30 years) of the use of toxicity tests in effluent control. Some examples of their use are:

(i)   Regular monitoring of acute toxicity of effluents to establish variability (in conjunction with chemical analysis to try to establish what is causing the toxic effect)

(ii)  Site investigations to establish toxic streams and/or constituents (again with chemical analysis to aid interpretation)

(iii) Acute toxicity studies to indicate what initial dilution must be achieved when designing effluent outfalls

(iv)  New plant discharges are studied in depth. Effect on the acute toxicity of the overall site effluent is established and, if there is evidence of the presence of conservative material, bioaccumulation studies are undertaken. Test species are selected that relate to the receiving environment.

c) Periodical monitoring of rivers to assess the integrity of their aquatic communities.

d) Surveys aimed at locating and identifying damage caused by accidental spills or discharges.

Although adequate coordination and well matched procedures are still lacking, more countries in Europe are adopting similar approaches for waste control.

## CONCLUSIONS

Protection of the quality of life is the most important parameter for the conservation of the environmental quality, and the procedures of "ecological surveillance" are the best way to establish such control. In particular, the feasibility of developing the use of bioassays and bioassessments in ecological surveillance is closely related to a host of factors including the following:

a) The development of basic research on the meaning of biological responses, particularly at the level of populations and communities

b) The correct definition and use of procedures of ecological surveillance can reduce the gap between estimated risk and real effects upon the environment

c) The standardizing of criteria for the collection, treatment, and interpretation of biological information

d) The definition of concrete and operational research methods stressing the centrality of biological responses in the assessment of environmental quality

e) The promotion of effective means for disseminating the concept of biological quality of the environment

f) The need for standard biological methods of water quality monitoring to be required by law, both nationally and internationally

g) The training of scientifically and ecologically aware research workers and operators who are capable of participating in, and significantly contributing to, programs of ecological surveillance. They are faced with an opportunity unprecedented in the history of environmental research to influence social policy decision on resource management

# REFERENCES

Amavis, R. and J. Smeets. 1975. *Principles and Methods for Determining Ecological Criteria on Hydrobiocenoses.* Pergamon Press, Oxford. 531 p.

Best, E. P. H. and J. Haeck. 1982. *Ecological Indicators for the Assessment of the Quality of Air, Water, Soil and Ecosystems.* D. Reidel Publishing Company, Boston. 283 p.

Blandin, P. 1986. Bioindicateurs et diagnostic des systèmes ecologiques. *Bulletin d'Ecologie* 17(4):1-308.

Cohn, F. 1875. Uber den Brunnenfaden (*Crenotix polyspora*) mit Bemerkungen uber die mikroskopische Analyse der Brunner wassers. *Beitrage zuer Biologie der Pflanzzen* 1:1-34.

Department of Civil Engineering. 1978. *Biological Indicators of Water Quality.* Proceeding of Symposium, University of Newcastle-upon-Tyne, UK. 546 p.

De Pauw, N. 1988. Biological assessment of surface water quality: the Belgium experience. *In: "La Qualita' delle Acque Superficiali."* Atti del Convegno Internazionale, Riva del Garda, Italy. Provincia Autonoma di Trento, Italy. 285 p.

De Pauw, N., P. F. Ghetti, P. Manzini and R. Spaggiari. In press. *Biological Assessment Methods for Running Waters.* International Conference on River Water Quality — Ecological Assessment and Control, C.E.C. (Brussels).

Farrington, B. 1953. *Greek Science.* Penguin Books Ltd. 118 p.

Garric, J. and G. Larbaigt. 1986. Aspects de l'experience francaise dans le control biologique des effluents et du milieu recepteur continentale. *In: "International Seminar on the Use of Biological Tests for Water Pollution Assessment and Control".* Joint Research Centre, O.E.C.D, Ispra. 235 p.

Ghetti, P. F. 1978. *Biological Water Assessment Methods.* Technical Report, Parma 8-13 October 1978. Report E.E.C., ENV/395/78 EN:1-42.

Ghetti, P. F. 1980. Biological indicators of the quality of running waters. *Boll. Zool.* 47:381-390.

Ghetti, P. F. 1985. I macroinvertebrati nell'analisi di qualita' dei corsi d'acqua. *Provincia Autonoma di Trento* (Italy) 1-111.

Hellawell, J. M. 1978. *Biological Surveillance of Rivers: a Biological Monitoring Handbook.* Water Research Center, Stevenage, England. 332 p.

Institute for Systems Engineering and Informatics. 1990. *Workshop on Indicators and Indices for Environmental Impact Assessment and Risk Analysis.* Proceedings, Joint Research Centre, CEC, Ispra. 333 p.

Knopp, H. 1954. Ein neuer Weg zur darstellung biologischer Vorfluterunter-suchungen, erlautert an einem Gutelangschnitt des Maines. *Wasserwirtschaft* 45:9-15.

Kolkwitz, R. and M. Marsson. 1902. Grundsatzliches die bensteilung des wassers nech seiner flora und fauna. *Mitt. der Kaiser. Prufan. fur Wass. und Abwass.* (Berlin-Dahlem) 1-33.

Nylander, W. 1866. Les Lichens du Jardin du Luxembourg. *Bull. Soc. Bot. Fr.* 13:364-372.

O.E.C.D., C.E.E. 1986. *International Seminar on the Use of Biological Tests for Water Pollution Assessment and Control.* Joint Research Centre, CEC, Ispra. 235 p.

Pantani, C., P. F. Ghetti, and A. Cavacini. 1989. Action of temperature and water hardness on the toxicity of exavalent chromium in *Gammarus italicus Goedm.* (Crustacea, Amphipoda). *Environ. Tech. Let.* 10:661-668.

Provincia Autonoma di Trento. 1988. *La Qualita' delle Acque Superficiali: Criteri per Una Metodologia Omogenea di Valutazione.* Atti del Convegno Internazionale, Riva del Garda, Italy. 285 p.

Ravera, O. 1979. Consideration on the effects of pollution at community and population level. *Experientia* 35:710-713.

Solbe, J. 1986. United Kingdom experience in biological examination for the control and assessment of water pollution. *In: International Seminar On the Use of Biological Tests of Water Pollution Assessment and Control.* Joint Research Centre, O.E.C.D., Ispra. 235 p.

Tittizer, T. 1975. Comparison of biological-ecological procedures for assessment of water quality. pp. 403-463. *In: Principles and Methods for Determining Ecological Criteria on Hydrobiocenoses.* Amavis and Smeets, Pergamon Press, Oxford. 531 p.

Woodiwiss, F. S. 1976. *Biological Water Assessment Methods.* Technical Seminar, *Nottingham.* Report, E.E.C., ENV/223/76 EN:1-53.

Woodiwiss, F. S. 1978. *Biological Monitoring of Surface Water Quality. Summary Report.* Commission of the European Communities. ENV/787/80-EN: 45 p.

# SECTION III

# Experimental Design

# CHAPTER 5

# The Statistical Validity
# of Biomonitoring Data

**John Cairns, Jr.**[1,2] and **Eric P. Smith**[1,3]
[1]University Center for Environmental and Hazardous Materials Studies,
[2]Department of Biology, and [3]Department of Statistics, Virginia Polytechnic
Institute and State University, Blacksburg, Virginia.

## INTRODUCTION

The term "biological monitoring" has been widely used to include almost any type of data gathered to assess the environmental impact of discharges. In our opinion, biological monitoring should be limited to a continuing collection of data to establish whether explicitly stated quality control conditions are being met. A concomitant implication is that, if these conditions are not being met, an immediate decision to take corrective action will be made. The close relationship between the information gathered and the decision to take action, or take no action, is a crucial one. The information must be collected and analyzed in time to make the action useful and to give confidence that any action undertaken will have biological benefits. This is feedback control.

Biological monitoring will quickly fall into disfavor if the action taken, based on the monitoring results, is inappropriate. The methodology for biological monitoring must be evaluated in terms of false positives and false negatives. A false positive is an indication that an excursion beyond previously established quality control conditions (i.e., unacceptable conditions) has occurred when, in fact, one has not. A false negative is an indication that conditions are acceptable when, in fact, they are not. Statistics must play a more important role in biological monitoring because they are capable of explicit statements of confidence in the biological monitoring results. With appropriate statistical evaluation of the data, professional judgment on whether to initiate immediate action or wait for more confirming data will be more objective and reliable.

In order to optimize the usefulness of biological monitoring, the selection of biological monitoring methodology cannot be based on the investigator's favorite organism or group of organisms. Neither can convenient methodologies espoused by regulatory agencies be a prime consideration. The selections must be based on the compatibility of data generated with the decision making process, including the statistical establishment of confidence in the result obtained.

## PURPOSES OF BIOLOGICAL MONITORING

Biological monitoring plans have several possible purposes:

1) To provide an early warning of a violation of quality control systems in time to avoid deleterious effects to ecosystems
2) To detect episodic events such as accidental spills, failure of predictive models, failure of early warning systems, or illegal disposal of wastes at night, etc.
3) To detect trends or cycles
4) To determine information redundancy
5) To evaluate environmental effects associated with the introduction of genetically engineered organisms into natural systems

These purposes and the statistical considerations entering into each type of application are discussed next.

## Monitoring to Provide Early Warning

The purpose of early warning systems is to alert decision makers to the impending entry of deleterious materials into natural ecosystems or impending harm to the biota before it occurs. As a consequence, these biological monitoring systems must be established at a point in the waste flow where management has several options when information suggests the presence of unacceptable concentrations of deleterious materials. Examples of possible actions include: (1) shunting the waste to a holding pond or other holding facility; (2) recycling the waste through the treatment system; and (3) shutting down production until the problem has been corrected. If the industry is part of a drainage basin network, there would be two additional management options: (1) augmenting flow from upstream reservoirs during the critical period in order to dilute the deleterious material sufficiently to avoid biological harm; and (2) reducing the amount of discharge from other sources. Since the ecosystem responds to the cumulative or aggregate effect of all the wastes in the system, other dischargers might hold back on their discharges during the emergency so that the net ecosystem effect remains stable. A cooperative effort would be justified since each industry or other discharger might well have similar emergencies at times and require similar assistance. A more extended discussion of these options may be found in Cairns (1975a,b). A compilation of the literature on the older early warning systems may be found in Cairns and van der Schalie (1980).

It is possible to use single species as early warning systems (early versions of this were the canary in the mine and the king's winetaster), but more recently developed complex multivariate systems, such as microcosms and mesocosms, have more environmental realism. A crucial characteristic of all early warning systems is the ability to predict effects in complex natural systems. If single species early warning systems are utilized (e.g., Cairns et al. 1970), the species chosen might well be either more sensitive or more tolerant to a particular chemical than the organisms in the receiving system. The assumption is that by choosing the most sensitive species available all other organisms in the ecosystems are protected. There are some serious flaws in this reasoning (Cairns 1986) because the relative tolerance of organisms to chemicals is not uniform. Mayer and Ellersieck (1986) have documented this point in considerable detail. Since any early warning system is intended to protect a more complex natural system with quite a variety of responses not measurable in the early warning system, it is important that the correspondence between the two types of responses be validated. This validation can be carried out in surrogates of natural systems such as microcosms or mesocosms. This particular point has been discussed in great detail by Cairns (1988). The purpose of the validation process is to establish the probability and frequency of false positives and false negatives. Validation is essential because false negatives could cause great harm to the ecosystem and false positives would result in expenditures in additional waste treatment that produce no additional ecological benefits. In our opinion, exclusive reliance on single species tests as early warning systems has not been well validated in ecological systems in this sense of identifying the frequency and seriousness of the consequences of false positives and false negatives. Some recent work by Kenneth Dickson and his colleagues (personal communication) suggests that the false positive and negative rates are not too high. This work compared association of impact using toxicity tests with assessment of impact using a variety of field measures. Validation in this study is an internal validation in which the similarity between the model and the real system is assessed. Also needed is an external validation that assesses how well the model will work in practice, i.e., predictive validation. Toxicity data are put into a mechanistic model and the output is a prediction of risk or probability of impact. The predictions from this model must be compared with reality.

Although mesocosms provide a test system with complexity and realism intermediate to conventional toxicity tests and the real world, they also have problems that must be resolved before they can be used as early warning systems. Along with greater complexity and realism, mesocosm tests typically have higher variability that can limit sensitivity. In addition, while the interlaboratory variability for toxicity tests is reasonable, it is higher for mesocosm studies. Clearly, work needs to be done in order to obtain standard methods for mesocosm studies that are reproducible among laboratories, and end points need to be chosen that exhibit better statistical properties.

Guidance for designing optimal early warning systems can be found in the statistical quality control literature. One approach is to view the system as stable, then some intervention causes a change in the system. The purpose of monitoring is to detect the change as quickly as possible. This is monitoring for compliance. Specifically, there is some pre-set standard that the end point is not to exceed. A variety of methods must be used in this context (Schaeffer et al. 1980, Loftis et al. 1987, Bailar 1987). A quality control approach is also used in compliance studies. The system is conceptualized as a process that is in "control" and a standard value is used to judge deviations from control. Graphical and testing methods can then be used to assess the system.

Statistical methods can also aid in designing the monitoring study (Price 1987). One approach is to use a fixed sampling interval and some measure of the stability of the system. For example, in a system using a fish species, the respiration of the fish might be an indication of overall quality. Based on baseline data, an interval or normal operating range (NOR) can be established for the end point. When the measure lies inside the interval, the system exhibits good quality. When it lies outside the interval, the problem requires attention or remediation. This approach is essentially a test that the next observation is different from the others so far collected. The evaluation of a single observation relative to the previous observations is common practice in statistical quality control and the data are analyzed using a Shewart chart, which displays limits and samples. A variation on this approach is to use a variable sampling interval (Arnold 1970). Some recent work (Reynolds et al. 1988) indicates that the approach can be improved by using two sets of intervals and two sampling rates. For example, when the fish respiration or other end point is close to the norm, sampling should be done infrequently. When the fish respiration is farther away from the norm, but is not great enough to signal trouble, sampling should be done more frequently. A flexible system, in which changes to what is sampled, as well as when it is sampled, can be made, may be the most cost-effective approach for detection of change. Several levels of testing at differing frequencies can be used in a comprehensive monitoring system. These techniques illustrate the importance of feedback control as a characteristic of effective biological monitoring systems. Analysis of previous data determines what is done next in an ongoing process. The information gathered is used to make decisions and is not just archived.

The quality control approach has been primarily used with chemical data rather than with biological data. Setting a standard (for example, drinking water standards have maximum contaminant levels) with chemical data is often easy. Ecological data are more difficult, and setting a standard may be impossible. Biological indices of environmental health will fluctuate naturally in time and space: over season, over succession, over substrate, over ecoregion (e.g., Hughes et al. 1994). While it is possible to allow a standard to fluctuate, the variance in the system may be quite large and the test for a violation may be weak. Thus far, simple measures that are accepted as related to overall system health, such as BOD or DO, have been used most frequently. These measures may be useful for signaling gross effects (such as fish kills) but may not be as useful as indicators of subtle deteriorations. It may be difficult to design a system of end points that are high in ecological complexity, yet also capable of providing early warnings. Appropriate choice of end points is essential but is subject to a great deal of controversy (see Suter 1989, Plafkin et al. 1989). In general, as the hierarchical complexity of the system increases and the physical scale of the end point increases, the ability to detect the change of interest will decrease. The time scale for fish community changes is considerably larger than that for diatoms (Patrick 1994). While diatoms may be useful for early warning systems, they may not be as good when interest is in long-term changes or changes in diet of fish or energy flows.

Two issues must be resolved in the selection of field end points of environmental health for early warning systems. The first consideration is the question of detection. A good system needs to be cost-effective and detect the change rapidly so that corrective action can be taken. Although field data are usually the most informative and relevant index of environmental effects, our experience suggests that ecosystem studies tend to have higher costs and variability than toxicity tests. However, Karr (1991) indicates that the Index of Biotic Integrity (IBI) can be cost competitive with toxicity testing. End points that are sensitive, inexpensive to measure, and have lower variance can be identified through continued evaluation and standardization. A second issue is whether the decision is correct, which is the problem of validity. We suspect that single species tests would signal quickly and efficiently but have high error rates when used as the sole indicators of ecosystem health. It is difficult to find a single measure of ecosystem health. Measures that combine several end points may be more accurate, but these indices tend to be subjective and no combination of variables enjoys wide support among investigators (Suter 1989, Plafkin et al. 1989).

In addition to monitoring for early warning of detrimental change, monitoring may also evaluate recovery. The quality control approach is also applicable to studies of recovery. In this application, the assumption is that the system is initially out of control. Monitoring follows a return to the predefined acceptable range. While the problem can still be formulated as a quality problem, the limitation is that the time required for testing recovery may be quite long.

## Monitoring to Detect Episodic Events

The Valdez oil spill in Alaska provides a good example of the value of biomonitoring data. One of us (Cairns) was asked to evaluate some of the evidence of the environmental consequences of the Valdez spill. However, no explicit and detailed data on the structure or function of the entire Prince William Sound ecosystem were sent. There are, of course, substantial data on commercially and recreationally valuable species and the like. However, despite repeated requests for biological monitoring information giving long-term trends and cycles of microorganisms, invertebrates, etc., and functional attributes such as energy flow and nutrient cycling, the evidence was not produced, presumably because it was not available. Having a pre-accident data base would have enabled a more definite determination of the degree of ecological disequilibrium caused by the oil spill and would then indicate what remedial measures would be most beneficial. Thus far, the "restoration" has been restricted to the cleanup of the oil itself rather than to true ecological restoration to predisturbance conditions.

Of course, the Exxon Valdez oil spill was a large-scale and well-documented event, with instant and widespread recognition. However, one of the primary purposes of ecosystem monitoring is to detect deleterious events that would otherwise go undetected. Most regulatory measures are focused on the materials entering the system rather than the condition of the system itself. For example, industrial waste discharges must fit certain specifications in concentrations of hazardous materials, volume, relationship to total flow, and the like. These specifications are based primarily on laboratory tests low in environmental realism. An assumption is made that if these conditions are met, the ecosystem will be unchanged in either structural or functional attributes. Unfortunately, this crucial assumption is rarely validated. In short, the accuracy of predictive models of ecosystem response, based primarily on laboratory evidence, has not been well established in natural systems. One purpose of ecosystem monitoring is to correct this deficiency and to provide an error control when predictions from laboratory tests are inaccurate or when cumulative or aggregate impact from a number of dischargers and other ecosystem stressors is greater than one would predict from the additive effects of the individual discharges into the ecosystem.

## Monitoring to Detect Trends or Cycles

Originally, biological monitoring was developed to detect effects from single point source discharges, or, perhaps, effects resulting from multiple discharges originating within one drainage basin. However, the studies on acid rain and, more recently, the prospects of global warming suggest that trends within an ecosystem may be driven by events geographically distant from the ecosystem itself. In the case of acid rain, this may be many hundreds of miles. Now it is evident that global effects, such as global warming, might cause considerable ecosystem change. If long-term monitoring is not carried out on a global scale, changes on a global scale might be confused with cumulative impacts within a more limited region. Methods of assessing change in regional and global monitoring studies are still being developed. Besides the question of sample size for these studies, concern should also focus on the scale (or size of sampling unit) needed to address the change and the correlation between units (Jeffers 1989). Scales that are too small may be too erratic, while a larger scale may not exhibit enough variation to detect the change (see for example, Carlile et al. 1989). Legendre et al. (1989) indicate that a two-year program to collect baseline data is needed to address the question of scale for the important processes in a bay. Larger scales require longer times to establish baselines. The problem of correlation is also important. Studies in space and time typically do not yield independent observations. It is important to know how far apart sampling units need to be in space and time in order to minimize the effects of the correlations.

## Monitoring to Determine Information Redundancy

Monitoring is costly and cannot be carried out in great detail at every level of biological organization—any population, community, ecosystem—globally. However, in the initial stages of development of a biological monitoring network, it is important to determine which types of data are most informative under each set of circumstances. Several benefits can be derived from this exercise. (1) The most informative and important end points can be selected from among an array of attributes. These attributes will have a high correspondence in response with other crucial attributes. (2) Attributes that are unique and have low redundancy with other attributes can be determined and their significance to the continuation of data gathering can be evaluated. (3) Consistency of trends from multiple lines of evidence can be determined in order to increase confidence in the implications of the changes being measured. The fact that certain types of information are redundant in a particular ecosystem does not mean that they will be so in every ecosystem. As a consequence, the types of information that are redundant must be determined and this correspondence must then be validated in a variety of ecosystems prior to their widespread use in biological monitoring. Out of this analysis should emerge the attributes of ecosystems that are worth measuring on a global basis because of their high correspondence, or even because they are unique representations of particular ecosystem response. Estimation of redundancy must be done carefully because the information that appears redundant for short-term monitoring may not be so for long-term monitoring to establish trends, cycles, and the like. Nevertheless, information redundancy analysis provides the surest way of reducing the volume of monitoring data generated, while at the same time ensuring that the quality remains high. Finally, uncertainty is reduced when there is high information correspondence from a diversity of methodologies and attributes. Therefore, even when redundancy is established, it may be well to continue gathering multiple lines of corresponding evidence, especially during the early stages of a monitoring program. An interesting statistical problem in this area is the problem of combining information from different sources. One approach is meta-analysis, in which a variety of tests (usually independent) are combined to provide a single test statistic for assessing an effect. This approach has been useful in combining medical studies, where sample sizes are not adequate, but it has not received much attention in ecological studies (however see Neuhold 1987 and Suter et al. 1983). A recent paper by Mann (1990) provides a good introduction and some examples of meta-analysis.

## Monitoring to Assess Effects of Genetically Engineered Organisms

All the forms of monitoring previously discussed are appropriate when genetically engineered organisms (GEMs) are introduced, either deliberately or accidentally, into a natural system. Any ecological disequilibrium these organisms might cause could presumably be determined in the ways previously described. Additionally, however, the persistence of these organisms should also be monitored, especially if they are microorganisms. Methodology for determining the presence of GEMs in extremely low numbers is available (e.g., Scanferlato et al. 1990). Determination of die-off curves in the natural systems should be compared to the estimate of die-off measured under laboratory conditions. These same techniques can be used to monitor the introduction of a naturally occurring species exotic to a particular ecosystem. For example, monitoring microorganisms in bioremediation of an oil spill can determine whether the microbial community returns to normal or whether the introduced organisms have displaced organisms naturally occurring in that particular ecosystem. The transport of GEMs or exotic organisms from the ecosystem where they were accidentally or deliberately introduced into ecologically different ecosystems adjacent to the area of introduction could also be monitored.

A statistical problem that occurs in the monitoring of GEMs is the issue of rareness. For the community to return to normal, is it required to have all the genetically engineered organisms below a certain density, or completely eliminated from the system? Claiming that no altered organisms are left is difficult using sampled data. The problem is one of sampling rare organisms, which is a rather difficult problem (see for example, Kalton and Anderson 1986). This problem is additionally complicated because genetic material may have been transferred so that the genome may persist even though the monitored organism does not.

## STATISTICAL MODELS AND THE DETECTION OF CHANGE

A variety of statistical techniques can guide the design and analysis of data from biological monitoring programs. Some specific considerations have been discussed above; however, many methods are applicable to biological monitoring programs for more than one purpose.

Episodic events are commonly associated with a type of effect called the step change or trend. The step change model occurs when data are collected before and after the start of the impact (or in some cases restoration) process. Thus, the data prior to the change are used to set the standard for the post change data. The step change is often thought of in terms of the hypothesis that the mean (or median or variance) of the measurements in the before period are the same as the mean (or variance) in the after period. Many statistical methods may be used to test for a step change (e.g., Green 1979, Gilbert 1987, Hirsch 1988).

The BACI (Before-After-Control-Impact) test (Stewart-Oaten et al. 1986) is a more sophisticated method that addresses some of the problems of comparing control and impact sites. Two or more sites are needed to use the BACI method, and measurements must be taken on each site, before and after the impact has started. Measurements are typically taken over time. Differences between the control and impact sites are calculated for each sampling time. The test to ensure that the event has not impacted the system is obtained by testing that the mean of the before differences is the same as the mean of the after differences. Nonparametric methods can also be used to carry out the test (see Stewart-Oaten et al. 1986 for more details).

Eberhardt (1976a, 1976b) and Thomas et al. (1978) have provided some of the best guidance and examples of statistical approaches for impact assessment studies (see also Eberhardt and Thomas 1991, for a general view). These papers describe a number of studies to assess the impact of power plants on ecosystems. More importantly, they describe a number of the problems that might be encountered in designing and implementing a biological impact assessment. For example, one of the difficult problems is the choice of adequate sample sizes to detect biologically important changes (Millard and Lettenmaier 1984). Another problem is the lack of adequate controls; for example, the upstream site may be primarily sand, while the downstream is primarily silt. Under these conditions, some statistical methods may lead to statistically significant results that are biologically irrelevant. Also, statistical models may necessarily be complex to account for all the sources of variation. This statistical complexity may lead to problems in the analysis and interpretation of the data.

When no control sites are available, but measurements are taken a number of times before and after restoration, changes can be evaluated (if there are a large enough number of times) using methods of time series analysis, specifically intervention analysis. Intervention analysis may be effectively used, for example, to assess the changes due to adding artificial aeration to a lake (Fast 1973, 1974), or in a monitoring context to signal the change in fish ventilation behavior (Thompson et al. 1982). Intervention analysis is useful for assessing changes in data collection over time and requires a fairly long series of observations. In intervention analysis, an intervention (for example, a spill, impact, or restoration change) is noted as occurring at some time. Data are available prior to the intervention and collected following the intervention. The data prior to the intervention are modeled using time series methods that allow for correlations between observations. The intervention is assumed to cause a change in this time series, such as a shift in the mean (step-change) or a trend. Assuming a time series model, a test for the intervention can be initiated. An example of the use of intervention analysis to test for changes in the chlorophyll *a* levels in sea water is given in Musters et al. (1988).

There are several difficulties in using an intervention model that may confound results. First is the length of the series, which needs to be long enough to characterize the time-series and estimate parameters. For hydrologic variables, such as flow, this may not be a problem, as they may be inexpensive to measure and a long series can be obtained. However, other variables, such as the abundance of a species, are more expensive to measure and cannot be measured as frequently. Second, it is preferable to have evenly spaced measurements. While unevenly spaced measurements will work, the approach is not as simple. Biological data are again problematic to work with because sampling is often (and should be) done in an irregular manner. For example, why sample aquatic macrophytes in the winter if they are not in the streams? Wetlands should be sampled more frequently during important hydrological events.

Another possible drawback to the use of the intervention model is the lack of power. As pointed out by Woodfield (1987) and Lettenmaier et al. (1978), the power of the intervention test for a step-change is typically controlled by the number of data points prior to the intervention. This number is fixed. Although the number of points after the intervention can be increased, the increase has very little effect on the power (once the number of points after is larger than the number in the before period). Unless the number of pre-intervention points is large, the power may be small. If, however, the effect of the intervention is a shift followed by a decay, then the emphasis should be placed on data collection in the post-intervention period. Thus, different types of interventions lead to different approaches to biomonitoring. This problem may be serious, especially when funds are limited and the expected change is unknown. Some other discussions on the design of time series studies are given by Smith (1984) and the references therein.

The literature on quality control provides some information that may be useful for the design of impact studies. Quality control methodologies stress the importance of understanding the process. There has been a recent revolution in quality control regarding the approach to understanding the process that we believe would benefit ecological assessment, especially from the point of view of restoration. The main contributions to the revolution in quality assurance are associated with Genichi Taguchi. As noted in Myers et al. (1992) and Kackar (1985), Taguchi suggests that a process could be viewed as being driven by two types of variables—noise variables and control variables. The noise variables would be environmental variables such as temperature, flow, etc. These variables could be controlled in a laboratory setting but not in most field settings. Other factors, such as substrate type, can be controlled or adjusted in field settings as well as in the laboratory. Taguchi noticed that as the process mean changed, so did the variability. Rather than trying to remove the heterogeneity of the variance (i.e., via transformation), Taguchi suggests incorporating the variability into the statistical model. Thus, the performance of the system depends not only on the mean of the system, but also the variance, and variance becomes a performance criteria. Among the control factors are those affecting the mean and those affecting variance. Taguchi suggests using the ones affecting the mean to move the system to target values and using the other factors to reduce the variability or make the system more robust. Thus, two points arise. First, the variance as well as the mean must be considered in the monitoring process (see also Green 1984, 1987). Second, knowledge of the relationships between the mean, variance, and the control factors can be more useful than relationships between just the mean and the control factors.

Another common approach to assessing change is to test for a trend. Frequently, this approach is used when there are no data prior to the restoration or if any prior data would be of no consequence. For example, in creating a wetland, prior data may not exist or information on certain attributes measured when the site was not a wetland may not provide usable information about the success of the wetland, only that it is different.

Trend assessment has been frequently reported in the environmental literature (Hirsch and Slack 1984, Hirsch et al. 1982). The methods frequently assume that the trend is linear, although nonlinear trends can also be used. Other models for analysis of change, such as analysis of covariance, can also be useful (Mathur et al. 1980). Trends can be incorporated into intervention analysis as described above.

Also related to trend assessment is the use of analysis of variance models to assess effects. This is perhaps the most common approach for assessing change. There are a variety of models that have been proposed, ranging from simple two-sample tests (parametric or nonparametric) to two-factor designs (Green 1979, Stewart-Oaten et al. 1986) to the more complex paired sampling approaches of Eberhardt (1976a, 1976b) and

Skalski and McKenzie (1982). An example of using this approach to design studies is given in Millard and Lettenmaier (1984).

Systems that are created or restored often undergo successional change. In some cases, the successional change may be toward a climax state. If the change toward climax is rapid, the climax state can be compared with stable control systems, using standard statistical approaches. In many cases, the system does not reach a climax and the evaluation process is more difficult. Loehle and Smith (1990) describe some methods that might be useful for evaluation of succession in this case.

The approaches discussed above imply a movement toward a goal. In many systems, the interest is more on obtaining a system that can persist through time. Natural wetlands are often in a state of flux. The state of the system depends strongly on external factors, primarily hydrologic. For such wetlands, stability is not the same as that for terrestrial systems but is a "pulsed stability" (Odum 1971). What is more important than reaching a climax state in a successional process is the ability for the system to respond to fluctuating environmental conditions and persist. Thus, the goal of interest is not to duplicate another system but to make the created wetland a persistent system (Carter 1986, Niering 1989).

There are two problems with applying statistical tests to the restoration problem. First is the philosophical problem that the tests used were designed for experimental situations and do not apply to data collected on recovery. This problem is referred to in the ecological literature as the pseudoreplication problem (Hurlbert 1984). In typical experimental design, a researcher has control over the way the treatments are allocated to the experimental units and the way that the experimental units are selected. In restoration problems, the researcher does not have this control. The treatment is the restoration process. There is typically one control site and one restoration or treatment site. One of Hurlbert's criticisms of this approach is that if more and more samples are taken at each site, then eventually differences will emerge because no two sites are identical. Use of multiple control sites is not feasible because of expense, and it is foolish to spend money on the controls when the recovering site is of much more interest. Replication when costs are high may cause reduction in frequency of sampling and a weakening of effort. Thus, the control site may be statistically different from the recovery site, but the difference may not be scientifically important. The limitations of the statistical method in this approach must be recognized.

As mentioned above, these methods primarily test for a change. The second problem with the testing approach is that the methods do not test for recovery. Testing for recovery requires a specific definition of recovery. As recovery is often dynamic and recovered systems are not fixed systems, it may be difficult to specify recovery in terms of the value of a single parameter or to define recovery simply. When this is the case, expert judgment is of great value. In much the same manner that a medical specialist reviews tests made on a subject and combines the results with experience, a restoration expert would view the results of the statistical analysis and interpret results based on the tests and the expert's interpretation of the system.

Besides the statistical issues discussed above, a number of other problems may arise in monitoring. Goals of monitoring must be clearly defined because any monitoring program has limitations as to the questions that can be answered. If the researchers have clear ideas as to why the monitoring program is needed, then a very powerful design of the monitoring program is possible. Powerful, as used here, means a minimum cost design is available that will answer the questions of interest with high reliability. Often the goals of the research change or the data collected are inadequate to answer the necessary questions. There may be little or no data on specific chemicals, important species, or important environmental processes for a region or time period during which important events are known to have occurred. However, difficulties in statistical analysis of existing data may persist. Environmental data are often complex and techniques may not be available to address the problems inherent in the data.

Changes in the monitoring program will often compromise the effort. The best intentions often run afoul and programs that are set up with adequate funding may in the future find the funding cut. The result is that the sampling plan changes, and, hence, the questions that can be answered change.

A related problem is one of compatibility of data. When data are collected over time, there are possible effects due to changing the method of collection or changing the researchers. For example, switching from collecting fish in nets to collecting fish using electrofishing can create considerable problems in assessing change. Changes in the detection levels for chemicals can have an effect on the inference about their impact. Improved accuracy of chemical measurements means that the variance of the measures is decreasing (which conflicts with the usual assumption of homogeneity of variance).

Multiple interventions are common in the real world, yet more difficult to assess. The simplest view for assessing change is that the change is caused by a single source, but many environmental effects are due to multiple sources. When the magnitude of the sources is small, it is not difficult to assign a single source as the cause, and these interventions can sometimes be dealt with using special programs. For example, if interest is in the effect of a manufacturing plant on the biota and a sewage treatment facility is also added in the vicinity of the plant, a design to separate the potential effects of the two potential impactors is needed.

Over the last 30 years, a number of environmental regulations have been issued. Environmental data on large scales should reflect the effects of the programs put into effect to meet the regulatory challenges. Current statistical methods do not accommodate these types of interventions. The biological aspects of multiple stresses are only recently receiving attention (Cairns and Niederlehner 1989). It thus becomes difficult to assess the effects of policies and regulations.

## CONCLUSION

The term "biological monitoring" has been widely used in this discussion to include almost any type of data gathered to assess the environmental impact of discharges. In our opinion, biological monitoring is limited to a continuing collection of data to establish whether explicitly stated quality control conditions are being met. If these conditions are not being met, there will be an immediate decision to take corrective action. Purposes of biological monitoring include providing early warnings of hazards, detecting spills, detecting environmental trends or cycles, determining the best and least redundant information for monitoring, and evaluating the environmental effects associated with the introduction of genetically engineered organisms into natural systems. One design will not serve each purpose, but, if the researchers have clearly defined goals for the monitoring program, powerful designs are possible.

# REFERENCES

Arnold, J. C. 1970. A Markovian sampling strategy applied to quality monitoring of streams. *Biometrics* 26:739-747.

Bailar, J. C., III. 1987. ASA/EPA Conference on Compliance Sampling, EPA-230-03-047. U. S. Environmental Protection Agency, Washington, D.C. 163 p.

Cairns, J., Jr. 1975a. Quality control systems. pp. 588-612. *In*: B. A. Whitten, ed., *River Ecology*. Blackwell Scientific Publications Ltd., London. 725 p.

Cairns, J., Jr. 1975b. Critical species, including man, within the biosphere. *Naturwissenschaften* 62(5):193-199.

Cairns, J., Jr. 1986. The myth of the most sensitive species. *BioScience* 36(10):670-672.

Cairns, J., Jr. 1988. Integrated resource management: The challenge for the next ten years. pp. 559-566. *In*: W. J. Adams, G. A. Chapman, and W. G. Landis, eds., *Aquatic Toxicology and Hazard Assessment: Tenth Volume*. American Society for Testing and Materials, Philadelphia, PA. 960 p.

Cairns, J., Jr. and B. R. Niederlehner. 1989. Adaptation and resistance of ecosystems to stress: A major knowledge gap in understanding anthropogenic perturbations. *Specul. Sci. Tech.* 12:23-40.

Cairns, J., Jr. and W. H. van der Schalie. 1980. Biological monitoring. Part I: Early warning systems. *Water Res.* 14:1179-1196.

Cairns, J., Jr., K. L. Dickson, R. E. Sparks, and W. T. Waller. 1970. A preliminary report on rapid biological information systems for water pollution control. *J. Water Poll. Control Fed.* 42(5):685-703.

Carlile, D. W., J. R. Skalski, J. E. Baker, J. M. Thomas, and V. I. Cullinan. 1989. Determination of ecological scale. *Landscape Ecol.* 2:203-213.

Carter, V. 1986. An overview of the hydrologic concerns related to wetlands in the United States. *Can. J. Bot.* 64:364-374.

Eberhardt, L. L. 1976a. Some quantitative issues in ecological impact evaluation. pp. 307-315. *In*: R. L. Sharma, J. D. Buffington, and J. T. McFadden, eds., *Proceedings of the Workshop on Biological Significance of Environmental Impacts*, NRC-CONF-002. U. S. Nuclear Regulatory Commission, Washington, D.C. 327 p.

Eberhardt, L. L. 1976b. Quantative ecology and impact assessment. *J. Environ. Manag.* 4:27-70.

Eberthardt, L. L. and J. M. Thomas. 1991. Designing environmental field studies. *Ecol. Monogr.* 61:185-199.

Fast, A. W. 1973. Effects of artificial aeration on primary production and zoobenthos of El Capitan Reservoir, California. *Water Res. Bull.* 9:607-623.

Fast, A. W. 1974. Restoration of eutrophic lakes by artifical hypolimnetic oxygenation. pp. 21-34. *In: Proceedings, Human Accelerated Eutrophication of Fresh-Water Lakes, December 1973.* Teatown Lake Reservation, Brooklyn Botanical Gardens, Ossining, NY.

Gilbert, R. O. 1987. *Statistical Methods for Environmental Pollution Monitoring.* van Nostrand, NY. 384 p.

Green, R. H. 1979. *Sampling Design and Statistical Methods for Environmental Biologists.* John Wiley and Sons, NY. 257 p.

Green, R. H. 1984. Statistical and nonstatistical considerations for environmental monitoring studies. *Environ. Mon. Assess.* 4:293-301.

Green, R. H. 1987. Statistical and mathematical aspects: Distinction between natural and induced variation. pp. 335-354. *In*: V. B. Bouk, G. C. Butler, A. C. Upton, D. V. Parker, and S. C. Asher, eds., *Methods for Assessing the Effects of Mixtures of Chemicals* (SCOPE). John Wiley and Sons, NY. 894 p.

Hirsch, R. M. 1988. Statistical methods and sampling design for estimating step trends in surface-water quality. *Water Res. Bull.* 24:493-503.

Hirsch, R. M. and J. R. Slack. 1984. A nonparametric trend test for seasonal data with serial dependence. *Water Res. Bull.* 20:727-732.

Hirsch, R. M., J. R. Slack, and R. A. Smith. 1982. Techniques for trend analysis for monthly water quality data. *Water Res. Bull.* 18:107-121.

Hughes, R. M., S. A. Heiskary, W. J. Matthews, and C. O. Yoder. 1994. Use of Ecoregions in Biological Monitoring. *In*: S. L. Loeb and A. Spacie, eds., *Biological Monitoring of Aquatic Systems.* Lewis Publishers, Boca Raton, FL.

Hurlbert, S. H. 1984. Pseudoreplication and the design of ecological field experiments. *Ecol. Monogr.* 54:187-211.

Jeffers, J. N. R. 1989. Statistical and Mathematical Approaches to Issues of Scale in Ecology. pp. 47-56. *In*: T. Rosswall, R. G. Woodmansee, and P. G. Risser, eds., *Scales and Gobal Change (SCOPE 35).* John Wiley and Sons, NY. 355 p.

Kackar, R. 1985. Off-line quality control, parameter design and the Taguchi methods. *J. Quality Tech.* 17:175-188.

Kalton, G. and D. W. Anderson. 1986. Sampling rare populations. *J. Royal Stat. Soc.* (Series A) 149:65-82.

Karr, J. R. 1991. Biological integrity: A long-neglected aspect of water resource management. *Ecol. Appl.* 1:66-85.

Legendre, P., M. Trousslier, V. Jarry, and M. Fortin. 1989. Design for simultaneous sampling of ecological variables: From concepts to numerical solutions. *Oikos* 55:30-42.

Lettenmaier, D. P., K. W. Hipel, and A. I. McLeod. 1978. Assessment of environmental impacts, part 2: Data collection. *Environ. Manage.* 6:537-554.

Loehle, C. and E. P. Smith. 1990. An assessment methodology for successional systems. II. Statistical tests and specific examples. *Environ. Manage.* 14:259-268.

Loftis, J. C., P. S. Porter, and G. Settenbre. 1987. Statistical analysis of industrial wastewater monitoring data. *J. Water Poll. Control Fed.* 59:145-151.

Mann, C. 1990. Meta-analysis in the breech. *Science* 249:476-480.

Mathur, D., T. W. Robbins, and E. J. Purdy, Jr. 1980. Assessment of thermal discharges on zooplankton in Conowingo Pond, Pennsylvania. *Can. J. Fish. Aquat. Sci.* 37:937-944.

Mayer, F. L., Jr. and M. R. Ellersieck. 1986. *Manual of Acute Toxicity: Interpretation and Data Base for 410 Chemicals and 66 Species of Freshwater Organisms.* Resource Publication 160, Fish and Wildlife Service, U. S. Department of the Interior, Washington, D.C. 506 p.

Millard, S. and D. P. Lettenmaier. 1984. Optimal design of biological sampling programs using analysis of variance. *Estuarine, Coastal Shelf Sci.* 22:637-656.

Musters, C. J. M., H. C. van Latesteijn, and W. J. ter Keurs. 1988. The effect of the dumping of waste-acid from the titanium-dioxide production on the cholorophyl-a concentration of North Sea water: A time *series analysis. Environ. Mon. Assess.* 10:181-203.

Myers, R. H., A. I. Khuri and G. Vining. 1992. Response surface alternatives to the Taguchi robust parameter design approach. *Am. Stat.* 46:131-139.

Neuhold, J. M. 1987. The relationship of life history attributes to toxicant tolerances in fishes. *Environ. Toxicol. Chem.* 6:709-716.

Niering, W. A. 1989. Vegetation dynamics in relation to wetland creation. *In*: J. A. Kusler, ed., *Wetland Creation and Restoration: The Status of the Science.* U.S. Environmental Protection Agency, Washington, D.C. 591 p.

Odum, E. P. 1971. *Fundamentals of Ecology.* W. B. Saunders Co., Philadelphia, PA.

Patrick, R. 1994. What are the Requirements for an Effective Biomonitor? *In*: S. L. Loeb and A. Spacie, ed., *Biological Monitoring of Aquatic Systems.* Lewis Publishers, Boca Raton, FL.

Plafkin, J. L., M. T. Barbour, K. D. Porter, S. K. Gross, and R. M. Hughes. 1989. *Rapid Bioassessment Protocols for Use in Streams and Rivers: Benthic Macroinvertebrates and Fish.* EPA 444/4-89-001. National Technical Information Service, Springfield, VA. 190 p.

Price, B. 1987. Quality control issues in testing compliance with a regulatory standard: Controlling statistical decision error rates. *In*: J. C. Bailar, III, ed., *ASA/EPA Conference on Compliance Sampling.* EPA-230-03-047. U. S. Environmental Protection Agency, Washington, D.C.

Reynolds, M. R., Jr., R. W. Amin, J. C. Arnold, and J. A. Nachlas. 1988. Xbar Charts with variable sampling intervals. *Technometrics* 30:181-192.

Scanferlato, V. S., G. H. Lacy, and J. Cairns, Jr. 1990. Persistence of genetically-engineered *Erwinia carotovora* in perturbed and unperturbed aquatic microcosms and effect on recovery of indigenous bacteria. *Microb. Ecol.* 20(1):11-20.

Schaeffer, D. J., K. G. Janardan, H. W. Kerster, and M. S. Shekar. 1980. Graphical effluent quality control for compliance monitoring: What is a violation? *Environ. Manage.* 4:241-245.

Skalski, J. R. and D. H. McKenzie. 1982. A design for aquatic monitoring programs. *J. Environ. Manage.* 14:237-251.

Smith, W. 1984. Design of efficient environmental surveys over time. pp. 90-97. *In*: S. M. Getz and M. D. London, eds., *Statistics in the Environmental Sciences*. American Society for Testing and Materials, Philadelphia, PA.

Stewart-Oaten, A., W. R. Murdoch, and K. R. Parker. 1986. Environmental impact assessment: "Pseudoreplication" in time? *Ecology* 67:929-940.

Suter, G. 1989. Ecological end points. pp. 2.1-2.26. *In*: W. Warren-Hicks, B. Parkhurst, and S. Baker, Jr., eds., *Ecological Assessment of Hazardous Waste Sites: A Field and Laboratory Reference*. EPA 600/3-89-013. National Technical Information Service, Springfield, VA. 282 p.

Suter, G. W., II, D. S. Vaughn and R. H. Gardner. 1983. Risk assessment by analysis of extrapolation error: A demonstration for effects of pollutants on fish. *Environ. Toxicol. and Chem.* 2:369-378.

Thomas, J. M., J. A. Mahaffey, K. L. Gore, and D. G. Watson. 1978. Statistical methods used to assess biological impact at nuclear power plants. *J. Environ. Manage.* 7:269-290.

Thompson, K. W., M. L. Deaton, R. V. Foutz, J. Cairns, Jr., and A. C. Hendricks. 1982. Application of time-series intervention analysis to fish ventilatory response data. *Can. J. Fish. Aquat. Sci.* 39:518-521.

Woodfield, T. J. 1987. Statistical power considerations in time series intervention analysis. pp. 102-106. *In*: *ASA Proceedings of the Statistical Computing Section*. American Statistical Association, Alexandria, VA. 360 p.

# CHAPTER 6

# Implementation of Large-Scale Stream Monitoring Efforts: Sampling Design and Data Analysis Issues

**Loveday L. Conquest**, Center for Quantitative Science; **Stephen C. Ralph**, Center for Streamside Studies; **Robert J. Naiman**, Center for Streamside Studies, University of Washington, Seattle

## INTRODUCTION

Throughout western North America there is a growing concern about the long-term ecological effects of broad scale land use impacts on fluvial aquatic systems (Naiman 1992). In the State of Washington, the focus of this concern is on industrial scale timber harvesting, which occurs on roughly 42 percent of the total land area and has been a primary land use activity for 135 years (Yates and Yates 1987). Forested areas that are subject to periodic timber harvesting activities include private, state, tribal, and federally owned lands, implying a variety of land management objectives. Resource managers and policymakers require reliable, quantifiable information about the status and trends of fish, wildlife and water quality, about their sensitivity to cumulative effects associated with harvesting activities, and about the potential of the ecological system to recover from watershed disturbances. Since these disturbances constitute large-scale perturbations, landscape-level information on aquatic communities, wildlife communities, and water quality is required to determine the effects of such disturbances.

Forest harvest practice within watersheds containing steep slopes, unstable soils, and abundant rainfall trigger disturbance events (such as hill-slope failures or debris dam-break floods) that dramatically affect streams (Benda et al. 1993). Other, less dramatic processes account for increased frequency of sediment inputs or changes to watershed hydrology. Repeat harvesting of riparian corridors over the past century has virtually eliminated the input of large woody debris into stream channels, resulting in diminished structural complexity of inchannel aquatic habitat (Master et al. 1988). Concomitant changes to the timing of snowmelt and rainfall runoff alter the magnitude and frequency of peak discharge events and otherwise change the hydrologic cycle in subtle, but significant ways (Naiman et al. 1992).

Management of fish, wildlife, and water quality resources in forested watersheds has traditionally been poor or nonexistent. Yet, there are strong state and federal legislative mandates, and public support, for ensuring that commercial harvesting does not adversely affect public resources (e.g., Federal Clean Water Act and Federal Endangered Species Act). Indian tribes holding treaty rights for fish harvesting, environmental groups, and state resource agencies demand that the integrity of the stream habitat supporting the resources not be adversely affected by forest management activities (e.g., harvesting, fertilization, road construction, and maintenance). However, there is a lack of reliable information, either qualitative or quantitative, on the status or trends of public natural resources found within these areas that are subject to land use impacts. The growing public perception, which is supported by a segment of the scientific community, is that disturbances associated with industrial scale forest management are posing significant risk to public resources (Halbert and Lee 1990). For its part, the timber industry is willing to apply its best management practices to provide mitigation for these effects, but the industry, along with state and federal resource regulators, needs reliable information on the effectivenss of those measures.

Monitoring is often required as a condition to permit a wide range of land-use activities, or to document the response of a resource to those activities. Increasing attention is being paid to the design and execution of intensive field monitoring in a forested landscape, due to natural spatial and temporal heterogeneity in landscape responses. Monitoring programs need to be well designed in order to (1) assess reliably the status of the resource of concern and (2) be of value in guiding future management decisions associated with formulating effective public policy.

The objectives of this paper are to outline key issues associated with monitoring to assess effects of large-scale disturbances to landscapes; to describe fundamental organizational approaches that have been used to address these issues; and to present an ongoing statewide monitoring project as a case study to assess effects of a particular type of large-scale perturbation, i.e., forest management practices. The specific program involved with this endeavor, the Washington State Ambient Monitoring Project, generalizes to any study conducting monitoring across a diverse landscape.

# A General Approach

A stream habitat monitoring program concerns the effects of land-use practices on stream habitat. The objective is to quantify the status and trends in physical instream fish habitat and channel condition in response to forest management practices. One program component is comprised of the response variables chosen for investigation; these are determined in part by the specific questions being asked. Once the set of response variables is chosen, the sampling scales are selected. A good sampling program that will withstand statistical scrutiny is designed before any data gathering takes place. Spatial and temporal sampling scales, in terms of the size of areal units and when measurements take place, are carefully considered. Both the nature of the response variables and the choice of sampling scales will determine, in part, the quantitative tools needed for analysis (see Figure 6.1).

## *Spatial and Temporal Considerations*

Implementing a stream monitoring project across a large land mass is complicated by the diverse combinations of climate, geology, vegetation, and land form. One is always faced with selecting representative sampling sites from a diverse landscape. Similarly, the statistical universe of inference that is reasonable from a particular study must be well-defined, both spatially and temporally. Admonitions about avoiding pseudoreplication in space and time (Hurlbert 1984, Stewart-Oaten and Murdoch 1986) must be heeded in order to ensure independent replicates. Also, defining the "sampling unit" and the "experimental unit", which are not necessarily the same thing (Steel and Torrie 1980, pp. 124-125), is an important consideration. In a landscape monitoring study, the experimental unit (the unit to which a "treatment" is applied) may be an entirely

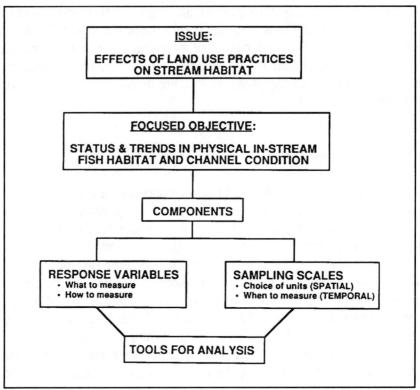

**Figure 6.1.** Process for implementation of a large scale habitat monitoring program.

different scale than the sampling unit. For example, timber harvesting activities may be applied to large portions of the landscape (the experimental units), but sampling may occur at the level of stream segments. It then becomes necessary to summarize information from various sampling units in an organized manner in order to make statements about processes at the larger landscape level.

Similarly, how to organize temporal units for sampling and for inference must be considered. One may record a given response variable instantaneously or combine information over different temporal periods. Different response variables lend themselves to different temporal units. Certain responses (e.g., flow) may change by the hour, while others (e.g., stream channel width) are much more stable. Some landscape processes may occur over years or decades. Whichever the time unit selected for sampling a particular response, the information must be organized and summarized so that conclusions may be drawn for longer periods of time, such as years or decades.

Consideration of scale in any landscape study is of fundamental importance. Definition of fundamental boundary characteristics is difficult, due to a lack of quantitative information, and is compounded by scale-dependent characteristics (Gosz 1991). Processes and parameters important at one scale may not be as important at another scale. This does not necessarily mean that it is impossible to characterize temporal or spatial changes in a landscape, but that there must be a keen awareness of both temporal and spatial scale when making inferences (Turner 1989). Frissell et al. (1986) offer examples of different scales of temporal and spatial resolution within a hierarchical classification system for coastal Oregon streams.

## Analytical Approaches to Landscape Classification

One way of organizing an inherently variable landscape is to employ a system of classification. The general intent of classification is to arrange units into meaningful groups in order to simplify sampling procedures and management strategies. When coupled with historic information about land use, disturbance events, and other key basic features (e.g., slope, soils, and geology), a classification-based monitoring program provides fundamental information for basin analysis of potential hazards. One may approach the concept of landscape classification and the creation of how to classify landscape units in several ways. Three familiar multivariate statistical techniques are principal components analysis, cluster analysis, and discriminant analysis. Principal components analysis is a technique used to shorten an otherwise long list of variables containing somewhat redundant information about landscape units. If most of the information in the data set can be expressed in terms of a few uncorrelated "factor scores" for each case (where the factor scores are linear combinations of the original longer list of variables), this becomes a statistically succinct way of handling information for subsequent analyses. For instance, Huang and Ferng (1990a) use principal component analysis to reduce an original list of 27 watershed management variables to a shorter list of 6 factor scores.

Cluster analysis (both hierarchical and non-hierarchical methods) has been used to create groupings of units. While the groups are initially formed by purely statistical rules (i.e., maximizing between group distance and minimizing without group distance), the final groups must make biological or physical sense to the researcher in order to be useful.

Researchers familiar with cluster analysis techniques know that the "final" cluster partitioning is not truly "final"—it may change according to the distance or similarity measure used, the number of clusters set, and the initial trial clustering in a non-hierarchical method (Afifi and Clark, 1984). Betters and Rubingh (1978) display maps showing how land type groups may change according to the desired level of homogeneity within each cluster. Cluster analysis is useful in assessing several criteria simultaneously and in coalescing multivariate information, but its use as an inferential tool must be tempered by common sense. For example, Howard and Howard (1981) claim to find no obvious discontinuities in the spatial distribution of units (100 km$^2$ map grid squares). Therefore, although the set of units may be partioned into statistical groups, the final partition chosen becomes somewhat arbitrary, reflecting optimization of certain statistical criteria.

Discriminant analysis has been used to confirm previous groupings of units. Sometimes the groups have been based, *a priori*, upon physical or biological definitions. Bailey (1984) uses discriminant analysis to demonstrate that two major ecoregions differ hydrologically, with monthly runoff as the response vector and hydrologic stations as the sampling units. Cupp (1989a) uses descriminant analysis to confirm a rule-based system defining nine valley segment types in the Gifford Pinchot National Forest, Washington. When the groups are not based upon prior definitions, but rather are based upon results from previous cluster analyses, discriminant analysis has more of a descriptive role than an inferential one. Since the clusters have been formed purposely to maximize between-group distance and minimize within-group distance, it is not appropriate to use F-tests and inferential procedures to "prove" that the groups are statistically different. However, it is useful to examine the makeup of the discriminant functions in order to identify which response variables account for most of the variation between groups, and in which order. Huang and Ferng (1990a) offer an example of this in summarizing characteristics for water quality management zones previously created by cluster analysis.

Appropriate choice of the landscape unit is another matter for consideration. Bailey (1984) uses hydrologic stations, a natural sampling unit where streamflow is the response variable. Other studies use map grid squares of arbitrarily chosen sizes. Problems may occur, of course, in defining the boundary of the squares, particularly when a natural landscape boundary runs through a square. Again, there is the problem of scale and its effect upon any results, as responses like "altitude range" or "stream channel density" may change according to the size of the grid used.

## Development of Classification Units

There are two major ways to develop classification units. The statistical methods discussed above depend upon having substantial site specific data that can be subjected to multivariate statistical analyses to derive patterns (grouping units with similar properties) and to suggest relationships. Whittier et al. (1988) term this the "*a posteriori*, inductive approach;" Huang and Ferng (1990a,b) refer to it as the "taxonomic approach." The results tend to reflect the uniqueness of that particular data set, requiring substantial data and not allowing extrapolation to other variables. Nevertheless, realistic regions can be produced from such an approach. Huang and Ferng (1990a) classify 89 watershed units into 5 watershed zones for land and water quality management purposes, using factor analysis to summarize response contributions to overall information, cluster analysis to form watershed zones, and finally, discriminant analysis to yield information on zone descriptions. Hence, if data are available and if the importance of certain response variables is understood, but the precise relationships between them is not readily apparent, such a data-intensive approach may prove helpful in revealing spatial patterns.

The second way to develop classification units may be thought of as a rule-based method, termed the "regionalization approach" by Haung and Ferng (1990a,b) and the "*a priori*, deductive approach" by Whittier et al. (1988). The deductive approach subdivides units hierarchically on the basis of identified variables and logical rules. While a site specific data set is not required, one must have enough information about the geographic area being classified to devise appropriate rules. Thus, one must search for patterns in land and climate characteristics that are known to be related to the ecosystem of interest (Warren 1979). For example, Omernik (1987) compiled maps of ecoregions of the U.S.; Omernick and Gallant (1986) compiled ecoregions of the Pacific Northwest. The rule-based approach is more commonly used in stream classification, partially because of the large investment of resources required to sample streams (e.g., for taxonomic data). Classification systems for streams in general have a long and complicated history and have used various spatial and temporal scales (see review by Naiman et al. 1991)

## A System of Hierarchical Landscape
## Classification for Monitoring Physical Habitat

A hierarchical landscape classification scheme offers a way to discriminate among features of the landscape at several scales of resolution. Such a system gives an organized view of spatial and temporal variation among and within stream systems. The hierarchical framework is useful because it provides for integration of data from diverse sources and at different levels of resolution, and it allows the scientist or manager to select the level of resolution most appropriate to the objectives. Also, it should provide a tool for comparing in-channel and stream habitat patterns among different basins or regions. Recent development of stream classification schemes has emphasized geomorphic templates and the relationship of a stream to its watershed across a wide range of scales (Lotspeich and Platts 1982, Rosgen 1985, Frissel et al. 1986, Naiman et al. 1991).

In the example of a current broad-scale monitoring effort discussed below, the focus is on monitoring the spatial and temporal characteristics of the aquatic physical habit rather than the instream biota. In Washington State, the primary instream biotic resources of concern are anadromous salmonid fishes. The seven species of salmon (*Onchorynchus*) and anadromous trout (*Salmo*) constitute a renewable resource that is integral to both the economy and identity of the Northwest. Once extremely abundant, the numbers of some species are now critically low, with some convincing evidence for their consideration for threatened or endangered listing under the Endangered Species Act (Nehlsen et al. 1991).

Physical habitats within rivers and streams are a primary base for production of anadromous salmon and trout. Adults return to spawn in their natal streams after having spent from two to five years feeding in the northern Pacific ocean. The early life stages of these fishes (from egg through smolt) residing in the stream environment depend upon specific habitat features to sustain them, and are particularly susceptible to conditions that diminish the suitability of these components (Bisson et al. 1981, Naiman et al. 1992). Disturbance events within the watershed that change the characteristics of the instream rearing habitat can render entire reaches unsuitable to support both anadromous and resident species for several years. Because it is difficult to reliably assess the relationship between essential instream habitat and the eventual survival to adulthood of anadromous salmonids, monitoring the physical attributes of habitats that support aquatic organisms is a fundamental first step in evaluating the link between the effects of timber harvesting and anadromous fishes. Further integration with biotic community assessment techniques are in progress (Beechie and Sibley 1990).

The stream valley classification system developed by Cupp (1989b) for drainage basins in forested lands of Washington is a *spatially nested hierarchy*, which includes:

1) Ecoregions,
2) Watersheds,
3) Basins,
4) Valley segments,
5) Habitat complexes, and
6) Habitat units.

Ecoregions are as definied by Omernik and Gallant (1986). For watersheds and their component sub-basins, there is no clear way to organize them except by stream order. Watershed characteristics suitable for classification can be highly variable between basins; timber and fisheries' management plans are often founded on basins and their associated stream systems. In our case, valley segments constitute the functional sampling unit (see Figure 6.2, illustrating its place within the hierarchy) for purposes of siting the actual riparian forest and fish habitat inventory procedures (Sullivan et al. 1987). Eighteen valley segment types (within five broad categories) have been identified to classify stream corridors in forested lands of Washington (Figure 6.3, after Cupp 1989b). The segment types have been chosen so that their classification characteristics do not change over long time periods (i.e., tens to hundreds of years). Attributes defining a segment are expected to remain temporally stable. Measured habitat responses may change, however, according to disturbance types. A smaller spatial scale might improve spatial resolution, but at the expense of temporal stability. Furthermore, maps at a scale of, for example, 1:2000 are not available for large areas of the landscape such as watersheds. Valley segments are distinguished by six diagnostic

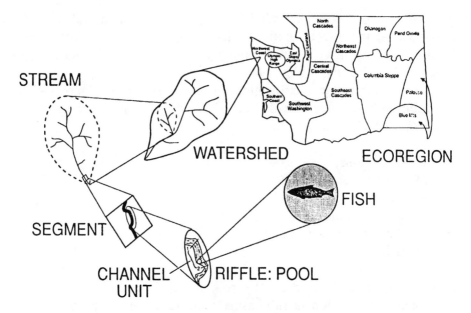

**Figure 6.2.** Spatially nested hierarchy for stream systems in Washington State by ecoregion, watershed, stream basin, valley segment, habitat unit, and biota.

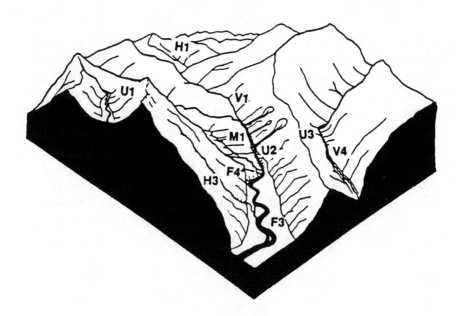

**Figure 6.3.** Oblique view of hypothetical watershed showing component valley segment types (after Cupp 1989b).

criteria, based on valley bottom and sideslope geomorphologic characteristics:

1) Position in the drainage network (i.e., stream order)
2) Valley sideslope gradient
3) Ratio of valley bottom width to active channel width
4) Channel gradient
5) Stream corridor geomorphic surface deposits
6) Channel pattern

This valley segment classification system uses primarily USGS topographic maps (1" = 24,000') and aerial photographs (1" = 24,000') to delineate initial apparent segment breaks. Final classifications must be confirmed by visual field inspection. Valley segment classification offers a potentially useful tool for assessing effects of various land management activities. With refined understanding of these relationships, they are then used to define management prescriptions to avoid or mitigate the hazards and risks to public resources. Differences in geomorphology may partially explain differing channel responses to effects of timber management and related activities on stream communities by linking stream geomorphology to physical habitat characteristics. The valley segment classification system provides a means to subsample from the available forested streams, and, because of the similarity in habitat character, to extrapolate results of habitat monitoring to areas with similar valley segment configuration. It offers a practical way of transferring information gained at sampled sites to the broader landscape, with a specified degree of reliability. When evaluating large scale environmental patterns, Green (1979) advocates breaking up larger areas into homogeneous subareas and allocating samples to each in proportion to the size of the subarea. The problems include that of sampling adequate numbers of segments to represent their proportional distribution across the landscape, and of adequately characterizing natural variation within a given segment class.

# Implementing a Statewide
# Monitoring Project: A Case Study

The Washington State Ambient Monitoring Project offers a case study where a system for landscape classification has been implemented and where one may address the issues discussed above with respect to large monitoring efforts. The Ambient Monitoring Project is designed to asssess status and trends (over time and space) of physical salmon and trout habitat within forested streams. The first component of the project includes identifying measurable characteristics of the physical habitat and developing a set of standardized methods to take the measurements. The second component involves stratifying the areas of the state by a system of hierarchical landscape classification based on combinations of features at differing levels of resolution. The physical habitat and channel characteristics of streams are related to the geomorphic setting and to the frequency and magnitude of disturbance events within the watershed.

There are a number of practical considerations in implementing such a monitoring program across a diverse landscape. Some, such as differing combinations of geology, climate, weather, and land use, can be dealt with through the use of a hierarchical classification system. The more fundamental aspects of implementation include successful completion of the field work and compilation of the resulting data in a fashion that allows ease of analysis. These aspects may be further broken down as follows:

1) **Variables**. One must choose appropriate response variables to measure; these depend, in part, upon the specific problem being addressed and how the effect is manifested at the hierarchical level of interest (e.g., habitat unit level). For example, amounts of woody debris serve to indicate the structural complexity of the inchannel aquatic habitat; estimation of coho salmon (*Onchorynchus kisutch*) spawning gravel area is a clear indicator of habitat suitability for reproduction.

2) **Methods**. Methods to measure the chosen variables may be taken from the literature (e.g., Platts et al. 1987, Bisson et al. 1981, Lisle 1987) and should be tailored to the site conditions and to the realities of time and manpower available. These issues are largely ones of resolution—matching the desired level of precision for the

measured variable to the available manpower, time, and site conditions. For instance, the suitability of instream fisheries' habitat for meeting the spawning requirements of salmon is, in large part, a function of the size of particles comprising the bed material found in riffles. Determining the status of this important instream attribute for streams within a basin could at first seem insurmountable. One technique, visual estimation of the apparent surface composition of bed material, is rapid, but its accuracy is low due to observer bias. Traditional sampling of bed material has been done through the use of coring devices, which give precise information on particle size distribution. These techniques are time consuming, involve the use of heavy equipment, and are difficult to apply when the scope of sampling is broad. A more suitable technique for sampling surface distribution of spawning gravels is the pebble count method (Wolman 1954), which gives accurate measurements of particles distributed throughout the channel. No equipment other than a ruler or caliper is required.

3) **Training**. Because of the extent and political context of our case study, there has been considerable diversity in the expertise of staff involved in field sampling. Limited funding for field staff has necessitated hiring temporary staff only. Generally, most of the field staff comprising the two-person field crews have been hired for the short (four month) field season and have little or no experience in evaluating streams and their component features.

In this situation it is essential to provide a field manual that describes in detail the procedures and the rationale for employing various methods. Without the manual the field staff may not understand how completion of a particular method contributes to the understanding of the whole program. Standardization of field methods is essential to ensuring reliable data, and no amount of detail should be overlooked when describing the elements that make up a particular method. Since the field observations and measurements employed are exclusively physical, it is important for everyone to use the same procedures to obtain desired information. They should be explicitly described to minimize subjectivity in application. Additionally, three-day field training sessions are conducted annually to provide initial exposure to the methods and their use under typical field conditions. This training has become an annual necessity since field staff turnover is nearly

100%. (One solution to this problem is for state agencies and tribes to cooperatively employ good field crews on a long-term basis in order to improve accuracy and precision in recording data from year to year.)

4) **Quality control**. In addition to the training sessions, follow-up in the field with individual crews is necessary to ensure that the methods are understood and consistently applied. Since the objective is to track the status and trends of instream habitat over time and space, confidence that our field methods are yielding reliable results is necessary. For example, changes detected during a second survey of a stream segment should be reliable measures of the change, not merely artifacts of observer bias or other errors attributable to field observations. Our quality assurance plan has looked at two sources of error: the consistency with which dimensions of habitat units can be visually estimated, and whether the spatial array of habitat units noted by one team correspond to those of another team, measuring the same stream segment at the same flow level. To our knowledge, few monitoring programs of this scale have been subjected to a critical examination of these quality control issues (see Bott et al. 1978).

5) **Data management**. One of the least conceptually interesting, but most challenging, aspects of mounting an extensive monitoring program is that of managing the large volume of data generated from the field work (Michener 1986). Field forms should be designed to accommodate rapid transcription into a database management program for later analysis. The database management system itself should be structured to allow ease of data formatting and production of various data summaries, while allowing for distribution to others needing information in a condensed form.

a) *Design of the field data form* – It was difficult to transcribe the open-ended field data efficiently from the field form to a computer based format because of the large number of people involved in field measurements and the volume of data. Field crews are surprisingly creative in their responses to data entry requirements; inconsistency of response was a problem when interpreting and transcribing data.

Optically scanned forms printed on "write-in-the rain" paper were designed for the second field season. These forms have several notable advantages over the traditional field forms. The data columns within which responses are entered can be more explicitly defined to minimize subjectivity when recording information. Data sheets are scanned on both sides simultaneously, using a Scantron 8200 optical scanner, and the resulting data is added to a unique computer field for that stream segment. Once the file is generated, an error checking program is applied to pinpoint data, recording errors and/or inconsistencies. Several hundred data sheets can be processed and corrected in less than an hour. Data are now compiled in a conventional PC-based language for ease of reporting and analysis.

b) *Data analysis* - The objectives of any monitoring project should clearly indicate the path for beginning the data analysis. However, when initiation of a monitoring program is in part driven by political expediency, these objectives and their attendant hypotheses may not be clearly articulated. Natural resource management agencies may not be able to describe their objectives in terms of testable hypotheses. It is first necessary to determine how the data derived from monitoring will be organized and manipulated. In our case example, despite the involvement of five state agencies, three state universities, 28 Indian tribes, and several dozen timber companies, there was no clear direction on the nature of the database management system to be used. Thus, it was decided to design a system to meet both short-term and anticipated long-term needs for flexibility in data manipulation, distribution, and reporting. We chose a PC-based application development software system using the widely used D-base language.

## Preliminary Results

The 1989 field season of the Washington State Ambient Monitoring Program yielded physical habitat information on 95 valley segments. This first field season may be considered as a pilot study to improve future sampling designs and to clarify objectives. For example, more sampling will be required in order to obtain a better representation of the distribution of the valley segment types found within watersheds across eco-regions of Washington State (Figure 6.4). Similarly, if it is desired to have certain valley segment types well represented in an experiment involving actual treatments (e.g., various methods of timber harvest), more sampling from the under-represented valley segment types will be needed.

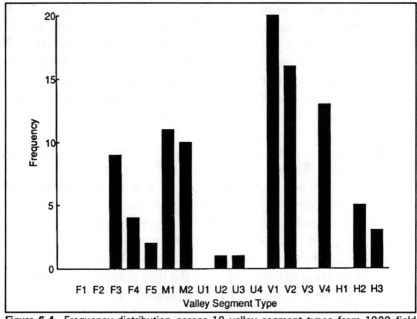

**Figure 6.4.** Frequency distribution across 18 valley segment types from 1989 field season, Washington State Ambient Monitoring Program.

Land-use information provided on the valley segments revealed that, of the 95 segments, 82 are at the extremes (0% or 100%) regarding timber harvest. When the segments are broken down by type, results are even more heavily weighted towards extremes. For example, among the sixteen V2 segments, twelve of them are 100% devoted to timber harvest (although the time since the last harvesting is not recorded), with the other four devoted 100% to riparian management zones. It would require more valley segments displaying a gradient of timber harvest use in order to investigate a statistical association between amount of disturbance and percentage of timber harvest use. Even if one wanted to establish the four riparian management zone (RMZ) segments (where logging has occurred adjacent to the stream, but trees have been spared to meet legal RMZ requirements) as "control" segments with respect to timber production, one would need to ensure that they fit the rules of appropriate regional reference sites outlined in Hughes et al. (1986). Hughes et al. (1986) suggest that regional reference sites be chosen according to low levels of human disturbance, stream size similar to that of affected areas, and stream channels typical of the region. Refuges (e.g., wilderness areas, wildlife refuges) should be considered as potential reference sites.

## System Testing

Any classification system should be tested before extensive field application occurs. Whittier et al. (1988) tested Omernik and Gallant's (1986) ecoregion classification system on a variety of ecosystem response variables in Oregon. They established a correspondence between eco-regions and spatial patterns in stream ecosystems and concluded that streams within an ecoregion were generally more homogeneous than streams between ecoregions. Beechie and Sibley (1990) tested certain aspects of the Cupp (1989b) classification system of valley segments, with both encouraging results and recommendations for system refinements. By pairing two of the segment types (F3 and M1) in five separate streams, they performed statistical tests on the percentage of summer pool space and concluded that F3 segments have a higher percentage of summer pool space than M1 segments. They also performed nonparametric testing on unpaired valley segment types, with regard to percentage of pool space and spawning gravel percentage, showing significant differences between means (parametric) or medians (nonparametric) for the two valley segment types. In general, they found that the segment types they evaluated (F3, M1, M2, H) distinguished the physical habitat characteristics of small streams with moderate success.

Univariate analyses of physical habitat variables (percentage of pools, spawning gravel percentages) on valley segment types show significant difference among means but some overlap of individual values (Beechie and Sibley 1990). Multivariate analyses (cluster analysis, discriminant analysis) identify valley segment groups more clearly. This is in keeping with properties of multivariate techniques for which separation is often much clearer when viewed in more than one dimension; when groups are projected onto individual variable axes, more overlap exists (Afifi and Clark 1984, p. 253).

## Future Sampling Considerations

More hypotheses may be formulated to focus sampling designs for future field seasons. Examples of specific hypotheses include:

$H_0$    Distributions of channel units are not significantly different for given valley segment types exposed to two distinct levels of management history (i.e., distinct management histories do not influence the distributions of channel units in a given segment type).

$H_0$    Spawning gravel percentages in a specific segment type are unaffected by land use (i.e., land use has no influence on this habitat in given segment types).

$H_0$    Distribution of large organic debris is independent of management history for given segment types.

Variability in channel unit distributions may be associated with stream discharge. Sampling recommendations include random sequences of sampling various segment types within a given flow regime (balanced against the actual costs of field crews to accomplish this). Care should be taken that individual segment types are not sampled under drastically different flow regimes; otherwise, variation due to segment type may be confounded with variation in flows.

# Further Issues Associated
# with a Large Scale Monitoring Program

In the context of timber harvesting, there is virtually no control of "treatments" (that occur both spatially and temporally across the sampling units) applied to watersheds or streams. Regional reference sites (as in Hughes et al. 1986; Hughes et al. 1994) that would allow sampling from discrete basins in their "steady state" are in short supply due to historic and current patterns of land uses. Since an upstream-downstream or before-and-after-impact approach is unlikely, streams with similar watershed features to the disturbed streams being monitored are required.

It is difficult to distinguish among natural inherent variability, variability associated with past disturbance events (e.g., fires, hydrologic, or climatic events), or variation due to events triggered by anthropogenic activities because there is so much unquantified variability across the landscape (Naiman et al. 1992). Apportioning these "observed changes" accordingly will be difficult, especially without reference sites providing some benchmark to judge that a change has occurred. We will analyze sites by accounting for their apparent nature and evidence of past disturbance events that have shaped their character.

For our case study, we have chosen a suite of response variables reflecting changes associated with inputs of sediment and hydrologic discharge. The inherent variation of some of these responses is unknown, although Platts et al. (1987), attempted to quantify this variation for some common aquatic responses. More understanding of this background variation for the responses chosen for monitoring should be revealed as more valley segments are sampled.

Although there is still great uncertainty at this stage of the program, real-world policy decisions must and are being made. However, additional research, coupled with broad-based educational training programs, will reduce that uncertainty in future years. Ultimately, there must be an ongoing, statewide program providing continuing information on Washington rivers, similar in nature to the New Zealand 100 Rivers Project (Biggs et al. 1990), which developed a national perspective on New Zealand rivers using consistent methodology. The ultimate goal is to provide managers with models for predicting the effects of forest practices on instream habitats and biota.

## ACKNOWLEDGMENTS

The authors wish to acknowledge the financial and conceptual support provided by the Ambient Monitoring Steering Committee of the Timber/-Fish/Wildlife Agreement, and the efforts and support of the Northwest Indian Fisheries Commission for providing field crews for the 1989 sampling season.

## REFERENCES

Afifi, A. A. and V. Clark. 1984. *Computer-Aided Multivariate Analysis.* Lifetime Learning Publications, Belmont, CA, 458 p.

Bailey, R. G. 1984. Testing an ecosystem regionalization. *Journal of Environmental Management* 19:239-248.

Beechie, T. J. and T. H. Sibley. 1990. *Evaluation of the TFW Stream Classification System: Stratification of Physical Habitat Area and Distribution.* Report for State of Washington Dept. of Natural Resources, EL-03, Olympia, WA. 85 p.

Benda, L. E., T. J. Beechie, R. C. Wissmar, A. Johnson. 1992. Morphology and evolution of salmonid habitats in a recently deglaciated river basin, Washington State. *Canadian Journal of Fisheries and Aquatic Sciences.* 49:1246-1256.

Betters, D. R. and J. L. Rubingh. 1978. Suitability analysis and wildland classification: an approach. *Journal of Environmental Management* 7:59-72.

Biggs, B. J., M. J. Duncan, I. G. Jowett, J. M. Quinn, C. W. Hickey, R. J. Davies-Colley, and M. E. Close. 1990. Ecological characterization, classification, and modeling of New Zealand rivers: an introduction and synthesis. *New Zealand Journal of Marine and Freshwater Research* 24:277-304.

Bisson, P. A., J. L. Nielson, R. A. Palmason, and L. E. Gore. 1981. A system of mapping habitat types in small streams, with examples of habitat utilization by salmonids during low stream flow. pp. 62-73. *In*: N. B. Armantrout, ed., *Acquisition and Utilization of Aquatic Habitat Information.* Western Div., American Fisheries Society. 376 p.

Bott, T. L., J. T. Brock, C. E. Cushing, S. V. Gregory, D. King, and R. C. Petersen. 1978. A comparison of methods for measuring primary productivity and community respiration in streams. *Hydrobiologica* 60:3-12.

Briggs, D. J. and J. France. 1983. Classifying landscapes and habitats for regional environmental planning. *Journal of Environmental Management* 17:249-261.

Bunce, R. G. H., S. K. Morrell, and H. E. Stel. 1975. The application of multivariate analysis to regional survey. *Journal of Environmental Management* 3:151-165.

Cupp, C. E. 1989a. *Identifying Spatial Variability of Stream Characteristics Through Classification.* M.S. Thesis, School of Fisheries, University of Washington, Seattle, WA. 92 p.

Cupp, C. E. 1989b. *Stream Corridor Classification for Forested Lands of Washington.* Report for Washington Forest Protection Association, Olympia, WA. 46 p.

Frissell, C. A., W. J. Liss, C. E. Warren, and M. D. Hurley. 1986. A hierarchical framework for stream habitat classification: viewing streams in a watershed content. *Environmental Management* 10:199-214.

Gosz, J. R. 1991. Fundamental ecological characteristics of landscape boundaries. pp. 8-30. *In*: M. Holland, R. J. Naiman, and P. G. Risser, eds., *Boundaries in a Changing Environment.* U.S. State Department. 142 p.

Green, R. H. 1979. *Sampling Design and Statistical Methods for Environmental Biologists.* John Wiley and Sons, NY. 257 p.

Halbert, C. L. and K. N. Lee. 1990. The Timber, Fish, and Wildlife Agreement: implementing alternative dispute resolution in Washington State. *The Northwest Environmental Journal* 6:139-175.

Howard, P. J. A. and D. M. Howard. 1981. Multivariate analysis of map data: a case study in classification and dissection. *Journal of Environmental Management* 13:23-40.

Huang, S. L. and J. J. Ferng. 1990a. Applied land classification for surface water quality management: I. Watershed classification. *Journal of Environmental Management* 31:107-126.

Huang, S. L. and J. J. Ferng. 1990b. Applied land classification for surface water quality management: II. Land process classification. *Journal of Environmental Management* 31:127-141.

Hughes, R. M., D. P. Larsen, and J. M. Omernik. 1986. Regional reference sites: A method for assessing stream potentials. *Environmental Management* 10:629-635.

Hughes, R. M., S. A. Heiskary, W. J. Matthews, and C. O. Yoder. 1994. Use of Ecoregions in Biological Monitoring. *In*: S. L. Loeb and A. Spacie, eds., *Biological Monitoring of Aquatic Systems*. Lewis Publishers, Boca Raton, FL.

Hurlbert, S. H. 1984. Pseudoreplication and the design of field experiments. *Ecological Monographs* 54:187-211.

Jasby, A. D. and T. M. Powell. 1990. Detecting changes in ecological time series. *Ecology* 71:2044-2052.

Lisle, T. E. 1987. *Using "Residual Depths" to Monitor Pool Depths Independently at Discharge*. USFS Pacific Southwest Forest and Range Experiment Station, Berkeley, CA. Research Note PSW-394. 18 p.

Lotspeich, F. and W. Platts. 1982. An integrated land-aquatic classification system. *American Journal of Fisheries Management* 2:138-149.

Master, C., R. F. Tarrant, J. M. Trappe, and J. F. Franklin, eds. 1988. *From the Forest to the Sea: A Story of Fallen Trees*. Technical Report PNW-GTR-229, Pacific Northwest Research Station, U.S. Dept. of Agriculture, Forest Service, Portland, OR. 153 p.

Michener, W. K., ed. 1986. *Research Data Management in the Ecological Sciences*. University of South Carolina Press, Columbia. 338 p.

Naiman, R. J. ed. 1992. *Watershed Management: Balancing Sustainability and Environmental Change*. Springer-Verlag, NY. 542 p.

Naiman, R. J., D. G. Lonzarich, T. J. Beechie, and S. C. Ralph. 1991. *Stream Classification and the Assessment of Conservation Potential*. Proceedings, Conference on the Conservation and Management of Rivers, York, England. 53 p.

Naiman, R. J., T. J. Beechie, L. E. Benda, D. R. Berg, P. A. Bisson, L. H. MacDonald, M. D. O'Connor, P. L. Olson, and E. A. Steel. 1992. Fundamental elements of ecologically healthy watersheds in the Pacific Northwest coastal ecoregion. *In*: R. J. Naiman ed., *Watershed Management: Balancing Sustainability and Environmental Change*. Springer-Verlag, NY. 542 p.

Nehlsen, W., J. E. Williams, and J. A. Lichatowich. 1991. Pacific salmon at the crossroads: stocks at risk from California, Oregon, Idaho, and Washington. *Fisheries*. 16(2):4-21.

Omernik, J. M. 1987. Ecoregions of the coterminous United States. *Annals of the Association of American Geographers* 77:118-125.

Omernik, J. M. and A. L. Gallant. 1986. *Ecoregions of the Pacific Northwest* (with map). EPA/600/3-96/033. U.S. Environmental Protection Agency, Corvallis, OR. 39 p.

Platts, W. S., C. Armour, G. D. Booth, M. Bryant, J. L. Bufford, P. Cuplin, S. Jensen, G. W. Lienkaemper, G. W. Minshall, S. B. Monsen, R. L. Nelson, J. R. Sedell, and J. S. Tuhy. 1987. *Methods for Evaluating Riparian Habitats with Applications to Management*. General Technical Report INT-221, Ogden, UT: U.S. Department of Agriculture, Forest Service, Intermountain Research Station. 177 p.

Rosgen, D. L. 1985. A stream classification. pp. 91-95. *In*: R. R. Johnson et al., eds., *Riparian Ecosystems and Their Management: Reconciling Conflicting Uses*. First North American Riparian Conference, Tucson, AZ. 218 p.

Steel, R. G. and J. H. Torrie. 1980. *Principles and Procedures of Statistics: A Biometrical Approach*. McGraw-Hill Book Company, NY. 633 p.

Stewart-Oaten, A. and W. Murdoch. 1986. Environmental impact assessment: "pseudoreplication" in time? *Ecology* 67:929-940.

Sullivan, K., T. Lisle, C. Dollof, G. Grant, and L. Reid. 1987. Stream channels: the link between forest and fisheries. pp. 39-97. *In*: E. Salo, and T. Cundy, eds., *Streamside Management: Forestry and Fisheries Interactions*. University of Washington Press, Seattle. 471 p.

Turner, M. G. 1989. Landscape ecology: the effect of pattern on process. *Ann. Rev. Ecol. Syst.* 20:171-197.

Warren, C. E. 1979. *Toward Classification and Rationale for Watershed Management and Stream Protection*. U.S. EPA Ecological Research Series EPA-600/3-79-059, 143 p.

Whittier, T. R., R. M. Hughes, and D. P. Larsen. 1988. Correspondence between ecoregions and spatial patterns in stream ecosystems in Oregon. *Can. J. Fish. Aquat. Sci.* 45:1264-1278.

Wolman, G. M. 1954. A method of sampling coarse riverbed material. *Transactions, American Geophysical Union* 35:951-956.

Yates, R. and C. Yates. 1987. *Washington State Year Book*. Information Press, Eugene, OR. 240 p.

# CHAPTER 7

# Spatial and Temporal Variation in Biomonitoring Data

**Arthur J. Stewart** and **James M. Loar**, Environmental Sciences Division, Oak Ridge National Laboratory, Oak Ridge, TN

## INTRODUCTION

Biological monitoring involves the use of biota to assess environmental conditions. Other papers in this book and in the ecological literature show that spatial or temporal variation in the abundances and activities of freshwater organisms and ecological processes is substantial. They also help identify sources of spatial and temporal variance and show that these sources of variation differ, depending on scale. Our objective in this paper is to assess variability in biological monitoring data acquired from an extensive biological monitoring program at U.S. Department of Energy (DOE) facilities in Oak Ridge, Tennessee.

In 1984, the U.S. Environmental Protection Agency (EPA) implemented a new national policy of water quality-based permit limitations for toxic pollutants designed to improve water quality through use of an "integrated strategy consisting of both conventional and biological methods to address toxic and nonconventional pollutants." This decision affected the three DOE facilities in Tennessee—the Y-12 Plant, the Oak Ridge National Laboratory (ORNL), and the Oak Ridge K-25 Site. Historically, these three facilities were built to fabricate nuclear weapons components (Y-12 Plant), conduct research needed to support the development and use of nuclear materials (ORNL), and isotopically enrich uranium for the purposes of national defense (K-25 Site). When the facilities were constructed, issues of national security and production were paramount. Siting of the facilities involved considerations of (1) remoteness from urban areas, (2) the availability of large supplies of electrical power and water, and (3) a low probability of disaster from natural phenomena such as earthquakes, hurricanes, and tornadoes. The Ridge and Valley Province of southeastern Tennessee provided conditions that met these needs. The geology of the Ridge and Valley Province is complex due to extensive folding, uplifting, and fracturing. The three facilities are located in separate valleys, but the numerous springs, seeps, and headwater streams found on the DOE Oak Ridge Reservation ultimately convey water to the Clinch River (Figure 7.1).

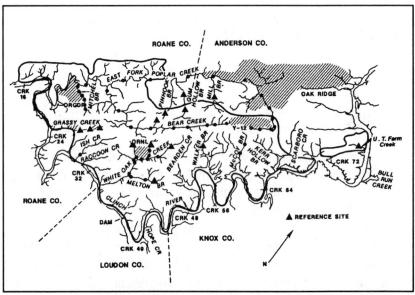

**Figure 7.1.** Locations of Department of Energy facilities (Oak Ridge National Laboratory, K-25 Site, and the Y-12 Plant) on the Oak Ridge Reservation in relation to receiving streams. Reference and non-reference biological monitoring sites on streams within and near the Oak Ridge Reservation are shown as triangles and circles, respectively.

Aspects of siting and history combined have made surface water quality problems inevitable. Early waste disposal practices, for example, included the use of unlined infiltration basins and well-intentioned, but technically flawed, attempts to immobilize high-level radioactive wastes by deep-well injection. Such situations, combined with an inevitable series of inadvertent releases of pollutants over the years, have contributed to surface and groundwater problems. By 1985, laws and environmental conditions collided. None of the three facilities was able to meet chemically based water-quality criteria; receiving streams were small and the volume of wastewater discharges was large.

In negotiations with the State of Tennessee prior to renewal of the National Pollutant Discharge Elimination System (NPDES) permit for each facility, biological monitoring was proposed to evaluate the adequacy of existing effluent limitations in protecting classified uses of waters of the State (e.g., the growth and propagation of fish and aquatic life), as designated by the Tennessee Department of Health and Environment. A secondary objective of biological monitoring was to document, for each facility, the ecological effects resulting from implementing a water pollution control program that included construction of several large wastewater treatment facilities.

Biological monitoring can be generally classified into two categories: ecological surveys and toxicity testing (Roop and Hunsaker 1985). Either or both types of biological monitoring can be incorporated directly into NPDES permits at the discretion of the regulatory agency. The NDPES permits for the three DOE facilities include both types of biological monitoring, including very specific requirements for effluent toxicity testing, and development of a plan for studies to evaluate the biological communities in designated receiving streams throughout the duration of the 5-year permit.

Because the three DOE facilities differ substantially with respect to operations, the biological monitoring plans differed in specific areas. However, the overall objectives are identical and the general areas of emphasis in the three biological monitoring programs are much more similar than they are different (Table 7.1). As part of the Toxicity Control and Monitoring Program for each facility, effluents are tested for acute and chronic toxicity according to NPDES requirements using fathead minnow larvae (*Pimephales promelas*) and *Ceriodaphnia*. These same test procedures are used to evaluate acute and chronic toxicity of ambient waters as part of the biological monitoring program for receiving streams.

**Table 7.1.** Tasks in the biological monitoring programs established for NPDES permit compliance for the Y-12 Plant, Oak Ridge National Laboratory (ORNL), and the K-25 Site. Effluent toxicity testing occurs for each facility under a separate program.

| Task/subtask | Organisms | Y-12 | ORNL | K-25 |
|---|---|:-:|:-:|:-:|
| Ambient toxicity | Fathead minnows | + | + | + |
| | *Ceriodaphnia* | + | + | + |
| | (others, ad hoc) | + | + | − |
| Bioaccumulation | Fish (PCBs, metals) | + | + | + |
| studies | Clams (organics) | + | + | − |
| | Pondweed (metals) | + | − | − |
| | Small mammals | − | + | − |
| Biological indicators | Fish | + | + | + |
| Ecological surveys | Fish | + | + | + |
| | Invertebrates | + | + | + |
| | Periphyton | + | + | − |
| Radioecological | Aquatic plants | − | + | − |
| studies | Fish | − | + | − |
| | Waterfowl | − | + | − |

# Spatial and Temporal Variation — Overview

To be useful, a biological monitoring program must accomplish one or more of the following objectives: (1) increase the understanding of the fate and ecological effects of contaminants, thus enhancing the ability to make accurate predictions about relationships between contaminants and ecological risk; (2) guide the implementation of cost-effective changes to improve environmental quality; and/or (3) communicate the value of improved water quality to the public. Biological monitoring can make important contributions to each of these areas.

An effective aquatic biological monitoring program, therefore, must simultaneously fulfill scientific, economic, and social objectives. Spatial and temporal variability in biological monitoring data influence the extent to which each of these objectives can be met. Because the objectives of biological monitoring differ greatly, the importance of variation in biological monitoring data differs. It is shaped by the perspective, and thus must be considered in several contexts.

Variation in the distribution and activity of aquatic organisms is evident at all spatial and temporal scales, but especially in streams, where biotic differences are often obvious from rock to rock within a reach, reach to reach within a watershed, and across watersheds. The distributions, abundances, and activities of aquatic biota vary conspicuously with time, too, over temporal scales ranging from seconds or minutes to years. These considerations affect exposure regimes of biota to contaminants and seriously complicate the task of relating ecological risk to chemical releases into the environment (see review by Spacie 1986). Examples from various components of the Oak Ridge biological monitoring programs will be used to demonstrate that neither spatial nor temporal variance is great enough to preclude meeting the stated objectives or the useful adoption of biological monitoring programs into the NPDES framework.

## Fish Community Studies

For biological monitoring purposes, everyone loves a fish (Cairns and Dickson 1980). Fish are known to the public, large enough to be analyzed for contaminant residues, contain prey and predatory species, tend to move through several trophic levels as they mature, can amplify concentrations of materials that bioaccumulate, occur abundantly in freshwater ecosystems, are sampled easily with simple equipment, and are commercially and recreationally important. It is not surprising that fish are emphasized in one of the first strongly-applied studies to be published in *Ecology* (Limburg and Schmidt 1990) and are used to quantitatively evaluate biological conditions in streams (the Index of Biotic Integrity, or IBI) (Karr 1981, 1991).

### Spatial Variance in Fish Data

The ecological literature provides substantial documentation of the factors that affect fish at small spatial scales (cf. Werner and Gilliam 1984, Power et al. 1985, Harvey 1991, Harvey et al. 1988). Within streams, departures from the acknowledged "deeper water—bigger fish" pattern may be indicative of a fish kill with insufficient time for recovery in numbers of large fish by growth and/or recolonization, or of a positive relationship between growth and mortality risk due to long-term exposure to some kinds of pollutants (Harvey and Stewart 1991) (Figure 7.2). These possibilities could be profitably explored.

At a larger spatial scale (e.g., within and between lakes), excellent examples of the propagation of fish effects and variance through top-down control of food web structures in lakes are given in a series of studies by Carpenter (e.g., Carpenter and Kitchell 1987, Carpenter 1989). Scale-dependent spatial correlations in predatory fish and their prey can also be detected over very large areas (e.g., 8 x 30 km study sites in oceanic areas) (Rose and Leggett 1990). It is not yet known how large-scale predator-prey correlations are affected by large-scale events (e.g., major oil-spills, habitat fragmentation, and deterioration of the ozone layer). Multivariate statistical techniques of ordination and classification can also be used to partition large geographic areas into smaller, more homogeneous "fish ecoregions" (Hawkes et al. 1986, Hughes and Larson 1988). Blocking by space is a simple example of variance control to meet biological economic objectives. This procedure reduces variance and makes it easier to statistically detect potential problems, which results in more effective management of water and fisheries resources.

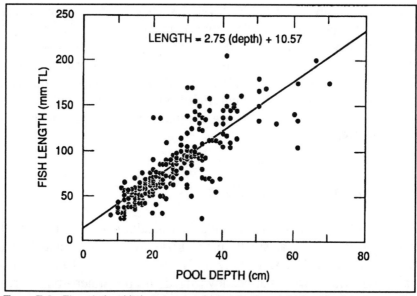

**Figure 7.2.** The relationship between maximum depth and the length of the largest fish in pools of Gum Hollow, Mill branch, and Ish Creek, small headwater streams on the Oak Ridge Reservation in eastern Tennessee (from Harvey and Stewart 1991).

## Temporal Variance in Fish Data

Carpenter and Kitchell (1987) convincingly showed that through empirical, experimental, and theoretical assessments of fish population dynamics that fish have highly variable recruitment. Strong year classes of predatory fish species can drive food web dynamics for years in lakes. Such results show that biotic forces are important in structuring food webs in lakes. Although cascading tropic effects are evident in streams (cf. Power et al. 1985, Power 1990), it might be reasonable to suppose that physical and biotic factors are about equally important in streams, and that large-scale geographic factors merely influence the relative importance of these two kinds of controls (Hubbs 1987, Stewart 1987).

At the Oak Ridge Y-12 Plant, data from ecological surveys of fish communities in Bear Creek were examined to determine if closure of four 0.37-ha waste infiltration basins located near the headwaters of Bear Creek (Jeter 1983, Southworth 1992; Figure 7.1) had detectable environmental benefits. This analysis involved seven sites, each of which was sampled eight times over a 42-month period (May 1984 through December 1987). Fish biomass ($g/m^2$) at the site nearest the basins was zero in three sampling periods and never exceeded 1.04 $g/m^2$; only 8% of the values at the other sites fell below 1.0 $g/m^2$. Density at the site closest to the basins was also very low (0.08 $\pm$ 0.10 $g/m^2$, versus $>2$ $g/m^2$ at all other sites). From these data, it was clear that the upstream site was inhospitable to fish, and statistical analysis easily demonstrated this finding.

The six sites farther downstream were catagorized into upstream and downstream reaches (3 sites in each). Analysis of variance (ANOVA) of biomass and density were conducted using either nontransformed or $log_{10}$-transformed data. Transformation did not markedly influence the outcomes: no significant time effects were detected for either parameter (e.g., $p>0.24$ for density, $log_{10}$ density, biomass, and $log_{10}$ biomass). The reach effect was significant for both parameters (e.g., $p<0.029$ for density, $p<0.013$ for biomass) and the time-reach interaction was not significant ($p=0.133$). A significant interaction could be taken as evidence for recovery of fish biomass in the upstream reach through time. In this analysis, the significant influence was that of reach (fish biomass in the upstream reach was, on average, only 68% of that in the downstream reach). This outcome could be expected simply because the downstream reach was deeper than the upstream reach, and deeper reaches can sustain more fish biomass per unit area than shallower reaches.

From March 1988 through October 1990, the site farthest upstream in Bear Creek was sampled six more times for fish. The mean density of fish at this site for the first three sampling events (March and November 1988, and March 1989) was $< 0.05$ fish/$m^2$; the mean density for the last three sampling events (October 1989, and March and October 1990) was $1.40 \pm 0.40$ fish/$m^2$ (M. G. Ryon, Environmental Sciences Division, ORNL, personal communication). Additionally, whereas the "community" at the upper site consisted, on average, of 1.5 species per sampling date prior to October 1989, the community during the last three sampling periods consisted of $3.7 \pm 0.6$ species per date. Thus, improvements at this site are now evident based both on species composition and fish abundance.

## Spatial and Temporal Variation in IBI Values

It is worth noting that IBI (Karr 1991) gains effectiveness by geographically restricting comparisons and by judicious use of numerical scores (5, 3, or 1) for up to 12 attributes (partitioned into three groups of metrics) of a fish community. The IBI is a good example of a biological monitoring analysis that essentially focuses on "relevant" variance and directs it to the advantage of the investigator or regulator. IBI values for large rivers tend to be robust temporally: within-year values differ by only about 8%, and between-year values for about 80% of the rivers surveyed by Steedman (1988) had IBI values that differed by $< 10\%$ (Karr 1991). These findings provide strong support for the IBI because they suggest that biological monitoring data obtained through time could be reasonably collapsed and still allow the generation of strong statements about the biological integrity of receiving waters. The sensitivity of the IBI to particular contaminants or conditions has not been determined.

At ORNL, the IBI was used to evaluate conditions at 16 sites in five streams in the White Oak Creek (WOC) watershed (Figure 7.3). Sampling was conducted four times over a 20-month period (Ryon and Schilling 1990). IBI scores for the 64 site-date combinations ranged from 15 to 30 (the possible range was from 12 to 60), with overall condition scores ranging only from poor to very poor. The major deficiency in fish communities in ORNL streams identified by the IBI was reduced species richness and composition components. Headwater sections of the streams used as reference sites lacked a number of fishes, such as *Notropis*, suckers, and darters, that were expected based on studies of fish communities in other headwater streams in the Tennessee Valley.

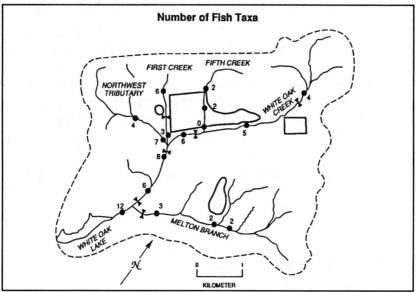

**Figure 7.3.** Number of fish taxa at biological monitoring sites in streams near the Oak Ridge National Laboratory (*from* Ryon 1990).

Two factors reduced the usefulness of the IBI on ORNL streams. First, reference sites for the study consisted of upstream areas of the impacted streams. The upstream reaches of the streams are small, nutrient-poor, and have extensive riparian cover; their invertebrate communities are strongly food-limited as well (Hill, Weber, and Stewart 1992). Thus, in the reference areas, "islands" (*sensu* MacArthur 1972) available to fish are small and the supply rate of energy needed to sustain higher trophic levels is low. It is not surprising that the fish communities at these sites are depauperate (Ryon and Schilling 1990, Harvey and Stewart 1991), consisting almost exclusively of blacknose dace (*Rhinichthys atratulus*), banded sculpin, creek chub (*Semotilus atromaculatus*), and a few stoneroller minnows (*Campostoma anomalum*). Extensive adjustments, needed to calibrate the procedure for reference streams that are naturally depauperate, increase uncertainty about the application method. Second, lower species richness in streams within the WOC watershed occurs due to the presence of permanent physical and intermittent chemical barriers that greatly restrict fish movements (e.g., a dam on WOC, weirs on WOC and a tributary, and acutely toxic reaches of some streams due to the presence of chlorine). Barriers of any type can be viewed as legitimate impacts that should be included in the IBI framework and may be a common problem when attempting to use the IBI in small- or intermediate-sized systems where rates of immigration or emigration are important in maintaining fish community structure (cf. Diamond 1975, Power et al. 1985). The problem of naturally depauperate fish communities in reference streams due to "excessively clean" conditions may be found to be more common than suspected with additional study. Presently, most IBI applications involve lower-gradient streams, which tend to be larger and more nutrient-rich than headwater streams. This is true simply because biological monitoring *per se* to date has been used primarily to evaluate conditions attributed to industrial and municipal activities, and industrial and municipal activities are more prevalent in valleys, on piedmont areas, or upon topographically flat areas than in mountainous areas.

## Benthic Invertebrate Studies

Like fish, the distributions and activities of benthic invertebrates vary greatly, both spatially and temporally, over virtually all scales in polluted and nonpolluted flowing-water systems (Hynes 1960, 1970, Cummins 1979). The use of aquatic invertebrates in biological monitoring studies is very likely to increase greatly for several reasons. Relative to fish, advantages for incorporating benthic invertebrates into biomonitoring programs include (1) their smaller size and limited mobility, which facilitates logistics of sampling and experiments, and (2) an acknowledged sensitivity of many invertebrate groups to diverse pollutants (Weber 1973, Platts et al. 1983). A disadvantage of invertebrates (compared with fish) is that they have a stronger seasonal cycle in abundance and/or activities. Invertebrates also tend to have shorter life cycles, and streams generally contain many more taxa of invertebates than fish (Table 2; see also Patrick 1975). The latter attributes provide a mixed blessing: shorter life cycles suggest more rapid responses at the community level but greater temporal variability, and more taxa per site means more effort must be expended on taxonomic identification. The benefit of this additional effort is more data for detecting differences through time or among sites.

Table 7.2. Number of fish and invertebrate taxa in four streams near Oak Ridge, TN. The White Oak Creek (White Oak) system contains contaminated segments and consists of White Oak Creek and four tributaries; Bear Creek and East Fork Poplar Creek (East Fork) are impacted near their headwaters by the Oak Ridge Y-12 Plant; and Brushy Fork is a reference stream draining agricultural and woodland areas north of Oak Ridge.

| Group | White Oak | Bear Creek | East Fork | Brushy Fork |
|---|---|---|---|---|
| Fish | 13 | 14 | 30 | 32 |
| Inverterates | 145 | 118 | 124 | 88 |

## Spatial Variance in Invertebrate Data

At small spatial scales, flow strongly affects the spacing patterns and foraging activities of invertebrates. For example, for blackfly larvae (*Simulium*), the spacing difference between competitive success or failure is on the order of millimeters (D. D. Hart, Philadelphia Academy of Natural Sciences, personal communication). Microhabitat flow patterns influence the distributions of hydropsychid caddisfly larvae at centimeter spatial scales as well (Osborne and Herricks 1987). Metabolic activity and/or feeding status (estimated through measurements of ammonia release rates) of individual pleurocerid snails (*Elimia clavaeformis*) collected from various substrate types in headwater streams differ by about a factor of two over 10- to 50-cm distances, and by about a factor of three from stream to stream (1- to 5-km spatial scales), even when water quality parameters, such as alkalinity, hardness, pH, and conductivity, show no stream-to-stream differences (Stewart, unpublished data). Presumably, small-scale variation in food quantity or quality on different substrates, and stream-to-stream variation in algal production, contribute to the differences in ammonia release rates. At whole-pool and within-pool spatial scales, species-level and ontogenetic shifts in behavior attributed to predation risk can strongly influence invertebrate communities (Gilliam et al. 1989). It is reasonable to suppose that as analytical procedures become more sensitive and field-portable, much more will be learned about the use of invertebrates for providing ecological information at very fine spatial scales. Behavior studies, such as those by Kohler and McPeek (1989), might be applied effectively to biological monitoring.

Over larger spatial scales, changes in invertebrate activities and abundances within streams and rivers can be large even in the absence of anthropogenic impacts. This expectation emerges naturally in consideration of the river continuum concept (Vannote et al. 1980). Major shifts in species and/or functional groups of aquatic insects occur in response to changes in substrates, temperature, chemical constituents, food supply, and predators, with increase in stream order. In a four-stream study in Texas, annual production of a predaceous aquatic insect (*Corydalus cornutus*) varied by more than 8-fold even though the streams were similar with respect to water quality factors. The variability was attributed to differences in stream size, flow, and temperature (Short et al. 1987). In a "uniform" 200-m reach of a noncontaminated stream near ORNL, the number of snails per square meter had a coefficient of variation (CV) of 39% (528.7 $\pm$ 207.0; mean $\pm$ SD; W. R. Hill, Environmental Sciences Division, ORNL, unpublished data). Substantial differences were also found in the size and lipid content of a diapausing caddisfly (*Neophylax etneri*) within a suite of 13 headwater streams in eastern Tennessee (Figure 7.4). In this case, much of the variance in *Neophylax* size and lipid content was attributed to competitive interactions. Snails (*Elimia* spp.) were present in some of the streams and absent in others, even though water quality was very similar. When *Elimia* was present, the mean size and lipid content of *Neophylax* was significantly lower (Hill, Weber, Stewart, 1992). In this study, the CV in number of snails per square meter for the six streams that contained snails was 49.3% (643.0 $\pm$ 316.9; mean $\pm$ SD). Thus, the variation in snail densities among sites within a stream were just about as great as it was among streams, and snails may competitively influence other stream invertebrates.

Annual community richness (the mean number of taxa per sample) for WOC headwater reference sites ranges from about 18.6 to 24.5 (J. G. Smith, Environmental Sciences Division, ORNL, personal communication), and declines substantially in ORNL streams affected by plant operations. The loss rate of taxa in affected streams reaches was 21.1 species per km for Fifth Creek, and 8.8 species per km in First Creek (Figure 7.5). Thus, large differences in community structure were readily detectable over relatively small spatial scales.

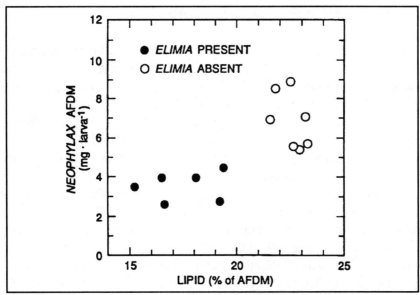

**Figure 7.4.** Lipid content and mass of *Neophylax* at diapause in streams that contain (solid circles) or lack (open circles) the pleurocerid snail *Elimia* (from Hill, Weber, Stewart, 1992).

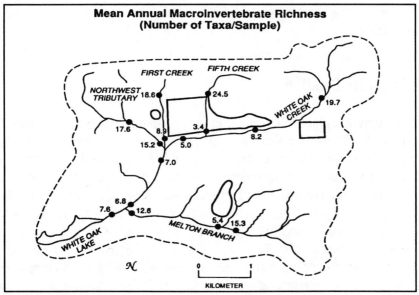

**Figure 7.5.** Mean annual number of macroinvertebrate taxa in biological monitoring sites in streams at ORNL.

## *Temporal Variance in Invertebrate Data*

Life-history patterns vary greatly among aquatic invertebrates, with life-cycle durations that range from less than two weeks to 10 to 20 years (Anderson and Wallace 1984, Resh and Solem 1984). This feature is an important source of temporal variation in biological monitoring data. Sampling and sample-processing procedures may be less effective for early life-history stages than for late instars or adults, for example, and individuals in diapause may be much less vulnerable to toxicants than actively-foraging specimens.

Month-to-month variation in the numbers of EPT taxa (insects belonging to orders Ephemeroptera, Plecoptera, and Trichoptera) at fixed sampling sites in selected streams near ORNL is considerably less than the variation among sites (Smith 1990, 1992; Figure 7.6). In this instance, the use of invertebrates as a biological monitoring parameter allows these sites to be analyzed at virtually any time of the year. In Bear Creek, the total number of EPT taxa showed a strong longitudinal gradient, and this pattern varied little year to year over a 3-year period (Figure 7.7). These findings suggest that the EPT component of aquatic invertebrate communities is relatively robust to seasonal and annual fluctuations that could compromise the usefulness of EPT analyses for biological monitoring assessments.

## Periphyton Studies

Periphytic algae are a ubiquitous and ecologically important component of many rivers and streams (Gregory 1983). The algal component of periphyton ultimately provides food needed to sustain high trophic levels, and the heterotrophic components of the periphyton (e.g., protozoans, bacteria and fungi) produce degradative enzymes (cellobiosidase, proteases, etc.) that promote the decomposition of organic matter. Both the autotrophic and heterotrophic components of the periphyton are consumed by biota in higher trophic levels. The use of periphyton as biomonitoring components has been reviewed recently by Steinman and McIntire (1990), Blanck and Wängberg (1988a,b) and Boston et al. (1991).

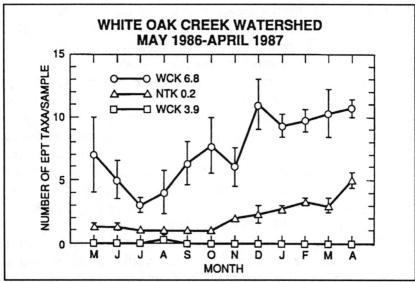

**Figure 7.6.** Month-to-month variation (mean ± SE) in number of EPT taxa (Ephemeroptera: Plecoptera and Trichoptera) in three stream sites at the Oak Ridge National Laboratory (ORNL). WCK 6.8 is a noncontaminated reference site in an upstream reach of White Oak Creek; NTK 0.2 is a moderately impacted site in Northwest Tributary (see Figure 1); and WCK 3.9 is a site on White Oak Creek that is impacted by various ORNL discharges (from Smith 1990).

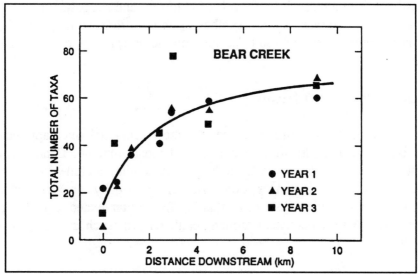

**Figure 7.7.** Year-to-year variation in total number of macroinvertebrate taxa at seven sites in Bear Creek.

Periphyton are effective bioaccumulators of various metals. In upper East Fork Poplar Creek (EFPC) within the Y-12 Plant, periphyton contained up to 607 ppm (dry wt) total mercury, but concentrations declined rapidly with distance downstream (Boston et al. 1991). The mercury content of periphyton from reference streams was 0.3 ppm (Boston et al. 1991). Periphyton in upper EFPC was also enriched in silver (concentrations up to 112 ppm dry wt), cadmium, copper, nickel, and zinc. Persistent contaminants such as these are easily transferred to higher trophic levels via grazing invertebrates or herbivorous fishes (e.g., stoneroller minnows). Reduction in food quality due to the presence of metal-contaminated aquatic vegetation has been suggested as a factor contributing to the reduced benthic species richness, especially Plecoptera and Ephemeroptera, in EFPC (Stewart et al. 1992).

Periphytic organisms have relatively short life cycles, so this community should reflect changes in water quality conditions quickly. Their rapid response to perturbation and their widespread occurrence in rivers and streams on logistically-tractable experimental units (cobbles and stones) are also advantages in biological monitoring. Short-term reciprocal transplant studies, for example, can be used to identify contaminant sources and estimate the impact of stream water quality on a key process, such as photosynthesis (Boston and Hill 1991, Hill and Boston 1991). These studies also showed that periphyton photosynthesis was more sensitive than biomass to low levels of chlorine, and suggested the use of biomass-adjusted rates of photosynthesis (mg $C/m^2/h$ per unit chlorophyll $a$) as a simple metric for use in periphyton biological monitoring studies.

## Bioaccumulation Studies

Bioaccumulation studies can be extremely beneficial when applied to receiving streams with toxic materials that bioaccumulate. Examples of contaminants that can be evaluated in such studies include PCBs and mercury. In some instances, advisories are issued by state regulatory agencies for receiving systems containing fish contaminated to unacceptable levels with a persistent human-health hazard (such as mercury or PCBs).

# Spatial and Temporal
# Variation in Bioaccumulation Studies

Redbreast sunfish (*Lepomis auritus*), bluegill (*L. machrochirus*), and carp (*Cyprinus carpio*) are collected from five sites in EFPC below the Oak Ridge Y-12 Plant for evaluations of mercury and PCBs. One purpose of this monitoring is to evaluate the effectiveness of remedial actions implemented within the Y-12 Plant, which is situated near the headwaters of the stream. The results of mercury analyses of redbreast sunfish fillets are summarized for 1984–1990 (8 fish per site-date combination) in Figure 7.8. A strong longitudinal gradient in mercury content was apparent on 11 of 12 sampling dates. Two major remedial actions associated with the control of mercury were conducted. A number of mercury-contaminated pipes were cleaned and relined from August 1986 – December 1987. The mercury content of the fish increased following this event, in association with increased aqueous mercury during the cleaning-relining process. The second action consisted of closing a settling basin that contained sediments with high levels of mercury. This action occurred in November 1988. Before and after closure, the mercury content of redbreast sunfish at the farthest downstream site (EFK 2.1) has not changed much. However, the slope of the longitudinal gradient in mercury content of redbreast sunfish throughout EFPC has declined by 38% since pond closure (Figure 7.8), suggesting a decline in point-source releases of Hg to upper EFPC. This study shows that multi-year biological monitoring can be important in evaluating the effectiveness of various remedial actions.

Even one-time synoptic surveys of PCBs in fish can provide strong evidence for large-scale point- or area-source contamination. Mitchell Branch (at the K-25 Site), upper EFPC (near the Y-12 Plant), and the WOC system at ORNL all show evidence of PCB contamination based on analyses of fish tissues (Figure 7.9). In terrestrial habitats, analyses of small mammals can be used to provide evidence for toxic materials that bioaccumulate, much as fish are used in aquatic habitats. This subject has been reviewed recently with respect to application to heavy metals, radionuclides, and organic chemicals (Talmage and Walton 1991).

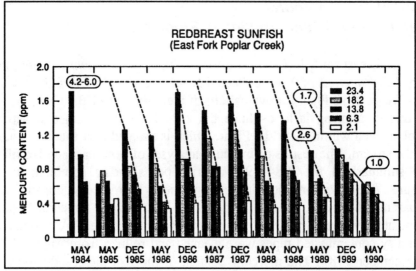

**Figure 7.8.** Mercury content (μ/sg dry weight) of redbreast sunfish from five sites in East Fork Poplar Creek (EFK). Site designations indicate distance upstream (in km) from the confluence of EFK with Poplar Creek (see Figure 7.1). The Y-12 Plant is closest to EFK 23.4. Dashed lines indicating slopes for the longitudinal gradient in mercury concentrations (dashed lines and their numerical values) were fit by eye. Mercury-contaminated pipe relining activities occurred during August 1986 – November 1987; a settling basin with mercury-contaminated sediments was closed in November 1988 (from Southworth and Peterson 1990).

## Toxicity Testing

The potential shortfalls of toxicity testing in assessing biotic integrity, and the recent shift in emphasis toward the incorporation of biological monitoring into a regulatory framework are reviewed by Karr (1991). Toxicity tests with *Ceriodaphnia* and fathead minnow larvae (*Pimephales promelas*) (EPA methods 1002.0 and 1000.0, respectively; Weber et al. 1989) are used to evaluate biological quality both of wastewaters and receiving streams at all three DOE facilities in Oak Ridge.

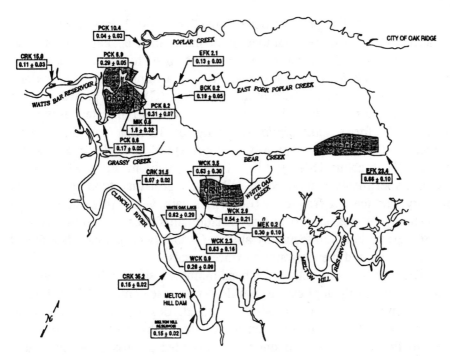

**Figure 7.9.** Mean (± SE) concentrations of PCBs (μg/g, set weight, N = 8) in bluegill (*Lepomis macrochirus*) collected in winter 1988–89 at sites on the Oak Ridge reservation and nearby reaches of the Clinch River. Rock bass (*Ambloplites rupestris*) and redbreast sunfish (*Lepomis auritus*) were substituted for bluegill at BCK 0.2 and MIK 0.6, respectively (from Southworth and Peterson 1990).

## Effluent Toxicity Testing

Effluent tests are used to assess the toxicity of various waste streams at the three DOE facilities. With repeated tests, these procedures can also be used to quantify wastewater variability through time. The 7-day *Ceriodaphnia* test appears more useful than the fathead minnow (*Pimephales promelas*) test in effluent testing. The *Ceriodaphnia* test has a statistical advantage over the minnow test in that it uses 10 replicates rather than four. In addition, *Ceriodaphnia* may be more sensitive than fathead minnow larvae to many common toxicants (e.g., chlorine, neutral salts, and various metals). Finally, the chronic endpoint of the *Ceriodaphnia* test (fecundity) can range from zero to about 230 offspring per female per 7-day test. The chronic endpoint of the fathead minnow test (growth), in contrast, can range from zero (or, in theory, some negative value) to about 0.6 mg dry wt per fish per 7-day test. It is easier to obtain an accurate count of the number of neonates in a beaker than it is to weigh very small dried fish on an electrobalance. It also takes less time to conduct a *Ceriodaphnia* test than it does a minnow test (Kszos and Stewart 1991). The coefficient of variation (CV) for growth in the minnow test is typically 5–15% among the control replicates, compared with a CV for *Ceriodaphnia* fecundity of 10-20% for the control replicates. Test-to-test variation (CV) in growth of fathead minnow larvae in controls is about 18% (0.43 $\pm$ 0.08 mg/fish; mean $\pm$ SD, for 58 tests). Experiments conducted at ORNL suggest that food quality and quantity, and feeding frequency (i.e., twice daily versus three times daily) importantly influences test-to-test variation in fathead minnow growth. Within-test variation in fecundity among *Ceriodaphnia* females is affected by water type and the presence of trace metals.

The usefulness of both the fathead minnow test and the *Ceriodaphnia* test is increased by application in a dose-response context. When five or six concentrations of an effluent are tested, dose-related reductions in survival, growth or fecundity clearly override within-treatment variation. Four to five tests of an effluent may be needed to provide a reasonable estimate of toxicity. In many instances, it is statistically, operationally, economically and scientifically advantageous to monitor effluent quality by testing more frequently (e.g., 6–12 times per year) with *Ceriodaphnia* than to test 3-6 times per year with both fathead minnow larvae and *Ceriodaphnia*.

## Ambient Toxicity Testing

Tests used to estimate acute and chronic toxicity of effluents can also be applied to receiving streams (Nimmo et al. 1990, Stewart et al. 1990, Kszos and Stewart 1991), but the "rules" of ambient toxicity testing are not identical to those of effluent testing. For example, receiving waters are rarely so toxic as to allow deriving a full dose-response relationship of the type often obtained from effluent tests. Also, whereas effluent tests are used to estimate the toxicity of an effluent, the objective of ambient testing is usually to determine which sites are better (or worse) than other sites (reviewed by Stewart et al. 1990).

Two variance-related findings associated with the use of ambient toxicity testing in biological monitoring programs are especially noteworthy. First, the results of ambient toxicity tests can be used to characterize stream water quality. Stream reaches at ORNL and the Y-12 Plant that are clearly degraded (based on other biological monitoring parameters, such as fish and invertebrate community structure, and population abundances) are frequently toxic to *Ceriodaphnia* and/or fathead minnow larvae. In such applications, the *frequency* with which a particular site fails a particular test (e.g., *Ceriodaphnia* survival was $\leq 50\%$ in 11 of 26 tests of water from site "L"; failure rate = 42%) can be used to effectively characterize site toxicity. Thus, ambient toxicity tests can provide useful information at a relatively low cost. Second, the fathead minnow test is prone to an interesting interference in some ambient applications. When the test is used to evaluate noncontaminated reference waters, pathogenic fungi can cause mortality in some replicates while not affecting fish in other replicates. In such situations, within-treatment variability in survival of the fish can be very large (e.g., 30%, 0%, 100%, 90% survival for four replicates). When a one-way ANOVA was used to compare among-replicate levels of variation (CV) for effluent and ambient tests with similar levels of mean survival (between 40% and 60%), the underlying variance structure of fathead minnow tests in these two applications differed greatly ($p < 0.001$, $DF_{1,25}$; Stewart, unpublished data). Thus, toxicity test variance and its interpretation depend in part upon whether the tests are being used to assess effluents or receiving streams.

## Variance in Biological Monitoring Data and the Importance of the Point of View

### Scientific Objectivies

To meet scientific objectives, sampling and analysis methods must be sound, and sufficient data should be obtained to provide enough statistical rigor to test the appropriate null model (cf. Green 1977, Cairns and Smith 1994): "There are no significant differences in parameter X among sites," for example, or "The number of fish per unit area in stream A was not detectably different from that in streams B, C, and D," with some specified level of certainty (typically $P \leq 0.05$). A typical approach for testing such statements involves an ANOVA of data on samples collected from multiple sites or streams. At a local scale, the statical power of this approach is affected by the number of replicates, the number of sites, and the types or numbers of comparisons that are made. The hypothesis is "objectively" (Berger and Berry 1988) rejected, or not rejected, based on the value of $P$. The failure to reject a hypothesis does not prove that the hypothesis is true (reviewed by Carpenter 1990 and Reckhow 1990), and alternatively, hierarchically-embedded hypotheses are inevitably possible (Pirsig 1974). Thus, in the context of meeting the scientific objectives of biological monitoring, variance must ultimately be viewed as enigmatic: it can comfortably be used to distinguish among alternatives, but iteratively drives a process akin to evolution. More data are always required because, like the Phoenix, new questions emerge from the ashes of the old questions.

The importance of "scientifically important" variance can also be affected by spatial scale. Invertebrate communities in WOC biological monitoring sites were examined at two spatial scales to demonstrate this point. Earlier, an example was given showing that the EPT taxonomic richness at some WOC sites was strongly impacted relative to others (Figure 6). Huston (1990) compared the EPT taxonomic richness of invertebrate communities in WOC biological monitoring sites to that in streams in the Tennessee Valley region using a regional database obtained from seven river systems and 107 sites. In this analysis, the relative taxonomic richness of combined EPT taxa in the WOC sites was much lower than would be expected (Figure 7.10), suggesting that the entire WOC system is relatively depauperate, relative to other Tennessee Valley stream systems of similar size. This analysis shows simply that the importance of "local variance" in biological monitoring data can change when viewed at a different spatial scale. Thus, results of biological monitoring data collected to guide operational changes or remediation activities can be useful at a local scale, and may also generate larger-scale questions about impacts at larger spatial scales.

## Economic Objectives

To meet purely economic objectives, variance in biological monitoring data is troublesome. Managers are understandably hesitant to make decisions about the expenditure of large sums of money unless the need has been demonstrated. Thus, uncertainty (variance) generates

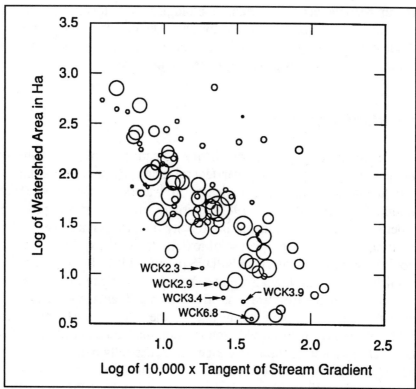

**Figure 7.10.** Relative benthic macroinvertebrate richness of EPT taxa at sites in the Tennessee Valley regional data base and White Oak Creek biomonitoring sites. Richness is the number of different taxa of each order that were found in a 0.09 m² sample of stream bottom. Circles are scaled in relation to the site with the highest density, so the density per site decreases proportionately with decreasing circle size (from Huston 1990).

discomfort, and the reduction in uncertainty can only be achieved by expenditure of funds. The request by scientists for more data is a tiresome burden, at best. Are conditions in the stream getting better, or worse? How many years of data will be required to convince regulators that a particular change in waste treatment operations is adequate? State-of-the-art approaches to remediation and water-quality improvements are not automatically more expensive than conventional approaches (Korte et al. 1989), and biological monitoring data can be used to identify monitoring procedures that optimize information per unit cost (cf. Jackson and Resh 1989, Kszos and Stewart 1991). It is also worth noting that quality assurance issues associated both with data acquisition and data interpretation can become very important if biological monitoring data address compliance issues that are driven by law. Quality assurance aspects of biological monitoring, and their associated costs, are likely to gain attention as biological monitoring is used more widely.

## Regulatory Objectives

At least three important regulatory objectives can be met through biological monitoring. First, clear federal mandates provide the impetus for reducing or eliminating toxicants in waters of the U.S. (Water Pollution Control Federation 1987). Initially, the most direct procedure for demonstrating a reduction in toxicity is met within the NPDES permitting system through the use of toxicity tests. Thus, the inclusion of biological monitoring criteria in NPDES permits fulfills both the letter and the spirit of federal law. Second, short-term toxicity tests presently recommended by the EPA are cost-effective and advantageous because (1) they can detect toxic conditions due to the presence of unsuspected constituents and (2) they provide information that can be easily translated into direct regulatory action. If toxicity is statistically detectable at some concentration using defined method "X", it can be set to automatically trigger confirmatory tests, chemical analyses, or more stringent operating limits. Third, and perhaps more importantly, biological monitoring in broad form provides a rational alternative to chemical limits. In this context, the type of biological monitoring requirements needed to confirm additional improvements in water quality can change as water quality improvements are made. Performance standards now commonly used to regulate specific activities will be replaced by impact standards, which require the achievement of a certain result (Courtemanch et al. 1989). Because various types and intensities of biological monitoring activities can be incorporated in NPDES permits to achieve impact standards, biological monitoring provides room for negotiations between the regulatory agency and the permittee.

## Public Sector Objectives

Time tables for demonstrable progress from remediation efforts may at times, span decades, and can involve very large sums of money. As money is spent to improve environmental quality, its expenditure must be justified to the general public. It is critical to show the public in very unambiguous terms that the expenditure of *their* money (through higher taxes, a lower standard of living, and/or increases in costs of goods and services), has resulted in demonstrable changes in water quality (Vining and Schroeder 1989). In this regard, it is far easier to communicate progress in fish units than in chemical units. A statement that the loading rates of cadmium from the plant to the stream has decreased by 69% over the past three years is not as easily comprehended as a statement that the number of fish in the stream has doubled over the past three years, and more sensitive species, such as darters, which were once absent, have now returned. As emphasized by Courtemanch et al. (1989), the conversion of an approach based on performance-based standards to impact-based standards shifts the emphasis from regulation to planning, which further requires effective communication with the public sector. This shift, which is already in progress, may well be both biological monitoring's greatest challenge and its strongest ally.

## The Web of Spatial and Temporal Variance

To meet all or most of these objectives, biological monitoring data, data analysis, and data interpretation must flow effectively among five parties: (1) Operators of waste treatment facilities; (2) plant mangers; (3) scientists conducting the biological monitoring studies; (4) regulators who must ensure that progress is being made; and (5) the public at large, who ultimately pays the bills. It is the propagation of variance in the interpretation of biological monitoring data through *this* web that controls the relationships between biological monitoring data and environmental progress. Studies that focus on the need to develop procedures that include management as a partner to science for effective resolution of resource issues have themselves become "cutting edge" science (cf. Walters and Holling 1990).

Overall, we conclude that variance in biological monitoring data does not appear great enough to prevent the more widespread use of biological monitoring in NPDES permits, resource management, or environmental restoration efforts. Biological monitoring programs very similar to those developed at ORNL for use at DOE facilities in Tennessee, for example, are now being used in Kentucky (Paducah Gaseous Diffusion Plant) and Ohio (Portsmouth Gaseous Diffusion Plant).

## Models in Biomonitoring

At ORNL, biological monitoring during the past several years has focused on obtaining a robust database that can be used to develop a deeper understanding of factors and processes that importantly influence the distribution and activities of biota in receiving streams. The Biological Monitoring and Abatement Programs at each facility are designed to include (1) a core of monitoring sites that are sampled systematically through time for parameters at various levels of organization (e.g., biomarkers that focus on the individual level, through communities; periphyton, invertebrates, and fish), and (2) an element of flexibility, wherein specific experiments or studies are conducted to identify and quantify mechanisms contributing to patterns identified by monitoring (cf. Loar 1990, Hinzman 1993). It is our expectation that these data will both foster the development of simulation models and allow more accurate parameterization of existing models.

### Disturbance

At large spatial scales, the spread of disturbance across aquatic or terrestrial landscapes, large or small, is a process that is influenced by spatial heterogeneity (Pickett and White 1986, Turner 1989). Disturbance is a concept that could unite "aquascape ecology" and biological monitoring simply because it explicitly includes processes, factors, and conditions that range from those that are natural (e.g., floods, lightning fires) to those that are anthropogenic (e.g., changes in urbanization and agricultural land-use patterns) (cf. Walker 1985, McDonnell and Pickett 1990, Limburg and Schmidt 1990, Odum 1989). It should be emphasized that episodic or continuous point- or area-source discharges of waste-waters and contaminants are factors important to regulators and applied ecologists, just as patch dynamics, nutrient flux rates, and shifting "top down versus bottom up" controls are important to more academic aquatic ecologists, who often seem to study only the most pristine systems.

## Spatial Heterogeneity

All habitats are spatially heterogeneous, and spatial heterogeneity is clearly one of the most important sources of variation in the distribution and activity of aquatic biota. However, spatial heterogeneity is not explicitly included in many simple population or food web models. The explicit inclusion of spatial heterogeneity into population models typically occurs in one of two ways: (1) simulation of the spatial spread of biota by diffusion, such that randomness is assumed at the level of the individual's movements; or (2) patch models, wherein populations are described stochastically in terms of the fraction of patches in a number of different states (cf. Pringle et al. 1988, Townsend 1989). The predictions of these two kinds of models are different because of the way variability is incorporated. The two kinds of models also apply to different combinations of temporal and spatial scales (reviewed by Hastings 1990). Because biological monitoring data will always be insufficient, models that can forecast contaminant fate, environmental effects, and environmental or human-health risk are needed to effectively protect our nation's water resources. Thus, additional studies on how spatial heterogeneity affects populations and communities in non-pristine waters would seem extremely advantageous.

## CONCLUSIONS

There is no great dearth of ecological detail of direct relevance to needs in applied ecology and biological monitoring. The ecological literature tells us in no uncertain terms that the adage "a fish is a fish" is naive. Even within species, adults differ importantly from juveniles by more than age: Factors such as diet, physiology, and level of activity differ substantially in the context of spatial heterogeneity, both for prey and predators. We know too, in more general terms, that ecological features, such as prey size versus prey-handling capabilities of predators, are influential in food webs, and that hydrodynamic considerations of flow and viscosity differentially affect the distributions and abundances of biota and their refugia and are key factors in how aquatic communities operate (Vogel 1981, Norwell and Jumars 1984, Statzner et al. 1988). These examples illustrate that spatial and temporal heterogeneity of the environment is ecologically important and generates variance in the distributions and activities of organisms. Presently there are fairly sound ecological data to show that aspects of spatial variability are critically important determinants of the long-term persistence of populations, communities, and ecosystems (cf. Hunsaker et al. 1990). Investigators are also learning to incorporate, albeit somewhat primitively, aspects of spatial and temporal variability into population and food web models. However, variance propagation through food webs and communities has been evaluated largely for relatively pristine systems, and it seems unlikely that the rules for how this occurs in impacted communities are identical to those for non-impacted communities. The acquisition of sound biological monitoring data should provide some scientifically interesting insights into this situation.

What may be needed most, with respect to variation in biological monitoring data, is simply more participation by ecologists and biologists in applied ecology (Slobodkin 1988). It is clear that we must learn how to effectively interpret variance from spatial and temporal sources in order to meet objectives that differ somewhat from those of more traditional scientific studies. The multiple objectives of biological monitoring must also be kept clearly in mind, by pure and applied ecologists, regulatory authorities, and the public at large, to maximize the effectiveness of biological monitoring.

## ACKNOWLEDGMENTS

This paper was improved through reviews by G. R. Southworth and S. W. Christensen, and comments by S. M. Adams, H. L. Boston, W. R. Hill, L. A. Kszos, M. G. Ryon, and J. G. Smith. Oak Ridge National Laboratory is managed for the U.S. Department of Energy by Martin Marietta Energy Systems, Inc., under contract DE-AC05-84OR21400.

## REFERENCES

Anderson, N. H. and J. B. Wallace. 1984. Habitat, life history, and behavioral adaptations of aquatic insects. pp. 38-58. *In*: R. W. Merritt, and K. W. Cummins, eds., *An Introduction to the Aquatic Insects* (second edition). Kendall/Hunt Publishing Company, Dubuque, IA. 398 p.

Berger, J. O. and D. A. Berry. 1988. Statistical analysis and the illusion of objectivity. *Amer. Scientist* 76:159-165.

Blanck, H. and S.-A. Wängberg. 1988a. Induced community tolerance in marine periphyton established under arsenate stress. *Can. J. Fish. Aquat. Sci.* 45:167-175.

Blanck, H. and S. A. Wängberg. 1988b. Validity of an ecotoxicological test system: short-term and long-term effects of arsenate on marine periphyton communities in laboratory systems. *Can. J. Fish. Aquat. Sci.* 45:1807-1815.

Boston, H. L. and W. R. Hill. 1991. Photosynthesis-light relations of stream periphyton communities. *Limnol. Oceanogr.* 36:644-656.

Boston, H. L., W. R. Hill, and A. J. Stewart. 1991. *Evaluating Direct Toxicity and Food-Chain Effects in Aquatic Systems Using Natural Periphyton Communities.* Second Symposium on the Use of Plants for Toxicity Assessments. ASTM STP 1115, American Society for Testing and Materials, Philadelphia, PA. 318 p.

Cairns, J., Jr. and K. L. Dickson. 1980. The ABCs of biological monitoring. pp. 1-31. *In*: C. H. Hocutt and J. R. Stauffer, Jr., eds., *Biological Monitoring of Fish.* Lexington Books, Lexington, MA. 416 p.

Cairns, J., Jr. and E. P. Smith. 1994. The statistical validity of biomonitoring data. *In:* S. L. Loeb, and A. Spacie, eds., *Biological Monitoring of Aquatic Systems*, Lewis Publishers, Boca Raton, FL.

Carpenter, S. R. and J. F. Kitchell. 1987. The temporal scale of variance in limnetic primary production. *Am. Nat.* 129:417-433.

Carpenter, S. R. 1989. Replication and treatment strength in whole-lake experiments. *Ecology* 70:453-463.

Carpenter, S. R. 1990. Large-scale perturbations: Opportunities for innovation. *Ecology* 71:2038-2043.

Courtemanch, D. L., S. P. Davies, and E. B. Laverty. 1989. Incorporation of biological information in water quality planning. *Environ. Manag.* 13:35-41.

Cummins, K. W. 1979. The natural stream ecosystem. pp. 7-24. *In*: Ward, J. V. and J. A. Stanford, eds., *The Ecology of Regulated Streams.* Plenum Press, NY. 398 p.

Diamond, J. M. 1975. Assembly of species communities. pp. 342-344. *In*: M. L. Cody and J. M. Diamond, eds., *Ecology and Evolution of Communities.* Belknap Press of Harvard University, Cambridge, MA. 838 p.

Gilliam, J. F., D. F. Fraser, and A. M. Sabat. 1989. Strong effects of foraging minnows on a stream benthic invertebrate community. *Ecology* 70:445-452.

Green, R. H. 1977. Some methods for hypothesis testing and analysis with biological monitoring data. pp. 200-211. *In*: J. Cairns, Jr., K. L. Dickson, and G. F. Westlake, eds., *Biological Monitoring of Water and Effluent Quality.* ASTM STP 607, American Society for Testing and Materials, Philadelphia, PA. 246 p.

Gregory, S. V. 1983. Plant-herbivore interactions in stream systems. pp. 157-189. *In*: J. R. Barnes, and G. W. Minshall, eds., *Stream Ecology: Application and Testing of General Ecological Theory.* Plenum Press, New York, NY. 399 p.

Harvey, B. C. 1991. Interactions among stream fishes: predator-induced habitat shifts and larval survival. *Oecologia* 87:29-36.

Harvey, B. C., R. C. Cashner, and W. J. Matthews. 1988. Differential effects of largemouth and smallmouth bass on habitat use by stoneroller minnows in stream pools. *J. Fish Biol.* 33:481-487.

Harvey, B. C., and A. J. Stewart. 1991. Fish Size and habitat depth relationships in headwater streams. *Oerologia* 87:336-342.

Hastings, A. 1990. Spatial heterogeneity and ecological models. *Ecology* 71:426-428.

Hawkes, C. I., D. L. Miller, and W. G. Layher. 1986. Fish ecoregions of Kansas: stream fish assemblage patterns and associated environmental correlates. *Environ. Biol. Fishes* 17:267-279.

Hill, W. R., S. C. Weber and A. J. Stewart. 1992. Food limitation of two lotic grazers: quantity, quality, and size-specificity. *J. N. Am. Benthol. Soc.* 11:420-432.

Hill, W. R. and H. L. Boston. 1991. Community development alters photosynthesis-light relations in stream periphyton. *Limnol. Oceanogr.* 36:1375-1389.

Hinzman, R. L., ed., 1993. *Second Report on the Y-12 Plant Biological Monitoring and Abatement Program for East Fork Poplar Creek.* Y/TS-888. Y-12 Plant, Oak Ridge, TN. 427 p.

Huston, M. A. 1990. Interpretation of biotic change. pp. 198-217. *In*: J. M. Loar, ed., *Fourth Annual Report on the ORNL Biological Monitoring and Abatement Program.* (Draft ORNL/TM-11544 report.) Oak Ridge National Laboratory, Oak Ridge, TN. 338 p.

Hubbs, C. 1987. Summary of the symposium. pp. 265-267. *In*: W. J. Matthews and D. C. Heins, eds., *Community and Evolutionary Ecology of North American Stream Fishes.* University of Oklahoma Press, Norman, OK. 310 p.

Hughes, R. M. and D. P. Larson. 1988. Ecoregions: an approach to surface water protection. *J. Water Poll. Contr. Fed.* 60:486-493.

Hunsaker, C. T., R. L. Graham, G. W. Suter, II, R. V. O'Neill, L. W. Barnthouse, and R. H. Gardner. 1990. Assessing ecological risk on a regional scale. *Environ. Manag.* 14:325-332.

Hynes, H. B. N. 1960. *The Biology of Polluted Waters.* University of Toronto Press, Toronto, Canada. 202 p.

Hynes, H. B. N. 1970. *The Ecology of Running Waters.* University of Toronto Press, Toronto, Canada. 555 p.

Jackson, J. K. and V. H. Resh. 1989. Sequential decision plans, benthic macroinvertebrates, and biological monitoring programs. *Environ. Manag.* 13:455-468.

Jeter, I. W. 1983. *The Chemical and Radiological Characterization of the S-3 Ponds.* Y/MA-6400. Union Carbide Corporation Nuclear Division. Y-12 Plant, Oak Ridge, TN. 17 p.

Karr, J. R. 1981. Assessment of biotic integrity using fish communities. *Fisheries* 6:21-27.

Karr, J. R. 1991. Biological integrity: a long-neglected aspect of water resource management. *Ecol. Applications* 1:66-85.

Kohler, S. L. and M. A. McPeek. 1989. Predation risk and the foraging behavior of competing stream insects. *Ecology* 70:1811-1825.

Korte, N., P. Kearl, and D. Smuin. 1989. State-of-the-art approach to hazardous waste site characterizations. *Environ. Management* 13:677-684.

Kszos, L. A. and A. J. Stewart. 1991. Effort-allocation analysis of the 7-d fathead minnow (*Pimephales promelas*) and *Ceriodaphnia dubia* toxicity tests. *Environ. Toxicol. Chem.* 10:67-72.

Limburg, K. E. and R. E. Schmidt. 1990. Patterns of fish spawning in Hudson River tributaries: response to an urban gradient. *Ecology* 71:1238-1245.

Loar, J. M., ed., 1990. *Fourth Annual Report on the ORNL Biological Monitoring and Abatement Program.* (Draft ORNL/TM-11544 report.) Oak Ridge National Laboratory, Oak Ridge, TN. 338 p.

MacArthur, R. H. 1972. *Geographical Ecology: Patterns in the Distribution of Species.* Harper & Row, New York, NY. 269 p.

McDonnell, M. J. and S. T. A. Pickett. 1990. Ecosystem structure and function along urban-rural gradients: an unexploited opportunity for ecology. *Ecology* 71:1231-1237.

Nimmo, D. R., M. H. Dodson, P. H. Davies, J. C. Green, and M. A. Kerr. 1990. Three studies using *Ceriodaphnia* to detect nonpoint sources of metals from mine drainage. *J. Water Poll. Contrl. Fed.* 62:7-15.

Norwell, A. R. M. and P. A. Jumars. 1984. Flow environments of aquatic benthos. *Ann. Rev. Ecol. Syst.* 15:303-328.

Odum, E. P. 1989. *Ecology and Our Endangered Life-Support Systems.* Sinauer Associates, Inc., Sunderland, MA. 283 p.

Osborne, L. L. and E. E. Herricks. 1987. Microhabitat characteristics of *Hydropsyche* (Trichoptera: Hydropsychidae) and the importance of body size. *J. N. Am. Benthol. Soc.* 6:115-124.

Patrick, R. 1975. Structure of stream communities. pp. 445-459. *In*: M. L. Cody and J. M. Diamond, eds., *Ecology and Evolution of Communities.* Belknap Press of Harvard University Press, Cambridge, MA. 838 p.

Pickett, S. T. A. and P. S. White, eds., 1986. *The Ecology of Natural Disturbance and Patch Dynamics.* Academic Press, Orlando, FL. 472 p.

Pirsig, R. M. 1974. *Zen and the Art of Motorcycle Maintenance: an Inquiry into Values*. Morrow Publishers, NY. 412 p.

Platts, W. S., W. F. Megahan, and G. W. Minshall. 1983. *Methods for Evaluating Stream, Riparian, and Biotic Conditions*. U.S. Department of Agriculture General Technical Report INT-138, Intermountain Forest and Range Experiment Station, Ogden, UT. 70 p.

Power, M. E., W. J. Matthews, and A. J. Stewart. 1985. Grazing minnows, piscivorous bass, and stream algae: dynamics of a strong interaction. *Ecology* 66:1448-1456.

Power, M. E. 1990. Effects of fish in river food webs. *Science* 250:811-814.

Pringle, C. M., R. J. Naiman, G. T. Bretchko, J. R. Karr, M. W. Oswood, J. R. Webster, R. L. Welcomme, and M. J. Winterbourn. 1988. Patch dynamics in lotic systems: the stream as a mosaic. *J. N. Amer. Benthol. Soc.* 7:503-524.

Reckhow, K. H. 1990. Bayesian inference in non-replicated ecological studies. *Ecology* 71:2053-2059.

Resh, V. H. and J. O. Solem. 1984. Phylogenetic relationships and evolutionary adaptations of aquatic insects. pp. 66-75. *In:* R. W. Merritt and K. W. Cummins, eds., *An Introduction to the Aquatic Insects* (second edition). Kendall/Hunt Publishing Company, Dubuque, IA. 722 p.

Roop, R. D. and C. T. Hunsaker. 1985. Biomonitoring for toxics control in NPDES permitting. *J. Water Poll. Control. Fed.* 57:271-277.

Rose, G. A. and W. C. Leggett. 1990. The importance of scale to predator-prey spatial correlations: an example of Atlantic Fishes. *Ecology* 71:33-43.

Ryon, M. G. and E. M. Schilling. 1990. Fishes. pp. 169-198. *In:* J. M. Loar, ed., *Fourth Annual Report on the ORNL Biological Monitoring and Abatement Program.* (Draft ORNL/TM-11544 report.) Oak Ridge National Laboratory, Oak Ridge, TN. 338 p.

Short, R. A., E. H. Stanley, J. W. Harrison, and C. R. Epperson. 1987. Production of *Corydalus cornutus* (Megaloptera) in four streams differing in size, flow, and temperature. *J. N. Amer. Benthol. Soc.* 6:105-114.

Slobodkin, L. B. 1988. Intellectual problems of applied ecology. *BioScience* 38:337-342.

Smith, J. G. 1990. Benthic macroinvertebrates. pp. 151-169. *In:* J. M. Loar, ed., *Fourth Annual Report on the ORNL Biological Monitoring and Abatement Program.* (Draft ORNL/TM-11544 report.) Oak Ridge National Laboratory, Oak Ridge, TN. 338 p.

Smith, J. G. 1992. Benthic macroinvertebrates. pp. 126-153. *In:* J. M. Loar, ed., *Second Annual Report on the ORNL Biological Monitoring and Abatement Program.* ORNL/TM-10804. Oak Ridge National Laboratory, Oak Ridge, TN. 392 p.

Southworth, G. R., ed., 1992. *Ecological Effects of Contaminants and Remedial Actions in Bear Creek.* ORNL/TM-11977. Oak Ridge National Laboratory, Oak Ridge, TN. 290 p.

Southworth, G. R. and M. J. Peterson. 1990. Bioaccumulation studies. pp. 91-126. *In*: J. M. Loar, ed., *Fourth Annual Report on the ORNL Biological Monitoring and Abatement Program.* (Draft ORNL/TM-11544 report.) Oak Ridge National Laboratory, Oak Ridge, TN. 338 p.

Spacie, A. 1986. Spatial and temporal distribution of biota and its role in exposure assessment. pp. 152-162. *In*: H. L. Bergman, R. A. Kimerle, and A. W. Maki, eds., *Environmental Hazard Assessment of Effluents.* Pergamon Press, Inc., NY. 366 p.

Statzner, B., J. A. Gore, and V. H. Resh. 1988. Hydraulic stream ecology: observed patterns and potential applications. *J. N. Amer. Benthol. Soc.* 7:307-360.

Steedman, R. J. 1988. Modification and assessment of an index of biotic integrity to quantify stream quality in Southern Ontario. *Can. J. Fish. Aquat. Sci.* 45:492-501.

Steinman, A. D. and C. D. McIntire. 1990. Recovery of lotic periphyton communities after disturbance. *Environ. Manag.* 14:589-604.

Stewart, A. J. 1987. Responses of stream algae to grazing minnows and nutrients: a field test for interactions. *Oecologia* 72:1-7.

Stewart, A. J., L. A. Kszos, B. C. Harvey, L. F. Wicker, G. J. Haynes, and R. D. Bailey. 1990. Ambient toxicity dynamics: Assessments using *Ceriodaphnia dubia* and fathead minnow (*Pimephales promelas*) larvae in short-term tests. *Environ. Toxicol. Chem.* 9:367-379.

Stewart, A. J., G. J. Haynes, and M. I. Martinez. 1992. Fate and biological effects of contaminated vegetation in a Tennessee stream. *Environ. Toxicol. Chem.* 11:653-664.

Talmage, S. S. and B. T. Walton. 1991. Small mammals as monitors of environmental contaminants. *Rev. Environ. Contamin. Toxicol.* 119:47-145.

Townsend, C. R. 1989. The patch dynamics concept of stream community ecology. *J. N. Amer. Benthol. Soc.* 8:36-50.

Turner, M. G. 1989. Landscape ecology: the effect of pattern on process. *Ann. Rev. Ecol. Syst.* 20:171-197.

Vannote, R. L., G. W. Minshall, K. W. Cummins, J. R. Sedell, and C. E. Cushing. 1980. The river continuum concept. *Can. J. Fish. Aquat. Sci.* 37:130-137.

Vining, J. and H. W. Schroeder. 1989. The effects of perceived conflict, resource scarcity, and information bias on emotions and environmental decisions. *Environ. Management* 13:199-206.

Vogel, S. 1981. *Life in Moving Fluids: The Physical Biology of Flow.* Princeton University Press, Princeton, NJ. 352 p.

Walker, R. D. 1985. Agricultural nonpoint source pollution in the Midwest. pp. 497-498. *In: Perspectives on Nonpoint Source Pollution: Proceedings of a National Conference*. EPA 440/5-85-001.

Walters, C. J. and C. S. Holling. 1990. Large-scale management experiments and learning by doing. *Ecology* 71:2060-2068.

Water Pollution Control Federation. 1987. *The Clean Water Act of 1987* (second edition). Water Pollution Control Federation, Alexandria, VA. 318 p.

Weber, C. I. 1973. *Biological Field and Laboratory Methods for Measuring the Quality of Surface Water and Effluents*. EPA-670/4-73-001. U.S. Environmental Protection Agency, National Environmental Research Center, Cincinnati, OH. 187 p.

Weber, C. I. et al. 1989. *Short-Term Methods For Estimating the Chronic Toxicity of Effluents and Receiving Waters to Freshwater Organisms*. EPA/600/4-89/001. U.S. Environmental Monitoring Systems Laboratory, Cincinnati, OH. 249 p.

Werner, E. E. and J. F. Gilliam. 1984. The ontogenetic niche and species interactions in size-structure populations. *Ann. Rev. Ecol. Syst.* 15:393-425.

# CHAPTER 8

# Use of Ecoregions in Biological Monitoring

**Robert M. Hughes**. ManTech Environmental Technology, Incorporated. U.S. EPA Environmental Research Laboratory, Corvallis, OR

**Steven A. Heiskary**. Nonpoint Source Section. Minnesota Pollution Control Agency, Saint Paul, MN

**William J. Matthews**. Biological Station. University of Oklahoma, Kingston, OK

**Chris O. Yoder**. Ecological Assessment Section. Ohio Environmental Protection Agency, Columbus, OH

## INTRODUCTION

In order to better manage populations of lakes and streams it is useful to have some form of lake and stream classification. Such classifications may be based on site-specific data or on some form of regionalization generated from those or other data. The goal of any classification is to stratify variance, and the greater the variance that a classification accounts for, the more useful it is. The process of classification is iterative; as we learn more about aquatic ecosystems through site-specific monitoring and theoretical advances, we can use that knowledge to improve our classification and thereby the quality of our science and management.

In biological monitoring programs, particularly at the state, regional, or national level, an appropriate geographic framework is useful for developing estimates of organisms likely to be collected and conditions likely to be encountered. Such a framework is also useful for setting biological criteria, for interpreting the relative health of the site, and for extrapolating the results of collections or samples to a population of water bodies. State agencies, in particular, need a method for breaking a large complex set of systems into rational units for management and for predicting the results of management actions. Traditionally, aquatic biologists have used river basins to determine these frameworks.

As assemblage and water quality data bases and statistical software have become available, they have been used to frame regions. When mapped, the ordination results often are interpreted from some preconceived landscape level pattern, such as river basins (Hocutt and Wiley 1986), catchments (Matthews and Robison 1988), or physiography (Pflieger 1971).

Others have compared aquatic ecosystem patterns with various environmental variables. Ross (1963) showed a strong association between terrestrial biomes and North American caddisfly species in small streams. Legendre and Legendre (1984) found climatic, geomorphic, and vegetation patterns more useful than river basins for explaining fish distribution patterns in Quebec. Minshall et al. (1985) concluded that regional patterns in climate, geology, and land use were necessary for appropriately applying the river continuum concept across regional scales. Jackson and Harvey (1989), in a study of available data from 286 lakes in the northern Great Lakes area of Ontario, found that fish faunas were significantly correlated with geographical proximity, but not with lake area, maximum depth, or pH. They proposed that the faunal patterns were a function of differing post-glacial dispersal routes and climatic regimes. Corkum (1989), examining benthic invertebrate data from 100 river sites in northwestern North America, determined that drainage basin, physiographic region, and bedrock geology were more useful for classifying the fauna than were environmental variables measured at the sites. In a study of five sites along each of three rivers in eastern Ontario, she concluded that macroinvertebrate assemblages corresponded more with land use than with season, site location along the rivers, or drainage (Corkum 1990).

Recently, aquatic ecoregions, developed from a combination of landscape characteristics, have been proposed as a framework for assisting water body managers (Omernik 1987, Hughes and Larsen 1988). Ecoregions are defined as mapped regions of relative homogeneity in land surface form, soil, potential natural vegetation, and general land use. Ecoregions group water bodies that would be naturally similar in the absence of permanent human settlements and thereby they stratify variability; such regions are substantially less diverse than an entire nation or state. They have been shown to correspond to statewide differences in fish and water quality in Ohio streams (Larsen et al. 1986, 1988); fish, water quality, and physical habitat in Arkansas streams (Rohm et al. 1987); fish, macroinvertebrates, physical habitat, and water quality in Oregon streams (Hughes et al. 1987, Whittier et al. 1988); water quality in Minnesota lakes (Heiskary et al. 1987); and fish in Wisconsin streams (Lyons 1989). Ecoregions have also been evaluated on a national scale. The New Zealand Department of Scientific and Industrial Research proposes to use ecoregions as "an objective basis for establishing regional criteria that are feasible and protective of aquatic systems" (Biggs et al. 1990). Their conclusion was based on a study of catchment characteristics, hydrology, water quality, periphyton, benthos, and fish in 144 catchments.

The purpose of this paper is to (1) compare fish faunal regions and ecoregions, (2) summarize the experiences of two states that use ecoregions as management units, and (3) discuss concerns about the use of ecoregions.

## CORRESPONDENCE BETWEEN FISH FAUNAL REGIONS AND ECOREGIONS

Pflieger (1971) and Pflieger et al. (1981), using fish assemblage data from over 1600 localities, described three fish faunal regions and a large river fauna for Missouri. Except for northwest Missouri, their regions and those of Omernik show considerable correspondence (Figure 8.1), possibly because both frameworks are partly based on physiographic patterns.

Analyzing data from 410 stream sites, Hawkes et al. (1986) distinguished six major fish regions delineated mostly by drainage in Kansas. These bear no resemblance to the six regions that Omernik (1987) mapped for the state (Figure 8.2). Hawkes et al. observe, however, that the Walnut Creek drainage (which is split by an Omernik ecoregion) divides into two fish regions.

Using discriminant analysis of fish data from 350 stream sites, Bazata (1991) found that five fish faunal regions were more appropriate than the seven ecoregions of Omernik. However, the reduced number of faunal regions were obtained by combining two small portions of Omernik ecoregions with two larger ecoregions (Figure 8.3).

Hughes et al. (1987) used available data from 9,100 collections in 85 catchments to evaluate the ability of ecoregions, drainage basins, and physiographic regions to form ichthyogeographic regions. They found that 2 of 10 physiographic provinces, 5 of 18 river basins, and 4 of 8 aquatic ecoregions could serve as fish

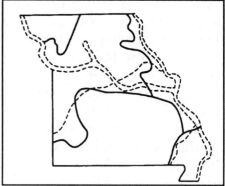

**Figure 8.1.** Comparison of fish faunal regions (dotted lines) of Pflieger et al. (1981) and ecoregions (solid lines) of Omernik (1987) in Missouri.

faunal regions in Oregon (Figure 8.4). This indicated that statewide, ecoregions offer the single most suitable system to classify fish assemblages, but that they alone are insufficient.

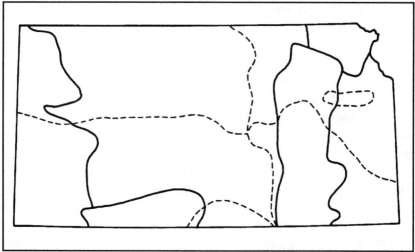

**Figure 8.2.** Comparison of fish faunal regions (dotted lines) of Hawkes et al. (1986) and ecoregions (solid lines) of Omernik (1987) in Kansas.

Rohm et al. (1987) and Matthews and Robison (1988) analyzed the fish fauna of Arkansas through use of an ecoregion and a data base/drainage units approach, respectively. Rohm et al. used

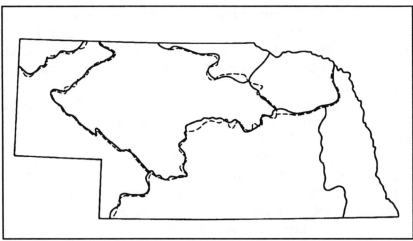

**Figure 8.3.** Fish faunal regions (dotted lines) of Bazata (1991) and ecoregions (solid lines) of Omernik (1987) for Nebraska.

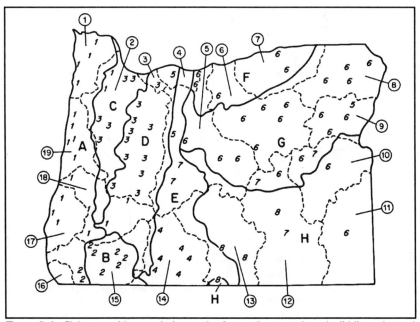

**Figure 8.4.** Fish assemblages relative to the 8 aquatic ecoregions (solid lines, letters) and 19 river basins (dotted lines, circled numbers) of Oregon (from Hughes et al. 1987). Uncircled numbers are from a cluster analysis.

collections from 22 streams in the six ecoregions of Omernik, whereas Matthews and Robison described five discontinuous fish faunal regions from over 2,000 collections, but mapped the regions as individual drainage units. In order to make comparisons at a similar scale for this paper, their 101 drainages were coded and plotted by the five major river basins and six ecoregions of Arkansas (Figure 8.5). Both Rohm et al. and Matthews and Robison distinguished two lowland regions and an Ozark region. However, Matthews and Robison's Ouachita-Ozark border region included Omernik's Ouachita and Boston Mountains regions, and part of his Ozark region. Matthews and Robison did not delineate a separate Arkansas River Valley region. The actual ordinations in both papers showed considerable differences between lowland and Ozark regions, but a gradual transition among the other regions.

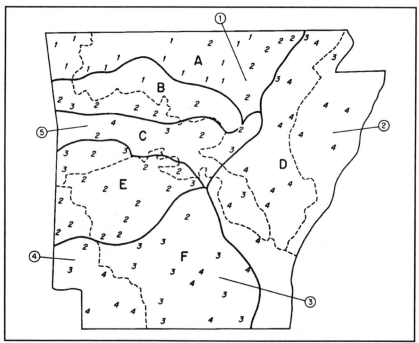

**Figure 8.5.** Fish assemblages relative to the six aquatic ecoregions (solid lines, letters) and five river basins (dotted lines, circled numbers) of Arkansas (from Matthews and Robison 1988). Uncircled numbers are from a detrended correspondence analysis.

These examples illustrate two inadequacies with river basins and Omernik's ecoregions as frameworks for fish monitoring. First, Pflieger and Matthews and Robison described a large river fauna that ecoregions and basins may miss. Second, available fish data offer more detailed information about fish presence and absence than ecoregions, and, consequently, should be examined while developing regional expectations for fish assemblages.

On the other hand, a focus on available data (fish or other assemblages) presents different problems. Single assemblage analyses are appropriate only for that assemblage or data base. It is unrealistic to expect state and federal agencies to develop individual regions for algae, benthos, fish, water quality, and physical habitat. Agencies must manage lakes and streams as aquatic ecosystems. It is theoretically possible to simultaneously ordinate a number of assemblages and physical and chemical habitat variables, but to our knowledge this has not been demonstrated.

Regardless of whether available water body data are used to draw ecological regions or whether Omernik's ecoregion map is used, there is a need for a hierarchical set of regions. Many state and federal monitoring agencies lack the resources to monitor and interpret data at a site-specific level, except at a small number of sites. Consequently, agencies tend to make screening assessments at a regional level. Regions at the scale of Omernik's ecoregions or Pflieger's fish faunal regions are appropriate for such purposes. Scientists concerned with cause and effect and compliance monitoring require less heterogeneous regions, perhaps at the habitat type scale. We are a long way from delineating such classifications at the state or national level. Intermediate regions at the scale of drainage units (Matthews and Robison 1988) or ecoregion subregions (Clarke et al. 1991, Gallant et. al. 1989) offer greater precision than basins or ecoregions at far less expense than habitat classification, but their usefulness is untested.

The research described above also reveals the wisdom of considering both drainages and landscape characteristics when delineating regions, regardless of scale. Clearly, the distributions of fishes and nonflying macroinvertebrates are restricted to historical and present water bodies. Additionally, it is clear that considerable heterogeneity exists within drainages. For example, Hughes and Gammon (1987) found distinctly different fish assemblages in four reaches of the mainstem Willamette River in Oregon. Omernik and Griffith (1991) were better able to stratify the heterogeneity of dissolved oxygen in Arkansas streams and the fish assemblages in the Calapooia River in Oregon, through the use of ecoregions than through drainages (Figure 8.6). Hawkes et al. (1986) indicated a substantial difference in Walnut Creek, Kansas, which is divided into two ecoregions (Figure 8.1). Smith et al. (1981) observed that fishes of the Raisin River in Michigan, were distributed in a nonrandom pattern; distributions of some species change near ecoregion boundaries (Figure 8.7).

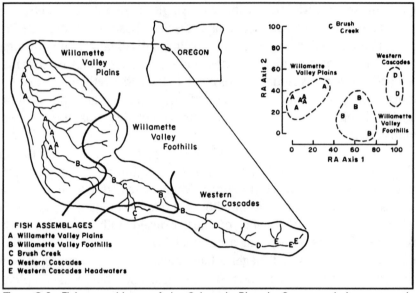

**Figure 8.6.** Fish assemblages of the Calapooia River in Oregon, relative to aquatic ecoregions (from Omernik and Griffith 1991). Letters are from a reciprocal averaging ordination.

In a study of 21 Tennessee River reservoirs, McDonough and Barr (1977) found considerable similarity among fish assemblages, and classified them by a combination of reservoir position and size, river basin, and elevation. However, the clusters do not distinguish the reservoirs in the Tennessee River from those in the Cumberland, French Broad, or Clinch Rivers, nor are the Little Tennessee and Hiwassee Rivers discriminated from each other. Reanalysis of their dendrogram reveals a clear (and simpler) relationship between clusters and ecoregions, with only two misclassifications (Figure 8.8).

Despite being classified by drainage unit, the results of Matthews and Robison reveal that ecoregions may better classify fish regions than do river basins, at least among lowlands. All the big rivers of Arkansas share a similar fauna. Within basins, there are regional differences as one moves across ecoregions; however, there are no obvious basin differences

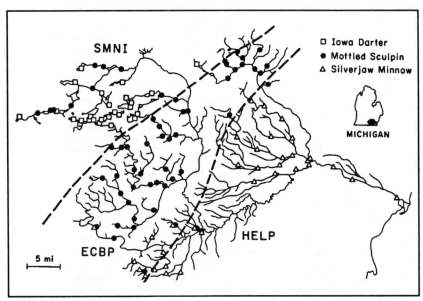

**Figure 8.7.** Three fish species distributions relative to ecoregions in the Raisin River basin of Michigan (from Smith et al. 1981). Southern Michigan/northern Indiana till plains (SMNI), eastern corn belt plains (ECBP), Huron/Erie lake plain (HELP).

within ecoregions. Although neither basins nor ecoregions classified fish assemblages accurately, ecoregions were consistently more accurate than basins (Figure 8.5, Table 8.1). It seems wise, then, to consider both ecoregions and drainage units when delineating fish faunal regions, regardless of the geographic scale of interest, the size of the water body, or whether the system is a lake or stream.

**Table 8.1.** Percent misclassifications* of Arkansas fish faunal regions by ecoregion and river basin.

| | % Misclassifications | |
| --- | --- | --- |
| Faunal Region | Ecoregion | River Basin |
| 1 | 34 | 64 |
| 2 | 61 | 74 |
| 3 | 67 | 71 |
| 4 | 46 | 83 |

*Misclassification is defined as a fish assemblage other than that predominating in that ecoregion or basin.

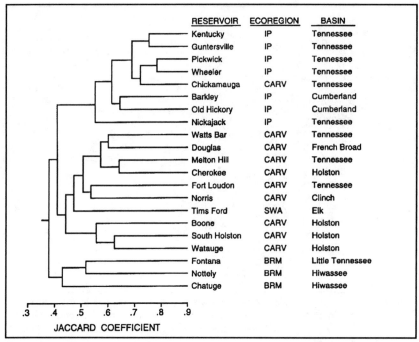

**Figure 8.8.** Cluster results for fish assemblages of Tennessee Valley Authority reservoirs (from McDonough and Barr 1977). Interior Plateau (IP), Central Appalachian Ridges and Valleys (CARV), Southwestern Appalachians (SWA), and Blue Ridge Mountains (BRM).

## MANAGEMENT APPLICATIONS OF ECOREGIONS

Although ecoregions and some other forms of water body classification are useful to researchers interested in large scale patterns, their greatest proponents are agency scientists charged with monitoring and assessing many waters across a large area. Two states, Minnesota and Ohio, have made extensive use of ecoregions, as well as of existing data.

In Minnesota, Moyle (1956) recognized regional patterns in lake productivity as a result of differences in geology, hydrology, vegetation, and land use. More recently, the Minnesota Pollution Control Agency (MPCA) has used Omernik's map to assess patterns in lake trophic state and morphometry and apply the results to lake management. Heiskary and his colleagues (Heiskary et al. 1987, Heiskary 1989, Heiskary and Wilson 1989) used trophic state variables from a 1,400-lake MPCA data base, existing fish data, and a data base from 90 minimally disturbed reference lakes to evaluate four lake-rich ecoregions. Marked ecoregional differences occurred for total phosphorus (Figure 8.9) and fish assemblages (Table 8.2). Heiskary and Wilson also found ecoregional differences in user perceptions about conditions suitable for swimming and for what constitutes a nuisance algal bloom. Persons in regions where lakes are typically clearer required lakes to be twice as clear as did people in regions characterized by more turbid lakes (Figure 8.10). Information such as this has been very useful for a staff of two to three limnologists in planning, goal setting, and communicating with citizens about the 12,000 lakes in the state. Ecoregions facilitated quantitative regional estimates of reasonable trophic state values and variability, and improved model predictions of trophic state variables and data interpretation among neighboring states.

**Table 8.2.** Ecological classification of all lakes greater than 60 hectares (approximately 1,800 lakes) within four Minnesota ecoregions (from Borchert et al. 1970 and Heiskary et al. 1987).

| | Ecoregion (%) | | | |
|---|---|---|---|---|
| Lake Class | NLF | CHF | WCP | NGP |
| Lake trout | 2 | — | — | — |
| Walleye | 20 | 5 | — | — |
| Bass-panfish-walleye | 48 | 37 | 13 | 7 |
| Bullhead-panfish | 4 | 6 | 14 | 4 |
| Winterkill-roughfish | 13 | 34 | 65 | 66 |
| No data/other | 13 | 18 | 8 | 23 |

NLF (northern lakes and forests), CHF (central hardwoods forest), WCP (western cornbelt plains), NGP (northern great plains).

Ohio, like Minnesota, was graced by early regional analyses conducted by an ichthyologist. Trautman (1957) found physiographic regions useful frameworks for fish faunas and he reviewed Omernik's draft Ohio ecoregion map. The Ohio Environmental Protection Agency

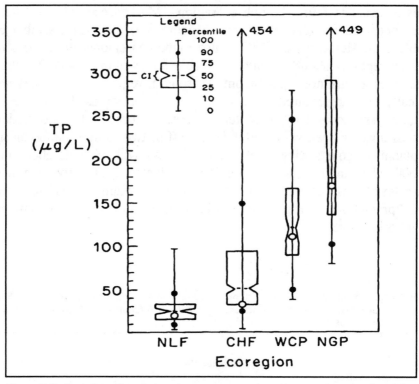

**Figure 8.9.** Box plots of total phosphorus concentrations by ecoregion—northern lakes and forests (NLF), central hardwoods forest (CHF), western cornbelt plains (WCP), northern great plains (NGP) (from Heiskary et al. 1987). Box width reflects number of lakes; CI is 95% confidence interval of the median; open circle is the reference lake median (from Heiskary 1989).

(OEPA) used Omernik's ecoregions and a fish and macroinvertebrate data base from over 300 minimally disturbed reference sites to develop biological criteria for the 45,000 miles of streams and rivers in Ohio (OEPA 1987, 1988, 1989a,b, Yoder 1989). The reference site data base was used to calibrate the index of well-being (Gammon et al. 1981), index of biotic integrity (Karr 1981, Karr et al. 1986), and an invertebrate community index (OEPA 1987). Index values determined from the regional reference sites were used to set regional criteria for ambient biological assemblages. Values for each index were plotted by ecoregion and warmwater criteria were set at the 25th percentile for streams of

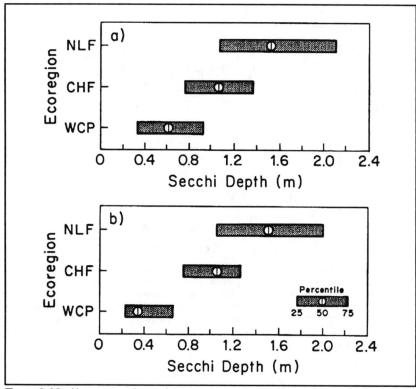

**Figure 8.10.** User perceptions of (a) recreational suitability (no swimming) and (b) physical appearance classes (high or severe algae). Ecoregions as in Figure 9 (from Heiskary and Wilson 1989).

three size ranges (headwater, wading, and boat) in each region for Figure 8.11 (Yoder 1989). Criteria for exceptional and physically modified sites were also developed. The ecoregional reference sites are systematically remonitored to adjust calibration curves and biological criteria and to evaluate background changes in biological integrity. Biological criteria are used by OEPA to demonstrate temporal and spatial trends in Figure 8.12 (Yoder 1989), detect impairment of aquatic life uses (Yoder 1991), provide biennial water resource summaries (OEPA 1990), and diagnose types of stressors. Quantitative biocriteria are not only a major improvement over chemical criteria, they also provide a more accurate measure of water resource quality than the more commonly used

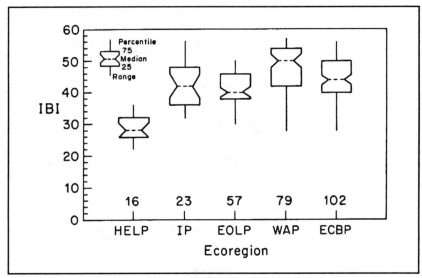

**Figure 8.11.** Box plots of Ohio reference sites for wading streams by ecoregion—Huron Erie lake plain (HELP), interior plateau (IP), Erie Ontario lake plain (EOLP), western Allegheny plateau (WAP), and eastern corn belt plain (ECBP) (from Yoder 1989). Box notch represents 95% confidence interval of the median; the number above the acronym is the number of reference sites per region.

narrative criteria. In a study of over 400 sites, OEPA found 61% of the sites attaining and 9% not attaining narrative biological criteria based on professional judgement. When numerical biological criteria based on macroinvertebrate assemblage scores at regional reference sites were applied to the same sites, 34% attained and 44% did not attain the criteria (Yoder 1991).

Use of regional reference lakes or stream reaches (Hughes et al. 1986) to develop water resource criteria is hindered by ecoregion heterogeneity. Regional reference sites are inappropriate benchmarks for sites that naturally differ from them, to a considerable degree. For example, sites with substantial natural difference in gradient, substrate, or water quality in the same ecoregion should not have the same set of reference sites. If such sites occur in a distinct geographic pattern, the ecoregion should be subregionalized (Gallant et al. 1989). If there is no pattern to the differences, reference sites for each type of natural gradient, substrate, or water quality should be selected.

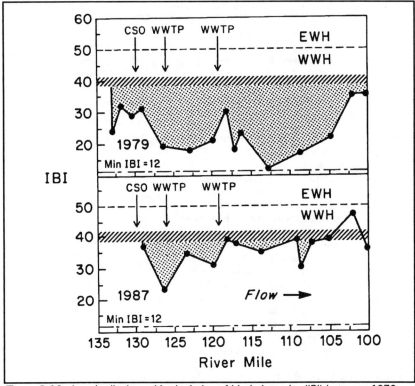

**Figure 8.12.** Longitudinal trend in the index of biotic integrity (IBI) between 1979 and 1987 for the Scioto River near Columbus, Ohio. Flow is from left to right, vertical arrows indicate major pollution sources, IBI of 40 equals biological criterion for warmwater habitat, and 50 is the criterion for exceptional warmwater habitat (from Yoder 1989).

Reference sites in extensively disturbed regions may be extremely difficult to locate and may represent unacceptable levels of degradation. However, use of least degraded or most desirable sites in such areas offers a realistic goal for improvement. Lake sediment cores may offer alternative references for assemblages preserved therein (Charles et al., 1994); however, criteria based on presettlement conditions are unrealistic without major changes in land use. Minimally disturbed catchments or sites can serve as references of the best we can expect, given current land use patterns, and they are certainly in better than average condition. Reference site selection involves both objective and subjective standards (Table 8.3) and many sites are necessary where regions are

**Table 8.3.** Steps in ecoregional reference site selection.

1)    Define area(s) of interest on maps
        Delineate ecoregions and subregions on 1:250,000 scale maps
        Delineate river basins and subbasins supporting different faunas
        Combine regions and basins to produce potential natural regions

2)    Define water body types, sizes, and classes of interest
        Evaluate their number and importance in the region

3)    Delineate candidate reference watersheds through use of maps, available data, air photos, and local experts
        Eliminate disturbed areas
                Point sources (atmospheric, aquatic)
                Hazardous waste sites, landfills
                Mines, oil fields
                Feedlots, poultry farms, hatcheries
                Urban, industrial, commercial, residential
                Channelization, dams
                Transportation and utility corridors
                Logged or burned forests
                Intensively grazed or cropped lands
        Seek minimally disturbed, typical areas or potential natural landscapes
                Agricultural or range oases
                Old growth forests, woodlots
                Roadless areas
                Preserves, refuges, exclosures

4)    Conduct field reconnaissance to locate potential natural sites
        Aerial observation, remote sensing
        Site inspection at ground level
                Extensive, old riparian vegetation
                Complex channel morphology
                Variable substrate, with large resistant objects and minimal sedimentation
                Considerable cover (overhanging vegetation, undercut banks, large woody
                        debris, deep pools, large turbulent riffles, macrophyte
                High water quality (clear, odorless)
                Vertebrates and macroinvertebrates present
                Minimal evidence of humans, livestock, and human activity

5)    Determine number of ecoregional reference sites needed
        Balance regional variability, resources, and study duration

6)    Monitor sites and evaluate their chemical, physical, and biological integrity. At least two assemblages should be evaluated.

heterogeneous or where numerous water body types occur in a region. Many of these issues will require additional federal and multi-state research and analysis in order to be resolved.

## CONCERNS ABOUT THE USE OF ECOREGIONS

**Is ecosystem health a value concept and consequently not determinable scientifically?** Ecosystem health certainly involves values; however, Frey 1977, Karr 1994 and Karr and Dudley (1981) offered a solution. They defined ecosystem health, or biological integrity, as "a balanced, integrated, adaptive community of organisms having a species composition, diversity, and functional organization comparable to that of natural habitat of the region." Ecoregions offer a regional framework for classifying natural habitat and for stratifying entire communities, as opposed to an individual assemblage such as fish or macroinvertebrates. Ecoregional reference sites represent comparable natural habitat against which the integrity of other sites in the ecoregion can be compared. Moyle 1994 emphasized that there is not one healthy community, but several. The set of ecoregional reference sites used by Ohio EPA and Minnesota PCA provide examples of such communities. Contrary to the implications of Conquest et al. 1994, we should not require only pristine systems as references. Doing so usually restricts us to no benchmarks for evaluating restoration results or deterioration. Also, many of our most pristine arctic and mountain waters are impacted by atmospheric deposition. Even if they were not, they would be inappropriate references for waters in other regions. Regional reference sites would certainly offer Stewart and Loar 1994 and others benchmarks for evaluating assemblages in streams with highly perturbed headwaters. Of course, such sites must be monitored regularly to evaluate the degree to which they change as a result of natural and anthropogenic modifications.

**Is any geographic framework ready for immediate adoption by state and federal management agencies?** Additional tests of ecoregions, basins, and other frameworks are needed in areas other than those already studied. Such tests should involve multiple assemblages to examine ecosystem or community differences and similarities, because we cannot afford to develop many issue- or assemblage-specific regions for large areas. Even if we could develop many assemblage-specific regional maps, state and federal management agencies would find them difficult to use.

The research would be even more useful if done at more than one scale. Existing lake chemistry and fish and macroinvertebrate data bases are promising sources of information for such analyses, particularly the fish data bases residing in many museums.

**Can we base ecoregions on geology or individual chemical, hydrological, and topographic variables?** Omernik's ecoregion map actually integrates these variables by using maps of landform, soil, and vegetation, in addition to other mapped information. These maps, in turn, were drawn from considerable site specific chemical and physical data.

**Do the 76 ecoregions of the U.S. incorporate too much hetero-geneity?** The answer to this question depends on their use. They are too detailed for some uses and users and not detailed enough for others. This is why a hierarchical set of regions would be appropriate. Omernik is now working with individual states, focusing on ecoregions that are particularly troublesome or important to managers. The Science Advisory Board of the U.S. Environmental Protection Agency (1991) recommended that the Agency fund subregionalization, along with further tests of ecoregions, but this has not yet occurred.

**Must we deal with individual lakes or stream reaches?** If our research is limited only to the system that we study, that leaves an enormous number of unstudied systems about which we are ignorant. It is very useful to study individual water bodies and be able to extrapolate the results to others. Ecoregions provide a useful framework for bounding our extrapolations. It is unwise to assume that structures and functions characteristic of a small northern mountain lake or stream typify all lakes and streams, or even all mountain waters. Water resource managers do not have the luxury of such research. Whether they take action or accept the status quo, they must extrapolate their knowledge of a relatively small proportion of waters. If we ever hope to understand and manage all lakes or streams, we must begin to assess and regulate them as populations, as well as individuals. Then we can use population-based information to help us make decisions about individual water bodies. Ecoregions are simply a way to classify landscapes and their watersheds so that we can generalize, much in the same manner that we use taxonomic classifications to generalize in biology.

**Might we be just as well, or better, off with a coarse classification of vegetation (say agriculture, grassland, deciduous forest, and coniferous forest) if we are interested in process or function rather than composition?** This is an interesting point because most of our examples involve species composition. Minshall et al. (1985) stated that climate, geology, and land use may explain a great deal of the process differences found across the northern U.S. However, ecoregions can be easily aggregated by color on Omernik's map, which will offer a biome level classification, and the map of Omernik and Gallant (1990) is at the biome scale of resolution. Note also that many ecoregions, such as those in Minnesota, subsume these major vegetation breaks and provide a useful framework for stratifying lake processes. As we learn more about stream processes and monitoring data become more available from more places, we may want greater detail than that provided by biomes. Presumably, we will also continue to be interested in species and guilds, for which greater regional detail is useful.

**Can we use river basins as the classification tool as is done in Britain?** We do use basins in this way in the U.S., but they are problematic. Basins frequently cross ecoregions that are drastically different (mountains, plains). Therefore, river basins and hydrologic units frequently do not correspond with patterns in aquatic variables, as this paper, Omernik and Griffith (1991), Hughes and Gammon (1987), and Smith et al. (1981) demonstrate. It is appropriate to evaluate both basins and ecoregions, especially for fishes.

**Is it possible to regionalize by stressor (such as acid rain) and lake or watershed size?** This is an effective approach where there is a single very important issue. Omernik and his colleagues developed alkalinity maps (Omernik and Griffith 1986, Omernik and Powers 1983) that were used to frame regions of sensitivity to acidification for EPA's National Surface Water Survey and they produced nutrient maps for evaluating patterns in eutrophication (Omernik 1977, Omernik et al. 1988). However, when agencies are concerned with multiple problems and multiple indicators, a general ecoregion map is more appropriate. The map need not and cannot be optimal for all purposes and users; it should be adequate, useful, logical, and able to explain much of the observed variance. Water body and catchment sizes should be considered however one regionalizes, but size does not appear to be a regional phenomenon.

**When will the mapping of ecoregions be complete?** It is important to realize that there are two national ecoregion maps developed by EPA to date (Omernik 1987, Omernik and Gallant 1990). Work is continuing at the subregion scale through cooperative work with a handful of states, but it receives insufficient research funds from EPA at present. The entire ecoregion project has been an evolving process leapfrogging among monitoring, research, and regulation. The lack of a national ecoregion research and mapping program to assist states in developing biological criteria means that some states will be inadequately mapped and that inconsistent maps may be produced by the states.

**What are the dangers of using ecoregions blindly?** As with any new idea, especially one that has an entire country as its subject, it is likely that ecoregions will be misused. They are presently being applied to develop biocriteria. We would be concerned if states ignore basin differences where they exist, overlook subregion differences when they are great, disregard different stream or lake types and sizes in the same region, assess too few reference sites to obtain a meaningful estimate of variability, use highly disturbed sites as reference sites, or use ecoregions when they are obviously inadequate or for purposes for which they were not intended. On the other hand, we would be concerned if academic researchers fail to use ecoregions to bound their lake or stream study areas or to examine regional patterns in other lakes or streams of that type.

**Given the resource shortages of state agencies and university researchers, why not involve more academics in ecological monitoring?** This is something that Societas Internationalis Limnologiae and other professional associations focusing on water issues could try to foster. It would be possible only for academics that have the time to conduct considerable monitoring for months at a time and through the use of standard, quality controlled methods. Agencies need funding mechanisms that allow easy transfer of research funds, as opposed to hiring staff. Both parties would require equal access to the data and it should be in easily accessible data bases, along with historical data. As the Environmental Monitoring and Assessment Program (EMAP, Paulsen and Linthurst, 1994) is implemented, there is likely to be considerable demand for academic biologists to conduct much of the field work, taxonomic identifications, and indicator development. EMAP and ecoregion researchers must also continue to seek and evaluate existing data bases and consult with state and local experts.

## CONCLUSIONS

Study and management of anything as complex as lake and stream ecosystems requires recognition of local, regional, and historical factors or filters (Ricklefs 1987, Tonn 1990). An ecoregion focus can be damaging if we ignore the local scale, but more frequently we tend to focus on the local factors and fail to see the broader picture.

We can also be misled by available museum data. Although greater use of such data is warranted, that use must be tempered by knowledge of how thoroughly the data were collected at the site, the quality of the proportionate abundance estimates, and the spatial intensity of the collections, both regionally and by water body type and size. Nonetheless, museum data, especially if compiled nationally or across a multi-state region, could be valuable for estimating species pools, for calibrating indices such as the index of biotic integrity (Karr 1981, Karr et al. 1986), and for determining locations where sub-ecoregions require delineating.

Fish faunal regions generated from museum data, together with ecoregions, should facilitate federal and interstate cooperation in biological monitoring by providing common biogeographic frameworks. Of course, this will still require much more federal leadership than currently exists to avoid generating inadequate, redundant, and conflicting state and federal biological monitoring programs.

We should be encouraged by renewed interest in biological monitoring and criteria evidenced by this workshop and others (McKenzie, et al. 1992, USEPA 1987, Yount and Niemi 1990), by the EPA's requirement for states to develop biological criteria (USEPA 1990), by large national biological monitoring programs (Gurtz, 1994, Paulsen and Linthurst, 1994), and by the growing number of states that are increasing their biological monitoring. However, much remains to be done. A very small part of state and federal budgets are spent on monitoring the biological resources that citizens assume we are protecting and little of that monitoring information is used in making management decisions. This is particularly discouraging as the resources we love and study disappear to human overpopulation and overconsumption before we even get to know them very well (Hughes and Noss 1992).

## ACKNOWLEDGMENTS

This paper was, in part, derived from a panel discussion held at a workshop, Biological Monitoring of Aquatic Systems, held at Purdue University, West Lafayette, Indiana, December 1990. We thank Stanford Loeb for inviting us to participate and the Environmental Monitoring and Assessment Program-Surface Waters Group of the U.S. EPA for travel expenses. An earlier draft of this manuscript was improved through reviews by Susan Christie, Deborah Coffey, John Giese, Phil Larsen, Stanford Loeb, Terry Maret, Jim Omernik, and Bill Platts. The research was partially funded by the U.S. Environmental Protection Agency through contract 68-C8-0006 to ManTech Environmental Technology, Incorporated. This chapter was partially prepared at the EPA Environmental Research Laboratory in Corvallis, Oregon, and subjected to Agency peer and administrative review and approved for publication.

# REFERENCES

Bazata, K. 1991. *Nebraska Stream Classification Study*. Nebraska Department of Environmental Control. Lincoln, NB. 342 p.

Biggs, B. J. F., M. J. Duncan, I. G. Jowett, J. M. Quinn, C. W. Hickey, R. J. Davies-Colley, and M. E. Close. 1990. Ecological characterization, classification, and modeling of New Zealand rivers: an introduction and synthesis. *New Zealand Journal of Marine and Freshwater Research* 24:277-304.

Borchert, J. R., G. W. Orning, J. Stinchfield, and L. Maki. 1970. *Minnesota's Lakeshore: Resources, Development, Policy Needs*. II. University of Minnesota Geography Department. Minneapolis, MN. 72 p.

Charles, D. F., J. P. Smol, and D. R. Engstrom. 1994. Paleoreconstruction of the Environmental Status of Aquatic Systems. *In*: S. L. Loeb and A. Spacie, eds., *Biological Monitoring of Aquatic Systems*. Lewis Publishers, Boca Raton, FL.

Clarke, S. E., D. White, and A. L. Schaedel. 1991. Oregon, USA, ecological regions and subregions for water quality management. *Environmental Management* 15:847-856.

Conquest, L. L., S. C. Ralph, and R. J. Naiman. 1994. Implementation of Large-Scale Stream Monitoring Efforts: Sampling Design and Data Analysis Issues. *In*: S. L. Loeb and A. Spacie, eds., *Biological Monitoring of Aquatic Systems*. Lewis Publishers, Boca Raton, FL.

Corkum, L. D. 1989. Patterns of benthic invertebrate assemblages in rivers of northwestern North America. *Freshwater Biology* 21:191-205.

Corkum, L. D. 1990. Intrabiome distributional patterns of lotic macroinvertebrate assemblages. *Canadian Journal of Fisheries and Aquatic Sciences* 47:2147-2157.

Frey, D. G. 1977. *Biological Integrity of Water — An Historical Approach* pp. 127-140. *In:* R. K. Ballentine and L. J. Guarraia, eds. *The Integrity of Water*. U.S. Environmental Protection Agency, Washington, D.C. 230 p.

Gallant, A. L., T. R. Whittier, D. P. Larsen, J. M. Omernik, and R. M. Hughes. 1989. *Regionalization as a Tool For Managing Environmental Resources.* EPA/600/3-89/060. U.S. Environmental Protection Agency, Corvallis, OR. 152 p.

Gammon, J. R., A. Spacie, J. L. Hamelink, and R. L. Kaesler. 1981. Role of electrofishing in assessing environmental quality of the Wabash River. pp. 307-324. *In:* J. M. Bates and C. I. Weber, eds., *Ecological Assessments of Effluent Impacts on Communities of Indigenous Aquatic Organisms.* American Society of Testing and Materials. STP 703. Philadelphia, PA. 333 p.

Gurtz, M. 1994. Design Considerations for Biological Components of the National Water-Quality Assessment (NAWQA) Program. *In:* S. L. Loeb and A. Spacie, eds., *Biological Monitoring of Aquatic Systems.* Lewis Publishers, Boca Raton, FL.

Hawkes, C. L, D. L. Miller, and W. G. Layher. 1986. Fish ecoregions of Kansas: stream fish assemblage patterns and associated environmental correlates. *Environmental Biology of Fishes* 17:267-279.

Heiskary, S. A. 1989. Lake assessment program: a cooperative lake study program. *Lake and Reservoir Management* 5:85-94.

Heiskary, S. A. and C. B. Wilson. 1989. The regional nature of lake water quality across Minnesota: an analysis for improving resource management. *Journal of the Minnesota Academy of Science* 55:71-77.

Heiskary, S. A., C. B. Wilson, and D. P. Larsen. 1987. Analysis of regional patterns in lake water quality: using ecoregions for lake management in Minnesota. *Lake and Reservoir Management* 3:337-344.

Hocutt, C. H. and E. O. Wiley. 1986. *The Zoogeography of North American Freshwater Fishes.* John Wiley & Sons, NY. 866 p.

Hughes, R. M. and J. R. Gammon. 1987. Longitudinal changes in fish assemblages and water quality in the Willamette River, Oregon. *Transactions of the American Fisheries Society* 116:196-209.

Hughes, R. M. and D. P. Larsen. 1988. Ecoregions: an approach to surface water protection. *Journal of the Water Pollution Control Federation* 60:486-493.

Hughes, R. M. and R. F. Noss. 1992. Biological diversity and biological integrity: current concerns for lakes and streams. *Fisheries* 17:11-19.

Hughes, R. M., D. P. Larsen, and J. M. Omernik. 1986. Regional reference sites: a method for assessing stream potentials. *Environmental Management* 10:629-635.

Hughes, R. M., E. Rexstad, and C. E. Bond. 1987. The relationship of aquatic ecoregions, river basins, and physiographic provinces to the ichthyogeographic regions of Oregon. *Copeia* 1987:423-432.

Jackson, D. A. and H. H. Harvey. 1989. Biogeographic associations in fish assemblages: local vs. regional processes. *Ecology* 70:1472-1484.

Karr, J. R. 1981. Assessment of biotic integrity using fish communities. *Fisheries* 6:21-27.

Karr, J. R. 1994. Biological Monitoring: Challenges for the Future. *In*: S. L. Loeb and A. Spacie, eds., *Biological Monitoring of Aquatic Systems*. Lewis Publishers, Boca Raton, FL.

Karr, J. R. and D. R. Dudley. 1981. Ecological perspective on water quality goals. *Environmental Management* 5:55-68.

Karr, J. R., K. D. Fausch, P. L. Angermeier, P. R. Yant, and I. J. Schlosser. 1986. Assessing biological integrity in running waters: a method and its rationale. *Illinois Natural History Survey Special Publication 5*. 28 p.

Larsen, D. P., J. M. Omernik, R. M. Hughes, C. M. Rohm, T. R. Whittier, A. J. Kinney, A. L. Gallant, and D. R. Dudley. 1986. Correspondence between spatial patterns in fish assemblages in Ohio streams and aquatic ecoregions. *Environmental Management* 10:815-828.

Larsen, D. P. D. R. Dudley, and R. M. Hughes. 1988. A regional approach for assessing attainable surface water quality: Ohio as a case study. *Journal of Soil and Water Conservation* 43:171-176.

Legendre, P., and V. Legendre. 1984. Postglacial dispersal of freshwater fishes in the Quebec peninsula. *Canadian Journal of Fisheries and Aquatic Sciences* 41:1781-1802.

Lyons, J. 1989. Correspondence between the distribution of fish assemblages in Wisconsin streams and Omernik's ecoregions. *American Midland Naturalist* 122:163-182.

Matthews, W. J. and H. W. Robison. 1988. The distribution of the fishes of Arkansas: a multivariate analysis. *Copeia* 1988:358-374.

McDonough, T. A. and W. C. Barr. 1977. *An analysis of fish associations in Tennessee and Cumberland drainage impoundments*. Proceedings of the Annual Conference of the Southeast Association of Fish and Wildlife Agencies 31:555-563.

McKenzie, D. H., D. E. Hyatt, and V. J. McDonald, eds, 1992. *Ecological Indicators*. Elsevier, NY. 1619 p.

Minshall, G. W., K. W. Cummins, R. C. Petersen, C. E. Cushing, D. A. Bruns, J. R. Sedell, and R. L. Vannote. 1985. Developments in stream ecosystem theory. *Canadian Journal of Fisheries and Aquatic Sciences* 42:1045-1055.

Moyle, J. B. 1956. Relationships between the chemistry of Minnesota surface waters and wildlife management. *Journal of Wildlife Management* 20:303-320.

Moyle, P. B. 1994. Biodiversity, Biomonitoring, and the Structure of Stream Fish Communities. *In*: S. L. Loeb and A. Spacie, eds., *Biological Monitoring of Aquatic Systems*. Lewis Publishers, Boca Raton, FL.

OEPA. 1987. *Biological Criteria for the Protection of Aquatic Life: Vol II: User's Manual for Biological Field Assessment of Ohio Surface Waters.* Ohio Environmental Protection Agency, Columbus, OH. 229 p.

OEPA. 1988. *Biological Criteria for the Protection of Aquatic Life: Vol. I: The Role of Biological Data in Water Quality Assessment.* Ohio Environmental Protection Agency, Columbus, OH. 44 p.

OEPA. 1989a. *Addendum to Biological Criteria for the Protection of Aquatic Life: Vol II: User's Manual for Biological Field Assessment of Ohio Surface Waters.* Ohio Environmental Protection Agency, Columbus, OH. 21 p.

OEPA. 1989b. *Biological Criteria for the Protection of Aquatic Life: Vol. III. Standardized Biological Field Sampling and Laboratory Methods for Assessing Fish and Macroinvertebrate Communities.* Ohio Environmental Protection Agency, Columbus, OH. 45 p.

OEPA. 1990. *Ohio Water Resource Inventory.* Ohio Environmental Protection Agency, Columbus, OH. 174 p.

Omernik, J. M. 1977. Nonpoint Source-Stream Nutrient Level Relationships: A Nationwide Study. EPA-600/3-77-105. U.S. Environmental Protection Agency, Corvallis, OR. 151 p.

Omernik, J. M. 1987. Ecoregions of the conterminous United States. *Annals of the Association of American Geographers* 77:118-125.

Omernik, J. M. and A. L. Gallant. 1990. Defining regions for evaluating environmental resources. pp. 936-947. *In:* H. G. Lund and G. Preto eds., *Global Natural Resource Monitoring and Assessments: Preparing for the 21st Century.* American Society of Photogrammetry and Remote Sensing, Bethesda, MD. 1495 p.

Omernik, J. M. and G. E. Griffith. 1986. Total alkalinity of surface waters: a map of the western region. *Journal of Soil and Water Conservation* 41:374-378.

Omernik, J. M. and G. E. Griffith. 1991. Ecological regions versus hydrologic units: frameworks for managing water quality. *Journal of Soil and Water Conservation* 46:334-340.

Omernik, J. M. and C. F. Powers. 1983. Total alkalinity of surface waters—a national map. *Annals of the Association of American Geographers* 73:133-136.

Omernik, J. M., D. P. Larsen, C. M. Rohm, and S. E. Clarke. 1988. Summer total phosphorus in lakes: a map of Minnesota, Wisconsin, and Michigan, USA. *Environmental Management* 12:815-825.

Paulsen, S. G. and R. A. Linthurst. 1994. Biological Monitoring in the Environmental Monitoring and Assessment Program. *In*: S. L. Loeb and A. Spacie, eds., *Biological Monitoring of Aquatic Systems.* Lewis Publishers, Boca Raton, FL.

Pflieger, W. L. 1971. A distributional study of Missouri fishes. *University of Kansas Publications Museum of Natural History* 20:225-570.

Pflieger, W. L., M. A. Schene, and P. S. Haverland. 1981. Techniques for the classification of stream habitats, with examples of their application in defining the stream habitats of Missouri. pp. 362-368. *In*: N.B. Armantrout, ed., *Acquisition and Utilization of Aquatic Habitat Inventory Information*. American Fisheries Society, Bethesda, MD. 376 p.

Ricklefs, R. E. 1987. Community diversity: relative roles of local and regional processes. *Science* 235:167-171.

Rohm, C. M., J. W. Giese, and C. C. Bennett. 1987. Evaluation of an aquatic ecoregion classification of streams in Arkansas. *Journal of Freshwater Ecology* 4:127-140.

Ross, H. H. 1963. Stream communities and terrestrial biomes. *Archives fur Hydrobiologie* 59:253-242.

Science Advisory Board. 1991. *Evaluation of the Ecoregion Concept.* EPA-SAB-EPEC-91-003. U.S. Environmental Protection Agency, Washington, D.C. 25 p.

Smith, G. R., J. N. Taylor, and T. W. Grimshaw. 1981. Ecological survey of fishes in the Raisin River drainage, Michigan. *Michigan Academician* 13:275-305.

Stewart, A. J. and J. M. Loar. 1994. Spatial and Temporal Variation in Biological Monitoring Data. *In*: S. L. Loeb and A. Spacie, eds., *Biological Monitoring of Aquatic Systems*. Lewis Publishers, Boca Raton, FL.

Tonn, W. M. 1990. Climate change and fish communities: a conceptual framework. *Transactions of the American Fisheries Society* 119:337-352.

Trautman, M. B. 1957. *The Fishes of Ohio*. Ohio State University Press, Columbus, OH. 782 p.

U.S. Environmental Protection Agency. 1987. *Report of the National Workshop on Instream Biological Monitoring and Criteria*. Office of Water, Washington, DC. 34 p.

U.S. Environmental Protection Agency. 1990. *Biological Criteria: National Program Guidance for Surface Waters*. EPA-440/5-90-004. Office of Water, Washington, D.C. 57 p.

Whittier, T. R., R. M. Hughes, and D. P. Larsen. 1988. The correspondence between ecoregions and spatial patterns in stream ecosystems in Oregon. *Canadian Journal of Fisheries and Aquatic Sciences* 45:1264-1278.

Yoder, C. O. 1989. The development and use of biological criteria for Ohio surface waters. pp. 139-146. *In: Water Quality Standards for the 21st Century*. U.S. Environmental Protection Agency. Office of Water, Washington, D.C. 263 p.

Yoder, C. O. 1991. Answering some concerns about biological criteria based on experiences in Ohio. pp. 95-104. *In: Water Quality Standards for the 21st Century.* U.S. Environmental Protection Agency. Office of Water, Washington, D.C. 251 p.

Yount, D. J. and G. J. Niemi, eds. 1990. Recovery of lotic communities and ecosystems following disturbance: theory and application. *Environmental Management* 14(5). 515-762 pp.

# SECTION IV

# Community Response

# CHAPTER 9

# Bioassessment and Analysis of Functional Organization of Running Water Ecosystems

**Kenneth W. Cummins**, South Florida Water Management District, West Palm Beach, FL

## INTRODUCTION

Based on the record of the past 30 years, one can legitimately ask, "Will the 90's, the 'Decade of the Environment,' see innovation or business as usual in the arena of bioassessment of running waters, that is, analysis of, and predictions about, lotic ecosystems?" We need not debate the importance of water, undoubtedly the ultimate "limiting resource", however, the importance of biological insight in the routine inventory (monitoring) and "management" of running waters apparently remains in question. The lack of efficiency with which the rich and robust body of basic research data, concepts, and paradigms have been transferred to the realm of application would never have been tolerated in the health professions. The level of understanding of lotic ecosystem structure and function seems to far outstrip the use of this knowledge in applied problem solving.

At the crux is the question, "What is the point of biological monitoring of running water ecosystems?" The goal of such investigations is usually to assess the present and continuing condition of a given lotic system in regard to a measured or implied standard, and to itself, over time, and to make predictions about future conditions so as to permit implementation of appropriate man-made changes. The former is the purview of monitoring, the latter management. It is likely that a team of trained lotic ecologists could easily accomplish these goals. However, there are not enough professionals available, nor time and money if there were, to take this approach. Thus, the question is what can we impart to non-professional lotic investigators, or those with incomplete levels of professional training, to enable them to accomplish these tasks of monitoring and management? It should be remembered that the notion of management of natural systems is really quite arrogant, and ability to monitor does not imply ability to manage.

The main thread of this paper is that what is missing in the monitoring-management endeavor is technological transfer, and that among the most effective transfers will be those concepts and methodologies that bear on system function. With regard to technological transfer, the health profession analogy alluded to above seems a valid one. Basic research results in medical science find their way into the clinical practice of medicine efficiently through the continuous and massive efforts of technological transfer groups. Workers in applied areas of lotic ecology charge that basic researchers are unwilling to develop and implement criteria needed for biological monitoring and management of running water ecosystems (e.g., Karr 1991). But why should they? The biochemist at the National Institutes of Health working on cell function is not expected to develop the drug and plan the vaccination strategy. On the other hand, the basic scientist, asking non-mission oriented questions about ecological systems, sees little effort on the part of applied scientists to implant known basic concepts and methodologies into their problem solving agendas. It seems clear that each transfer would yield big dividends.

As for the targets for technological transfer, among the most promising, as stated above, should be the knowledge base relevant to system function. In this area, I believe that a strong case can be made for the use of invertebrate associations as a "window" through which to view the functional organization of running water ecosystems.

# FUNCTIONAL ANALYSIS

## Structure vs Process

In the deliberations that led to the National Science Foundation sponsored Long-Term Ecological Research (LTER) program, very few process-type measurements could be identified that were judged suitable for long-term data collection with a view to chronicling ecosystem changes over long time spans. In the biological realm, taxonomic composition of communities as a system measured over time received most support. That is, the biological structure of an ecosystem was assumed to be embodied in its taxonomic arrays. However, problems with the definition of a species and conflicting views about species diversity and system stability have limited the scope of application of taxonomic analyses. Given the perpetually incomplete status of the taxonomy of most groups of lotic organisms, the requirement that species identifications be the basis for all studies is rarely achieved and stifles ecological inquiry.

With regard to the diversity issue and the emphasis on species identifications, it would be quite useful to examine the role of intraspecific genetic diversity in species "packing" in a given ecosystem. For example, the same level of diversity of function, such as the feeding categories discussed below, might be realized either through a large number of "true" (reproductively isolated) species, each with low genetic diversity, or the reverse—fewer numbers of true species, but each with high genetic diversity (Figure 9.1). Further, it seems clear that information about the balance between inter-specific reproductive isolation (species) and intra-specific genetic diversity will be important for deciding what to monitor and how to manage.

## The Case for Invertebrates

The need for rapid, accurate biological monitoring of stream and river environments has grown dramatically in the last decade as the demand for competing water uses continues to increase. There is a need for techniques that enable field crews, having varying levels of expertise, to rapidly assess the biological characteristics and conditions of fresh

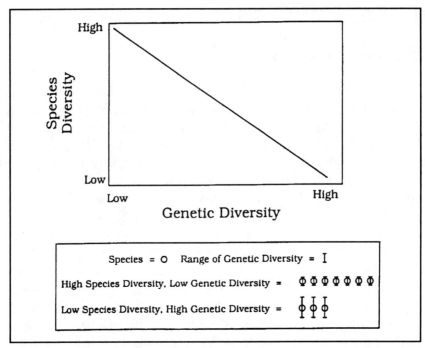

**Figure 9.1.** Proposed relationship between species diversity and intra-specific genetic diversity. (Model proposed by K. W. Cummins, F. J. Gottlieb, and W. P. Coffman, University of Pitsburgh, Department of Biological Sciences.)

water ecosystems. However, methods that adequately meet the required criteria of reliability, speed, and ease of use remain limited (e.g., Hilsenhoff 1987, Karr 1987).

Historically, assessment of disturbance or stress to the condition of freshwater ecosystems has been accomplished by instantaneous measurements of selected chemical contaminant concentrations (e.g., Averett and McKnight 1987) or the relative dominance of organisms considered to be indicators of either healthy or polluted conditions (e.g., Lenat 1988). Biological measures have often been preferred because of their time-integrative nature; for example the "biological memory" (loss of diversity, etc.), which can be detected long after short-term pulsed chemical inputs to a given environment have passed. Comparisons are usually made between disturbed systems (stressed by point or non-point alterations) and reference ecosystems for a given ecoregion (Omernik 1987, Hughes et al. 1994). These comparisons of systems that differ in levels of anthropogenic disturbance must be made within the context of natural fluctuations, both seasonal and long-term (e.g., Cummins et al. 1983, 1984).

Biological monitoring efforts have focused on the species diversity of invertebrates, fish, and occasionally, plants. There have been several major flaws in this approach. For example, low natural species diversity can be found in some pristine aquatic ecosystems, such as headwater streams. In many Western streams and rivers, fish species diversity is naturally quite low, in some cases only one or two in total. In fact, in many streams an increase in fish diversity, brought about through the introduction of exotic species, can be legitimately considered a stress. In addition, the importance of the balance between species and genetic diversity needs to be clarified before such analyses can really be predictive.

Assessments using macroinvertebrate species diversity require a high level of expertise and may be extremely time consuming to complete, even if sufficiently trained technical personnel are available. As suggested previously, the various techniques used to determine macroinvertebrate diversity frequently suffer from grossly incomplete taxonomy. The best example is the lumping of the dipteran Chironomidae into, at most, several taxa (e.g., subfamilies). This, despite the fact that the midges are almost invariably the most numerous and diverse group present in aquatic ecosystems, outnumbering all other groups combined. In spite of these taxonomic and other problems, macroinvertebrates have always been attractive subjects for the evaluation of aquatic ecosystems. They are universally abundant, easily collected, and large enough to be examined in the field with the unaided eye or a simple hand lens. From an ecosystem perspective, macroinvertebrates are ideal integrators between the rapid turnover microorganisms (e.g., bacteria, fungi, and algae) responsible for the majority of nutrient cycling, and the large, slow turnover, quantitatively trivial, fish, which are often qualitatively most significant for economic and social reasons.

## Functional Feeding Group Method

With a view to evaluating (monitoring) processes in running water ecosystems, the functional organization of the invertebrate associations has proven effective in providing a "window" to system function. A requirement for rapid lotic ecosystem evaluation and a need to address fundamental ecological questions about these aquatic ecosystems, without the prerequisite of high level taxonomic resolution, led to the development of a functional approach to macroinvertebrate analysis 17 years ago (Cummins 1973). This functional feeding group technique for the characterization of running water macroinvertebrate communities is based on a limited set of morphological-behavioral adaptations used by taxonomally diverse groups in acquiring their nutritional resources. The resource categories are coarse particulate organic matter (CPOM), such as leaf litter derived from the terrestrial riparian (stream-side) zone, fine particulate organic matter (FPOM), originating from a variety of sources including fragmentation of CPOM, periphyton, primarily attached algae, and prey animals (Table 9.1, Figure 9.2). The timing of the invertebrate functional group sampling of the populations is important, being adjusted to avoid periods of general inactivity or absence from the aquatic habitat (e.g., Cummins et al. 1989).

**Table 9.1.** Functional group categorization and food resources (modified from Merritt and Cummins 1984).

| Functional Groups | Feeding Mechanisms | Dominant Food Resources | Particle Size Range of Food (mm) |
|---|---|---|---|
| Shredders | Chew conditioned or live vascular plant tissue, or gouge wood | CPOM-decomposing (or living hydrophyte) vascular plan | >1.9 |
| Filtering Collectors | Suspension feeders— filter particles with sediment or brush loose surface deposits | FPOM-decomposing detrital particles; algae and bacteria | 0.01–1.0 |
| Gathering Collectors | Deposit feeders—ingest sediment or brush loose surface deposits | FPOM-decomposing detrital particles; algae and bacteria | 0.05–1.0 |
| Scrapers | Graze mineral and organic surfaces | Periphyton-attached algae and associated detritus and micro flora and fauna | 0.01–1.0 |
| Predators | Capture and engulf prey or ingest body fluids | Prey-living animal tissue | >0.5 |
| Plant Piercers | Pierce cells and such fluids | Macroalgae | 0.01–1.0 |

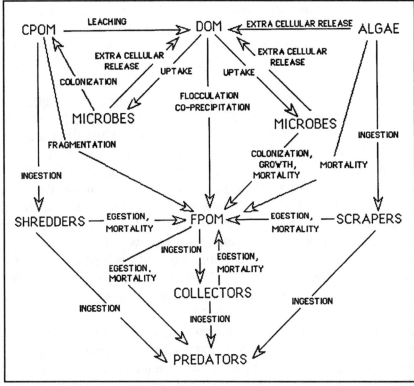

**Figure 9.2.** Simplified summary diagram of nutrient resource pools, invertebrate functional feeding groups, and microbe state variables, and the transfer functions between them. CPOM = Coarse Particulate Organic Matter (>1 mm); FPOM = Fine Particulate Organic Matter (<1 mm >0.45 μm); Algae = primary producers.

In the ensuing years since it was first proposed (Cummins 1973), the functional group technique has been refined (e.g., Cummins 1974, Cummins and Klug 1979, Merritt and Cummins 1984, Cummins and Wilzbach 1985, and Cummins 1988) and has been used in a number of basic research studies (e.g., Short and Maslin 1977, Cummins et al. 1981, Wallace et al. 1982, Webster and Benfield 1986). The procedure formed a primary element in the River Continuum Concept, one of the major running water ecosystem paradigms developed in the last decade (e.g., Vannote et al. 1980, Cummins et al. 1981, Minshall et al. 1983, 1985, Cummins et al. 1984). The linkage that exists between the taxonomically diverse but limited set of functional feeding groups and their food resource categories provides a sensitive and generalized measure of natural and anthropogenically driven changes in aquatic ecosystems.

However, despite the extensive world-wide use of functional group analysis in basic stream/river research, it has been used very little in repetitive monitoring or the analysis of impaired aquatic systems, even though it has obvious potential for rapid bioassessment (e.g., Plafkin et al. 1989).

The general characterics of the six functional groups, i.e., shredders, gathering collectors, filtering collectors, scrapers, predators, and plant piercers, are summarized in Table 9.1. Separations, which often transcend plant broad taxonomic categories, are based largely on morphological adaptations of the feeding structures (mouth parts, legs) and the behavior that drives these structures (e.g., modes of attachment that allow individuals positioned in the stream current to manipulate a filtering structure, and the construction of capture nets). The general characteristics used in functional group separation at a high level of efficiency may be either related or unrelated to these morphological-behavioral adaptations for food acquisitions. An example of a related generality would be all the families of net-spinning Trichoptera, which are filtering collectors. Unrelated generalities are exemplified by the presence of a color pattern on the dorsum of most nymphs belonging to predator plecopteran families, or the organic cases constructed by most shredder trichopteran genera in the family Limnephilidae. The basic procedure for separation of functional feeding groups, using both related and unrelated characteristics, is summarized in Figure 9.3. This flow diagram is for a first level of resolution. That is, the separation of functional feeding groups can be made at several levels of efficiency, depending upon the sophistication of the taxonomy that is applied (Figure 9.4). The first level, which requires only the most basic facility with invertebrate taxonomy, operates at an 80–85% efficiency of separation. By further categorizing the macroinvertebrates that either do not seem to fit easily into the groupings at the first level of resolution, or that are most likely to be misclassified, an additional 5–10% efficiency can be achieved. This is the second level of resolution. Another 1–5% efficiency of categorization can be attained if a third level of resolution is employed. Resolution at this level usually requires a microscope and taxonomic separations made at the generic or the species level.

At the first level of resolution, similarities resulting from convergent evolution in morphology and behavior that place taxa in the same functional group usually reside at the ordinal or familial levels. For examples, all Odonata and Megaloptera are predators, and filipalpian

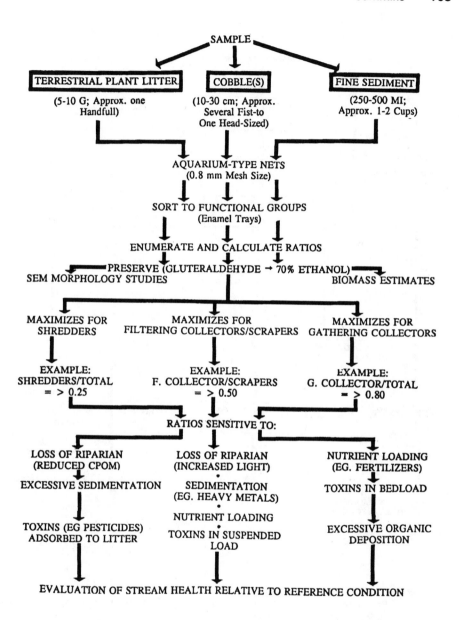

**Figure 9.3.** Sampling method for running water invertebrate functional feeding group analysis. Examples of the ranges for some group ratios are given for undisturbed (reference) headwater streams (orders 1-3). Examples of some factors most likely to affect the ratios are also indicated.

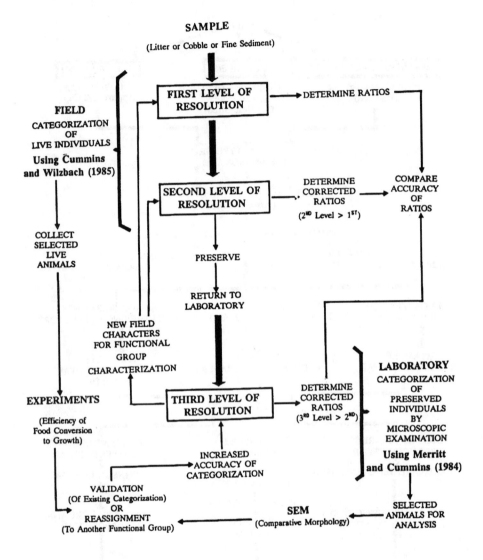

**Figure 9.4.** Macroinvertebrate functional group analysis at three levels of resolution. Refinement of categorizations by application of the results from morphological examinations and growth experiments.

plecopteran families are shredders (i.e., Peltoperlidae, Pteronarcidae, Capniidae, Leuctridae, and Nemouridae) (e.g., see Merritt and Cummins 1984, Cummins and Wilzbach 1985). The usual situation for the second level of resolution is that taxonomic separations are required at the generic level to accurately designate the appropriate functional group. For example, the genera *Tipula* and *Holorusia*, in the dipteran family Tipulidae, are shredders, while the genera *Hexatoma* and *Eriocera* are predators. At the third level of resolution, separation of species may be required to achieve accurate separation. For example, most species in the trichopteran genus *Rhyacophila*, are predators, but one is primarily a scraper.

Thus, the morphological and behavioral adaptations that control the acquisition of food resources are the basis for the macroinvertebrate functional group separations. However, the most efficient conversion of food to growth matches the best linkage between the acquisition system and the resource (Cummins and Klug 1979). In summary, the macroinvertebrate functional group-food resource associations (Table 9.1) are:

1) Shredders, with chewing mouth parts, feeding on coarse particulate organic matter (CPOM) with its attendant microbial flora, especially aquatic hyphomycete fungi (shredders - CPOM)

2) Gathering collectors, with generalized structures and strategies, feeding on sedimentary (depositional) fine particulate organic matter (FPOM, generally less than 1 mm in size) with its attendant microbial flora, primarily bacteria (gathering collectors - deposited FPOM)

3) Filtering collectors, with morphological-behavioral adaptations for filtering, feeding on water column (transported) FPOM (Filtering collectors - transported FPOM)

4) Scrapers, with adaptations for removing attached algae and associated items, the periphyton, from surfaces in fast flowing water (scrapers - periphyton)

5) Predators, specialized to locate, capture, and consume live prey by engulfing them or extracting their body fluids (predators - prey)

6) Plant piercers, with adaptations for sucking the fluids from individual cells of macroalgae (plant piercers - macroalgal cell fluids)

In accordance with the requirements of a rapid bioassessment, the functional group method can be accomplished quickly (e.g., two hours or less), primarily on site, and by field crews that have varying degrees of expertise in taxonomy, usually not beyond a level described as intermediate, and often less. The functional feeding group method is relatively independent of sample size, since the results are usually presented as dimensionless ratios; its use requires minimal equipment, and is accomplished at a level of resolution that can be chosen to be appropriate to the expertise of those performing the analysis. Because the method is responsive to changes in food resource base, it is sensitive to both localized, site-specific, and general, riparian-watershed (land use) influences. The localized input of a toxic effluent in particulate form might be an example of the former, while increased sediment or reduced litter inputs resulting from altered land use patterns would be examples of the latter (Figure 9.3).

An example of a stream functional group analysis is given in Table 9.2 for samples of cobble (see categories in Figure 9.3) substrates having similar surface areas. The counts were made by 21 "novices," i.e., undergraduate and graduate students with little or no experience in stream invertebrate taxonomy, and three "professionals," i.e., a trained

**Table 9.2.** Comparison of functional group analysis at first level of resolution. Counts from similar size rocks (Boulder/Cobble), Sugar Cr. PA. Novices, N = 21; Professional, N = 3.

| Functional Group | Novice Mean (SD) | Professional Mean (SD) |
|---|---|---|
| **Number** | | |
| Shredders | 1.6 (3.2) | 1.6 (3.1) |
| Collectors | | |
| Gathering | 48.9 (28.2) | 54.6 (19.7) |
| Filtering | 102.4 (44.9) | 127.0 (37.6) |
| Scrapers | 14.8 (10.7) | 3.9 (2.6) |
| Predators | 4.3 (4.2) | 28.0 (20.0) |
| **Ratios** | | |
| Shredders/Total | 0.01 (.02) | 0.06 (.04) |
| Collectors/Total | 0.87 (.08) | 0.84 (.17) |
| Scrapers/Total | 0.42 (.15)* | 0.02 (.02) |
| Filt Coll/Gath Coll | 3.21 (2.84) | 2.58 (1.84) |
| Shred + Coll/Scrape | 24.9 (37.9) | 119.3 (104.4) |

*Overestimate of scrapers (Ephemerellidae) at first level

technician and experienced graduate students. The similarity of the results, at the first level of resolution (Figure 9.4), between the counts made by the inexperienced and the experienced workers indicates the suitability of the method for use by crews having limited training.

A protocol for the refinement of the functional group method by adding further levels of resolution is summarized in Figure 9.4. The additional methods involve detailed morphological examination of potential feeding structures (mouth parts, legs, prolegs) and experiments conducted in laboratory culture or field enclosures. The experiments are designed to determine the efficiency with which food in the four resource categories (CPOM, FPOM, periphyton, and prey) is converted to growth (new biomass). This is based on the maximization of the match between the morphology behavior of food acquisiton, i.e., functional group, and the food resource category (Cummins and Klug 1979). The goal of these more detailed studies on invertebrates of unknown functional group association is to determine obvious characteristics of morphology and/or behavior, not necessarily directly related to food acquisition, that can be used in the field to make correct functional group assignments.

## CONCLUSIONS

The arguments presented above are that because of ineffective or non-existent technological transfer, there has been little incorporation of the extensive body of basic research data on lotic ecosystems into monitoring protocols or management strategies; and the most effective concepts and methodologies for transfer are likely those directed at evaluating functional characteristics of the systems. There are unresolved problems that bring into question the present almost total emphasis on taxonomically defined ecosystem structure. The case for using invertebrates as biological entities to serve as "windows" into system function is particularly valid for rapid, in-the-field evaluations. The invertebrate functional feeding group method is an example of a rapid field method that measures structure in functional categories. That is, although not a process measurement *per se*, the functional group procedure focuses on the dynamic linkages between co- and convergently evolved modes of food acquisition and a limited array of resource categories available to running water invertebrates. Furthermore, these resource categories are readily modified through changes in land use patterns and a variety of anthropogenic influences. The functional group approach is only one example of an extensive array of concepts and methodologies rooted in basic research that await technological transfer to the agendas of those charged with monitoring and management of freshwater ecosystems.

## REFERENCES

Averett, R. C. and D. M. McKnight, eds. 1987. *Chemical Quality of Water and the Hydrologic Cycle.* Lewis Publishers, MI. 381 p.

Cummins, K. W. 1973. Trophic relations of aquatic insects. *Ann. Rev. Ent.* 18:183-206.

Cummins, K. W. 1974. Structure and function of stream ecosystems. *BioScience* 24:631-641.

Cummins, K. W. 1988. The study of stream ecosystems: a functional view. pp. 247-262. *In*: L. R. Pomeroy and J. J. Alberts, eds., *Concepts of Ecosystem Ecology.* Springer-Verlag, NY. 384 p.

Cummins, K. W. and M. J. Klug. 1979. Feeding ecology of stream invertebrates. *Ann. Rev. Ecol. Syst.* 10:147-172.

Cummins, K. W., M. J. Klug, G. M. Ward, G. L. Spengler, R. W. Speaker, R. W. Ovink, D. C. Mahan, and R. C. Petersen. 1981. Trends in particulate organic matter fluxes, community processes, and macroinvertebrate functional groups along a Great Lakes drainage basin river continuum. *Verh. Int. Verein. Limnol.* 21:841-849.

Cummins, K. W., J. R. Sedell, F. J. Swanson, G. W. Minshall, S. G. Fisher, C. E. Cushing, R. C. Petersen, and R. L. Vannote. 1983. Organic matter budgets for stream ecosystems: problems in their evaluation. pp. 299-253. *In*: J. R. Barnes and G. W. Minshall, eds., *Stream Ecology.* Plenum Press, NY. 399 p.

Cummins, K. W., G. W. Minshall, J. R. Sedell, C. E. Cushing, and R. C. Petersen. 1984. Stream ecosystem theory. *Verh. Internat. Verein. Limnol.* 22:1818-1827.

Cummins, K. W. and M. A. Wilzbach. 1985. *Field Procedures for the Analysis of Functional Feeding Groups in Stream Ecosystems.* Pymatuning Laboratory of Ecology, University of Pittsburgh, Linesville, PA 16424. 18 p.

Cummins, K. W., M. A. Wilzbach, D. M. Gates, J. B. Perry, and W. B. Taliaferro. 1989. Shredders and riparian vegetation. *BioScience* 39:24-30.

Hilsenhoff, W. L. 1987. An improved biotic index of organic stream pollution. *Great Lakes Ent.* 20:31-39.

Hughes, R. M., S. A. Heiskary, W. J. Matthews and C. O. Yoder. 1994. Use of Ecoregions in Biological Monitoring. *In*: S. L. Loeb and A. Spacie, ed., *Biological Monitoring of Aquatic Systems*. Lewis Publishers, Boca Raton, FL.

Karr, J. R. 1987. Biological monitoring and environmental assessment: a conceptual framework. *Environmental Management*. 11:249-256.

Karr, J. R. 1991. Biological integrity: a long neglected aspect of water resource management. *Ecol. Applications* 1:66-85.

Lenat, D. R. 1988. Water quality assessment of streams using a qualitative collection method for benthic macroinvertebrates. *J. North Amer. Benthological Soc.* 7:222-253.

Merritt, R. W. and K. W. Cummins, eds. 1984. *An Introduction to the Aquatic Insects of North America*. Kendall/Hunt, Dubuque, 722 p.

Minshall, G. W., R. C. Petersen, K. W. Cummins, T. L. Bott, J. R. Sedell, C. E. Cushing, and R. L. Vannote. 1983. Interbiome comparison of stream ecosystem dynamics. *Ecol. Monogr.* 53:1-25.

Minshall, G. W., K. W. Cummins, R. C. Petersen, C. E. Cushing, D. A. Bruins, J. R. Sedell, and R. L. Vannote. 1985. Developments in stream ecology. *Can. J. Fish. Aquat. Sci.* 42:1045-1055.

Omernik, J. M. 1987. Ecoregions of the conterminous United States. *Ann. Assoc. Amer. Geogr.* 77:118-125.

Plafkin, J. L., M. T. Barbour, K. D. Porter, and S. K. Gross. 1989. *Rapid Bioassessment Protocols for Use in Streams and Rivers. Benthic macroinvertebrates and fish*. U.S. EPA EPA/444/4-89-001. p. 165.

Short, R. A. and P. E. Maslin. 1977. Processing of leaf litter by a stream detritivore: effect on nutrient availability to collectors. *Ecology* 58:935-938.

Vannote, R. L., G. W. Minshall, K. W. Cummins, J. R. Sedell, C. E. Cushing. 1980. The river continuum concept. *Can. J. Fish. Aquat. Sci.* 37:130-137.

Wallace, J. B., J. R. Webster, and T. F. Cuffney. 1982. Stream detritus dynamics: regulation by invertebrate consumers. *Oecologia* 53:197-200.

Webster, J. R. and E. F. Benfield. 1986. Vascular plant breakdown in freshwater ecosystems. *Ann. Rev. Ecol. Syst.* 176:567-594.

# CHAPTER 10

# Biodiversity, Biomonitoring, and the Structure of Stream Fish Communities

**Peter B. Moyle,** Department of Wildlife and Fisheries Biology, University of California, Davis, CA

## INTRODUCTION

The appalling loss of biodiversity that is now occurring worldwide is the foremost tragedy of our era because it is permanently altering our environment. This problem is not confined to tropical rainforests and third world countries but is, in fact, probably worse in temperate "developed" nations. Countries such as the United States, while having fewer species to lose, are losing higher proportions of them, despite having the knowledge and ability to save most of them. The problem is particularly acute in freshwater environments because they are downstream from most human activities, they are the foci of our urban areas, and they contain a substance in increasingly high demand—*fresh* water (Moyle and Leidy 1991). For example, 64% of 113 native fish taxa in California are extinct, endangered, or in need of special protection (Moyle and Williams 1990). This figure may seem on the high side compared to other regions without California's aridity and rapidly expanding human population; it surprised even me when it emerged from a study on fish "species of special concern" commissioned by the California Department of Fish and Game (Moyle et al. 1989). I suspect that similar surprises await biologists in other states who evaluate the status of their fish fauna taxon by taxon.

The magnitude of the problem of species decline and loss is so great and growing so rapidly that calls to save endangered species are starting to be replaced by calls to save endangered ecosystems. This sensible, if politically difficult, approach requires a great deal of biological knowledge in order to be effective. Protecting ecosystems requires the development of ways to monitor ecosystem health; the most effective ways seem to revolve around monitoring assemblages of organisms within the ecosystem. In freshwater systems this usually means fish (e.g., Karr 1991), aquatic invertebrates (e.g., Cummins 1994), or diatoms (Patrick 1994). Implicit in the idea of using assemblages of organisms (a.k.a. "communities") to monitor ecosystem health is that they are predictable; either their composition is fairly constant through time or they vary in a predictable manner in response to environmental changes. This in turn implies that the organisms making up the community evolved together and that biotic disputes had been settled in the distant past, leaving behind an assemblage of organisms well-adapted to each other and to their environment. This "Peaceable Kingdom" scenario, is, of course, more fiction than fact, especially in regard to fish communities, although some fish assemblages in tropical streams may fit this scenario surprisingly well (e.g., Wikramanayake 1990). In this paper I discuss why stream fish communities are so variable in composition, but why they nevertheless can be used to monitor ecosystem health. Most of my comments probably apply to other groups of stream organisms as well.

The problems with using stream fish communities to monitor ecosystem health stem from many factors that seem to promote variability in composition, usually in concert with one another. First, there is the background of natural events, especially the long-term influence of Pleistocene events and the short-term influence of natural variability in physical and chemical factors, particularly flow regime. Overlaid on this background is the influence of humans, resulting in most streams being altered in one way or another and containing introduced species. Distinguishing between natural and human induced causes of variability is very difficult (Grossman et al. 1990). Superimposed on these direct causes of biotic variability are difficulties in the *ways* in which we study fish communities. All these factors, which increase the variability in structure in stream fish communities (discussed in detail in the following sections), are countered in part by the considerable flexibility that temperate stream fishes have in their interactions with other species and with their environment. This flexibility allows fishes to rather quickly

develop fairly predictable community configurations following disturbances that cause changes in species composition.

## Pleistocene Events

During the Pleistocene, northern North America and Eurasia were
largely covered with glaciers and the rest of the continents had different
climatic regimes than they have today. The end of this era is usually and
arbitrarily designated as being about 10,000 years ago, but Pleistocene
events were, and *are*, a major factor shaping modern biotic communities,
a fact often not appreciated by ecologists studying them (Ricklefs 1987).
For example, in the western United States, late Cenozoic climates were
much wetter and the fish faunas were more diverse than they are today
(Smith 1981). As the region became drier, extinctions of fish species
occurred fairly rapidly in the many isolated basins of the region. Moyle
and Herbold (1987) argue that the remarkable similarity in the morph-
ologies and life histories of western North American freshwater fishes to
European freshwater fishes is the result of selective extinctions in the late
Pleistocene in response to changing conditions. Among other things, this
means that modern fishes in these regions probably evolved as part of
more complex faunas and coexisted with species that are now extinct
(Moyle and Vondracek 1985).

In contrast, in eastern North America the vast Mississippi drainage
served as a refuge for fishes and it is likely that late Cenozoic events
actually promoted speciation, helping to account for the astonishing
complexity of the fish fauna (Smith 1981, Moyle and Herbold 1987). The
northern parts of the Mississippi drainage were covered by glaciers and
had to be reinvaded by fishes after the glaciers melted, a process that may
still be taking place.

Overall, there is no reason to assume that modern temperate fish
communities were established under stable or benign conditions or that the
present distribution patterns of species reflect a full recovery from
Pleistocene events. Instead, natural invasions (or reinvasions) of species
into regions should be expected, as should the presence of "unsettled"
communities, where biotic interactions are changing community structure.
In fact, with the advent of accelerated global climate change, "natural"
shifts in fish communities should be occurring more rapidly than ever.

## Natural Variability

In a paper that sparked a lengthy debate in the literature, Grossman et al. (1982) argued that repeated and unpredictable flood events kept the fish community of an Indiana stream from developing a consistent, predictable structure. The responses to this paper almost universally found persistence and predictability in the fish communities of the Indiana stream as well as in other streams of eastern North America (e.g., Ross et al. 1985, Matthews et al. 1988). However, the picture that emerges from these studies is that stream fish communities are not nearly as neatly structured as had been previously assumed (Moyle and Li 1979), especially in streams with highly variable flow regimes (Matthews et al. 1988, Grossman et al. 1990).

Poff and Ward (1989) have created an excellent framework for understanding the effects of variable conditions on the structure of fish and invertebrate communities in streams by analyzing the flow regimes of 78 streams scattered around the USA. They found that the streams fell into nine clusters (types) according to various combinations of degree of intermittency, flood frequency, flood predictability, and flow predictability. They suggest that the relative importance of abiotic and biotic factors in determining community structure is related to the strength and interactions of these four factors. Thus, community predictability and complexity should increase with the permanence and predictability of the flow regime. Community predictability should also be high in streams characterized by extreme variability in flow regimes because such conditions would strongly select for a species-poor assemblage of fishes adapted to predictably severe conditions (Baltz and Moyle 1993).

The effects of short-term environmental fluctuations on aquatic communities is the subject of other papers in this volume and so will not be covered further here.

## Human Modification

Poff and Ward (1989) deliberately chose streams for their study that had unimpaired flow regimes, recognizing that streams with flows regulated by dams often had diminished faunas (Bain et al. 1988). Dams, however, are only the most conspicuous human factor that alter streams and their biotic communities. There are few streams anywhere that have not been changed by drainage-wide activities of farming, grazing of livestock, logging, and urbanization, usually in combination. Often the changes wrought by these activities are so pervasive and so old (150+ years) that we hardly know what the original streams looked like, much less what their fish communities were like. For example, Karr et al. (1985) document that streams of the Midwest are so extensively modified that presumably natural asemblages of fish exist primarily in isolated pockets of habitat in wooded areas, which are affected by both upstream and downstream stream alterations and pollution. In general, the least disturbed streams are smaller headwaters, which usually contain a small coldwater-adapted subset of the regional fauna that is nearly universal across geographic regions (*Oncorhynchus/Salvelinus* sp., *Cottus* sp., *Catostomus* sp. and *Rhinichthys* sp., Moyle and Herbold 1987). Even in these streams, the presence of introduced salmonids is common.

Despite the high degree of modification, many smaller Midwestern streams contain surprisingly diverse fish faunas that tend to have the characteristics expected of "natural" assemblages (e.g. Schlosser 1982a,b, Ross et al. 1985, Steedman 1988). It is likely that even such faunas contain more species now than they did 100-200 years ago, as the result of introductions (many unrecognized as such) and expansion of ranges of species adapted to the altered conditions. For example, Horwitz (1982) documented the spread of *Notropis spilopterus* in the lower Delaware River drainage in response to the creation of suitable habitats through human development. By and large, *N. spilopterus* failed to replace a very similar (ecologically and morphologically) species, *N. analostanus*, which showed signs of expanding its range as well. Both species were native to the drainage.

Obviously, altered habitats contain altered fish faunas. How altered a particular fish fauna is depends on the degree to which the habitat has been altered and on the nature of the original fish fauna. In the Mississippi drainage, there are many highly altered streams that appear to contain a high proportion of their original fish fauna. This is presumably due to a combination of reasonably high water quality, numerous habitat refuges, and the adaptability of the fishes to environmental change. For example, two of the most intensively studied streams in eastern North America, Jordan Creek, Illinois, and Brier Creek, Oklahoma, have drainages that have been extensively modified by agriculture. Nevertheless, their fish communities appear to be relatively predictable (Schlosser 1982a,b, Matthews et al. 1988).

More drastic faunal shifts are seen in altered streams of the western United States as a result of greater alteration (including the removal of water) of the streams and of a native fish fauna poorly adapted to the altered conditions (Moyle and Williams 1990). The altered habitats are thus more vulnerable to invasion by introduced fishes; the native fishes have largely been replaced by non-native fishes, mostly introduced from the Mississippi drainage. The reason for this seems to be that the introduced species have reproductive strategies more suited for the altered habitats and eliminate native fishes through predation and competition. In unaltered streams, introduced fishes are eliminated by extreme seasonal flood events that characterize the streams (Minckley and Meffe 1987). The altered streams often contain more species than were in the original native fauna but there is no indication of community persistence; the dominant species tend to vary from year to year.

## Introduced Species

Introduced species and altered habitats go together. Altered habitats are easier to invade than pristine habitats if the invaders are better adapted to the new conditions created, or if populations of native predators and competitors have been reduced or destabilized. Unfortunately, there are few streams today without introduced species, a reflection of the scarcity of pristine streams and of high frequency with which fishes and aquatic invertebrates are being introduced into new areas, either by anglers, by agencies, or by accident (Moyle 1986). Particularly insidious are "bait bucket" introductions by anglers because they are frequently fishes collected from nearby drainages and therefore hard to detect as introduced species. If an introduction is successful, it may disrupt the original fish community until an apparent "steady state" is established, which includes the introduced species. The degree to which an established fish community is altered by an introduced species varies with the species introduced but, in general, piscivorous fishes seem most likely to cause major alterations of a stream fish community while detritivores or omnivores seem least likely to cause major disruptions of a stream fish community.

## Piscivores

The importance of piscivorous fishes in structuring assemblages of aquatic organisms is just beginning to be appreciated (Power et al. 1985, Kerfoot and Sih 1987). When a newly introduced predator becomes abundant, it can have effects on the existing fish community that are major and rapid. The most spectacular examples of this have been invasions of lakes, where invasions of predators have severely altered native fish communities in short periods of time (e.g., Hughes 1986). The effects of introduced predators in streams are generally less severe. For example introduced Sacramento squawfish (*Ptychocheilus grandis*) spread throughout the Eel River drainage, a large coastal drainage in California, in less than 10 years without causing the immediate extinctions of any native fishes. As soon as they invaded a new area, there was an instantaneous shift in the use of space by the resident fish species, making competition among them more likely (Brown and Moyle, 1991). However, because threespine stickleback (*Gasterosteus aculeatus*) seem to be unable to avoid predation by large squawfish (L. Brown, unpublished data), it is likely that this species will be largely eliminated from the drainage. This is especially likely to occur after the stickleback populations have been reduced as the result of high stream flows, as happened in the spring of 1990. Likewise, Lemley (1985) found that introduced predatory green sunfish (*Lepomis cyanellus*) altered the distribution and abundance of native fishes in small North Carolina streams.

In Martis Creek, California, brown trout (*Salmo trutta*) apparently suppress native fish populations through predation (Moyle and Vondracek 1985), but their ability to do so is regulated by the flow regime of the stream in relation to spawning of both predator and prey. All the native fishes are spring spawners while the introduced brown trout is a fall spawner. As a result, if there are winter high flow events 2 – 3 years in a row, survival of brown trout eggs and young will be greatly reduced. With reduced brown trout populations, more of the native fishes (e.g., Tahoe sucker, *Catostomus tahoensis*) will reach maturity and spawn successfully. A model contructed to show the interactions between the suckers and trout indicated that if the suckers built up a large population of adults too large in size to be preyed upon by the trout, their high reproductive rates would allow them to keep ahead of trout predation (Strange et al. 1993). If, however, there are several springs in a row with conditions unfavorable for sucker reproduction, then brown trout would once again be able to keep suckers at low numbers. In this predator-prey system, at least two states ("equilibria") are possible, each capable of sustaining itself until the right series of environmental conditions change it.

## Omnivores

Small omnivorous fishes are common in streams (Vadas 1990) and they seem to be fairly flexible in their interactions with other fishes. Presumably, they also have a food supply that is so abundant that it rarely has a chance to limit population growth before other natural factors do such as flood and predators. Thus, their introduction seldom seems to alter the structure of fish communities into which they are introduced. In the Eel River, prior to the introduction of squawfish, California roach (*Lavinia symmetricus*) were introduced and became widespread. Although this small omnivore is now the most abundant fish in many parts of the drainage, it does not appear to have any impact on the native fishes (L. Brown, unpubl. data). However, it has had a major impact on invertebrates and algae in the river (Power 1990).

In the eastern United States, introduced omnivorous cyprinids are abundant in many drainages, apparently without noticeable effects on native omnivores; this has been used as evidence of "faunal undersaturation" (Hocutt et al. 1986). In Sri Lanka, the transfer of five omnivorous or herbivorous species to a depauperate drainage nearby seemed to have little effect on the native fishes, although at least one of the introduced species developed an expanded niche (Wikramanayake and Moyle 1989).

The results of introductions make it clear that monitoring programs need to take into account the effects of introduced species. They also make it clear that understanding how a community is structured is important because in some systems more than one structure may be indicative of the desired conditions.

## Problems of Study Design

A major problem faced by researchers in community ecology is how to design a study that truly demonstrates the nature of a stream fish community, or any other ecological system. This same problem faces anyone who is trying to design a community-oriented monitoring study as well (Conquest et al. 1994). Part of the problem is defining the community. Ideally, a community should be "an ensemble of species in some area whose limits are determined by the practical extent of energy flow (Drake 1990)." Although the boundaries of a community thus defined should be identifiable in a non-arbitrary manner, doing it in a practical way (compatible with a monitoring program) is not possible, given our present level of knowledge. Thus, we have to assume that one or more easily studied taxonomic assemblages (fish, invertebrates, diatoms) can be surrogates for the "real" community.

Grossman et al. (1990) discuss many of the problems of study design for stream fish communities, so here I will discuss just one as an example of the difficulties faced in proper design: the problem of determining at what scale sampling should take place. Will a single site do, or do we have to sample an entire drainage in order to understand what is going on with the fish communities? The study area used by Grossman et al. (1982), for example, was criticized for being both too small (Ross et al. 1985) and too large (Herbold 1984). Unfortunately, assemblages of stream fishes are rarely discrete; they merge imperceptibly with other assemblages both upstream and downstream (Meffe and Sheldon 1990). In practice, assemblages are defined as groups of fishes co-occurring within a broad habitat type or stream reach, with the area occupied being large enough so that most of the members of the assemblage can spend their entire life there. This means that probably several stations need to be sampled within each reach in order to obtain an adequate idea of the community being sampled. Grossman et al. (1990) concluded that even once the proper scale of study had been resolved, most studies showed that stream fish assemblages varied considerably in persistence and predictability and urged caution in applying the results of community studies of stream fishes to stream management.

## Can We Use Fish Communities for Monitoring?

There are many factors that can create unpredictability in stream fish communities, making them difficult to monitor or to use for monitoring. Essentially, each community is the result of a regional fish fauna having passed through a series of environmental filters, which results in the particular assemblage observed (Figure 10.1). One of the broadest filters, the effects of Pleistocene events, caused changes over broad geographic and temporal scales. Below this are a series of filters which reflect environmental variability, both spatially and temporally. In addition, most streams have been significantly altered by human activity and have been invaded by non-native fish species, altering the communities (although these changes are often poorly recognized).

Despite these problems, there is growing evidence that most stream fish communties are fairly predictable over at least short (2 - 10 yr) periods of time or that the direction of change can be predicted in response to environmental change. For example, Meffe and Sheldon (1990) showed that small South Carolina stream sites that had been experimentally defaunated recovered very rapidly to their original communities, while Matthews (1986) demonstrated the rapid recovery of the fish communities of an Arkansas stream following a catastrophic flood. The changes observed by Brown and Moyle (1991) in fish communities in response to the invasion of predatory squawfish were predictable, based on studies in streams where squawfish and other species naturally co-occurred. The situation in Martis Creek, where introduced brown trout prey on native fishes, is more complicated because at least two "equilibria" seem to exist, but predictable communities are nevertheless present (Strange et al. 1993).

In fact, it is likely that multiple solutions to the problem of community structure are possible in any collection of temperate fishes, accounting for some of the variability in structure noted by Grossman et al. (1990). Most assemblages of temperate stream fishes show resource partitioning in axes of both food and space. Yet most temperate fishes are not highly specialized and they are typically members of many different assemblages, as species composition of streams varies from place to place, region to region. The niche dimensions of many species, especially small cyprinids, are consequently likely to vary from place to place. This flexibility is not accidental because these fishes have made it through

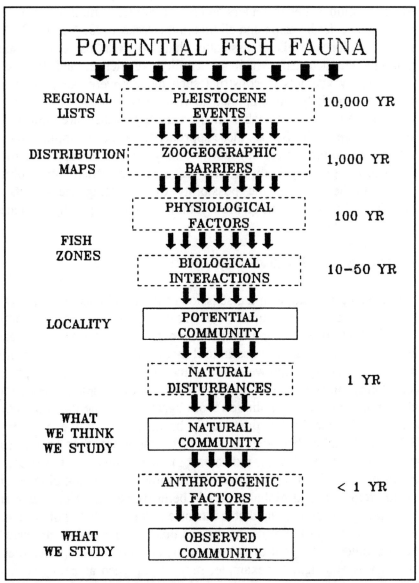

**Figure 10.1.** Fish fauna passing through a series of environmental filters that results in a particular assemblage.

the exceptional environmental turbulence of the Pleistocene. In such circumstances, extreme specialization is unlikely to develop (Leigh 1990).

To really understand how communities change through time in response to both natural and human-induced environmental change, we need to develop a better understanding of how communities are constructed by species invasion and elimination (Drake 1990, Nee 1990). We need to develop models of assembly that will help us to understand community characteristics such as stability, resilience, and food web patterns. Early attempts to construct such models were not particularly successful, especially when applied to fish communities (Li and Moyle 1981) but new, more tractable theoretical approaches are now being developed (Nee 1990). So far, these approaches have been applied mainly to laboratory communities, but the time may be ripe for applying these theories to the growing body of knowledge of stream fish communities. If they can be applied, the possibility of predicting the effects of environmental perturbations on the communities may be greatly enhanced.

## Monitoring and Biodiversity

I opened this essay with a discussion of the rapid loss of biodiversity that we are experiencing in aquatic systems because protecting biodiversity is closely tied to effective biological monitoring. At one level, each species can be regarded as a useful tool that may be particularly sensitive to a substance or condition we desire to monitor, or as an integral part of a community, useful for monitoring the effects of complex phenomena, such as non-point source pollution from agriculture. The usefulness of a community for monitoring presumably becomes less as the number of species it contains becomes less. More important than monitoring communities or species as indicators of water quality, however, is monitoring them for their own sake. Increasingly, protecting biodiversity at local, regional, and global scales is being regarded as our duty and responsibility and something that is intrinsic to the well-being of humanity (Leopold 1949, Nash 1989). This means that not just *any* assemblage of species should be the goal of management; an introduced species that is part of an assemblage, for example, may be regarded as being as much of a pollutant as a pesticide or chemical discharge from a factory. Thus, while we have recorded two equilibria in the fish community of Martis Creek, the one dominated completely by the introduced brown trout should be regarded as less acceptable than the one dominated by native fishes. We cannot get rid of the brown trouts of the world but we can manage the environment to favor the native species. I hope this philosophy will be part of any biological monitoring program in the future.

## ACKNOWLEDGMENTS

Comments by W. Bennett, L. Brown, L. Meng, E. Strange, and R. White on an early draft were extremely helpful.

## REFERENCES

Bain, M. B., J. T. Finn, and H. E. Booke. 1988. Streamflow regulation and fish community structure. *Ecology* 69:382-391.

Baltz, D. M. and P. B. Moyle. 1993. Invasion resistance to introduced species by a native assemblage of California stream fishes. *Ecological Applications*. 3:246-255.

Brown, L. R. and P. B. Moyle. 1991. Changes in habitat and microhabit partitioning with an assemblage of stream fishes in response to predation by Sacramento squawfish (*Ptychocheilus grandis*). *Canadian Journal of Fisheries and Aquatic Sciences*.

Conquest, L. L., S. C. Ralph and R. J. Naiman. 1994. Implementation of Large-Scale Stream Monitoring Efforts: Sampling Design and Data Analysis Issues. *In*: S. L. Loeb and A. Spacie, eds., *Biological Monitoring of Aquatic Systems*. Lewis Publishers, Boca Raton, FL.

Cummins, K. W. 1994. Bioassessment and Analysis of Functional Organization of Running Water Ecosystems. *In*: S. L. Loeb and A. Spacie, eds., *Biological Monitoring of Aquatic Systems*. Lewis Publishers, Boca Raton, FL.

Drake, J. A. 1990. Communities as assembled structures: do rules govern pattern? *Trends in Ecology and Evolution* 5:159-164.

Grossman, G. D., J. F. Dowd, and M. Crawford. 1990. Assemblage stability in stream fishes: a review. *Environmental Mangement* 14:661-671.

Grossman, G. D., P. B. Moyle, and J. O. Whitaker, Jr. 1982. Stochasticity in the structural and functional characteristics of an Indiana stream fish assemblage: a test of community structure. *American Naturalist* 120:423-454.

Herbold, B. 1984. Structure of an Indiana stream fish association: chosing an appropriate model. *American Naturalist* 124:561-572.

Hocutt, C. H., R. E. Jenkins, and J. R. Stauffer, Jr. 1986. Zoogeography of the fishes of the central Appalachians and central Atlantic coastal plain. pp. 161-212. *In*: C. H. Hocutt and E. O. Wiley, eds., *The Zoogeography of North American Freshwater Fishes*. John Wiley, NY. 866 p.

Horwitz, R. J. 1982. The range and co-occurrence of the shiners *Notropis analostanus* and *N. spilopterus* in southeastern Pennsylvania. *Proceedings of Academy of Natural Sciences of Philadelphia* 134:178-193.

Hughes, N. F. 1986. Changes in the feeding biology of Nile perch, *Lates nilotica* (L.) (Pisces: Centropomidae) in Lake Victoria, East Africa, since its introduction in 1960, and its impact on the native fish community of Nyasa Gulf. *Journal of Fish Biology* 29:521-548.

Karr, J. R., L. A. Toth, and D. R. Dudley. 1985. Fish communities of midwestern rivers: a history of degradation. *BioScience* 35:90-95.

Kerfoot, C. and A. Sih, eds. 1987. *Predation Direct and Indirect Impacts on Aquatic Communities*. University Press of New England, Hanover, NH. 386 p.

Leigh, E. G., Jr. 1990. Community diversity and environmental stability: a re-examination. *Trends in Ecology and Evolution* 5:340-344.

Lemly, A. D. 1985. Suppression of native fish populations by green sunfish in first-order streams of Piedmont, North Carolina. *Transactions of American Fisheries Society* 114:705-712.

Leopold, A. 1949. *Sand County Almanac*. Oxford University Press, NY. 228 p.

Li, H. W. and P. B. Moyle. 1981. Ecological analysis of species introductions into aquatic systems. *Transactions of American Fisheries Society* 110:772-782.

Matthews, W. J. 1986. Fish faunal structure in an Ozark stream: stability, persistence and a catastrophic flood. *Copeia* 1986:388-397.

Matthews, W. J., R. C. Cashner, and F. P. Gelwick. 1988. Stability and persistence of fish faunas and assemblages in three midwestern streams. *Copeia* 1988:945-955.

Meffe, G. K. and A. L. Sheldon. 1990. Post-defaunation recovery of fish assemblages in southeastern blackwater streams. *Ecology* 71:657-667.

Minckley, W. L. and G. K. Meffe. 1987. Differential selection by flooding in the stream communities of the arid American Southeast. pp. 93-104. *In*: W. J. Matthews and D. C. Heins, eds., *Evolutionary and Community Ecology of North American Stream Fishes*. University of Oklahoma Press, Norman, OK. 248 p.

Moyle, P. B. 1986. Fish introductions into North America: patterns and ecological impact. pp. 27-43. *In*: H. A. Mooney and J. A. Drake, eds., *Ecology of Biological Invasions of North America and Hawaii*. Springer-Verlag, NY. 321 p.

Moyle, P. B. and B. Herbold. 1987. Life history patterns and community structure in stream fishes of western North America: comparisons with eastern North America and Europe. pp. 25-32. *In*: W. J. Matthews and D. C. Heins, eds., *Evolutionary and Community Ecology of North American Stream Fishes*. University of Oklahoma Press, Norman, OK. 248 p.

Moyle, P. B. and R. L. Leidy. In press. Loss of biodiversity in aquatic ecosystems: evidence from fish faunas. *In*: P. L. Feidler and S. K. Jain, eds., *Conservation Biology: The Theory and Practice of Nature Conservation, Preservation, and Management*. Chapman and Hall, NY. 507 p.

Moyle, P. B. and B. Vondracek. 1985. Persistence and structure of the fish assemblage of a small California stream. *Ecology* 66:1-13.

Moyle, P. B. and J. E. Williams. 1990. Biodiversity loss in the temperate zone: decline of the native fish fauna of California. *Conservation Biology* 4:275-284.

Moyle, P. B., J. E. Williams, and E. D. Wikramanayake. 1989. *Fish species of special concern of California*. California Department of Fish and Game, Sacramento, CA. 222 p.

Nash, R. F. 1989. *The Rights of Nature*. University of Wisconsin Press, Madison. 180 p.

Nee, S. 1990. Community Construction. *Trends in Ecology and Evolution* 5:337-340.

Patrick, R. 1994. What are the Requirements for an Effective Biomonitor? *In*: S. L. Loeb and A. Spacie, eds., *Biological Monitoring of Aquatic Systems*. Lewis Publishers, Boca Raton, FL.

Poff, N. L. and J. V. Ward. 1989. Implications of streamflow variability and predictability for lotic community structure: a regional analysis of streamflow patterns. *Canadian Journal of Fisheries and Aquatic Sciences* 46:1805-1818.

Power, M. E. 1990. Effects of fish in river food webs. *Science* 250:811-814.

Power, M. E., W. J. Matthews, and A. J. Stewart. 1985. Grazing minnows, piscivorous bass, and stream algae: dynamics of a strong interaction. *Ecology* 66:1448-1456.

Ricklefs, R. E. 1987. Community diversity: relative roles of local and regional processes. *Science* 235:167-171.

Ross, S. T., W. J. Matthews, and A. A. Echelle. 1985. Persistence of stream fish assemblages: effects of environmental change. *American Naturalist* 126:24-40.

Schlosser, I. J. 1982a. Fish community structure and function along two habitat gradients in a headwater stream. *Ecological Monographs* 52:395-414.

Schlosser, I. J. 1982b. Trophic structure, reproductive success, and growth rate of fishes in a natural and modified headwater stream. *Canadian Journal of Fisheries and Aquatic Sciences* 39:968-978.

Smith, G. R. 1981. Late Cenozoic freshwater fishes of North America. *Annual Review of Ecology and Systematics* 12:61-93.

Steedman, R. J. 1988. Modification and assessment of an index of biotic integrity to quantify stream quality in southern Ontario. *Canadian Journal of Fisheries and Aquatic Sciences* 45:492-501.

Strange, E. M., P. B. Moyle, and T. C. Foin. 1992. Interactions between stochastic and deterministic processes in stream community assembly. *Environmental Biology of Fishes*. 36:1-15.

Vadas, R. L., Jr. 1990. The importance of omnivory and predator regulation of prey in freshwater fish assemblages of North America. *Environmental Biology of Fishes* 27:285-302.

Wikramanayake, E. D. 1990. Ecomorphology and biogeography of a tropical fish assemblage: evolution of assemblage structure. *Ecology* 71:1756-1764.

Wikramanayake, E. D. and P. B. Moyle. 1989. Ecological structure of tropical fish assemblages in wet-zone streams of Sri Lanka. *Journal of Zoology (London)* 218:503-526.

# CHAPTER 11

# Using Benthic Macroinvertebrate Community Structure for Rapid, Cost-Effective, Water Quality Monitoring: Rapid Bioassessment

**David R. Lenat,** NC Division Environmental Management, Water Quality Section, Raleigh, NC
**Michael T. Barbour,** Tetra Tech. Inc., Owings Mill, MD

## INTRODUCTION

The scientific literature relating to water pollution biology frequently describes very time-consuming and labor-intensive, surveys. Such surveys, however, may not represent the most typical kind of biological monitoring effort. Most state agencies are responsible for water quality monitoring on thousands of streams; this situation is also typical of many areas outside the United States. To expend large amounts of time and money on a single stream is equivalent to ignoring water quality problems in many other streams.

The development of more cost-effective biological monitoring strategies has come to be known as "Rapid Bioassessment". The term "Community Assessment Approach" has also been used in this context because of a focus on the evaluation of community structure and function. The emphasis in such monitoring is on obtaining "rapid" results in order to expedite both assessment of water quality problems and any subsequent management decisions. Specifically, the goal is to expend the minimum amount of effort required to get reproducible, scientifically valid results. Most rapid bioassessment programs are designed to go from field collections (3 to 5 sites) to a report in five working days. Rapid bioassessment also implies some shortcut techniques relative to traditional collections. This shortcut usually involves qualitative/semiquantitative sampling, or processing a targeted number of organisms/site.

One of the charges of the Societas Internationalis Limnologiae's Biological Monitoring Working Group involves education and sharing of information on water quality assessment methods. Many organizations with no prior biological monitoring experience will be looking to adopt rapid bioassessment methods; other agencies will be interested in adding rapid bioassessment to their existing program. It should be taken as part of the Working Group's charge to think about what will work in parts of the world where resources, personnel, and taxonomic knowledge are minimal.

This paper addresses the many choices to be made in setting up a rapid bioassessment program, with discussion of the advantages/disadvantages of each choice. We have not included all of the approaches used in North America, but we have reviewed a representative cross-section of "rapid" water quality assessment techniques. Procedural recommendations are based on the type of habitat sampled, the type of data required, available sampling equipment, and available taxonomic skill. The most important decisions involve the level of taxonomy and the general sampling method.

This is an "opinion" paper and reflects our biases. It borrows from many recent publications, including Plafkin et al. (1989), Lenat (1988, 1990), and Resh and Jackson (1993). Other important sources of information include bioassessment programs in Maine (Susan Davies/David Courtemanch), Ohio (Jeff DeShon), Arkansas (Bruce Shackelford) and New York (Robert Bode). Discussion will be limited primarily to stream/river benthos and will focus on North American work.

## CHOOSING A RAPID BIOASSESSMENT METHOD

### General Considerations

Before establishing a rapid bioassessment program, some general decisions must be made about what portion of the aquatic community will be collected, how to locate individual samples within a stream reach, and quality assurance methods. To some extent, these desicions may be influenced by existing programs in each geographic area.

## Which aquatic assemblage?

The organisms most commonly used in stream water quality monitoring are periphyton, fish, and benthic macroinvertebrates. Both Plafkin et al. (1989) and Karr (1991) discuss the use of different aquatic groups and this topic is not treated in detail here. Ideally, a bioassessment agency should have the capability of using all of these groups in biological monitoring, as each additional group will contribute further useful information towards an assessment of water quality. The Kentucky Division of Water is one of the few biological monitoring agencies which routinely uses periphyton, fish, and macroinvertebrate collections; they have found that such integrated monitoring yields results superior to the usual single-group collections (Mike Mills, personal communication). The U.S. Environmental Protection Agency (E.P.A.) also has proposed using all three groups for lake and stream monitoring (Paulsen et al. 1994) Periphyton collections are particularly useful because of the many indicator assemblages (Patrick 1994): enrichment, brine, etc. Fish are especially useful in relating changes in habitat quality and showing changes in the abundance of game fish. The Index of Biotic Integrity (IBI: Karr et al. 1986) has been very successful in the Midwest for relating fish community structure to water quality. The success of the IBI approach is being copied by investigators working with benthic fauna (Ohio EPA 1989, Shackleford 1988).

Each group of aquatic organisims has been commonly used for water quality monitoring. Further discussion in this paper focuses on benthic macroinvertebrates, but does not imply that invertebrates are always the best choice. The selection of a group of aquatic organisms is most strongly influenced by an agency's focus, the known sensitivity of the group, available taxonomic skills, and comparability of existing data bases.

## Location of individual samples.

Some biological monitoring agencies collect macroinvertebrate samples from what appears to be a representative habitat, while other agencies sample either the "most productive habitat" (Plafkin et al. 1989) or any high current habitat with "structure" (Lenat 1988). The latter approach usually means riffle sampling, although snags and root mats will be sampled in naturally sandy streams. The choice of sampling location affects the ability of an investigator to separate habitat effects from water quality effects.

When attempting to measure water quality impacts, most investigators attempt to minimize between-site differences in habitat. For this reason, benthic macroinvertebrate sampling must always be accompanied by some measurement of habitat characteristics. Changes in habitat quality, however, often cannot be avoided in some studies, particularly investigations of nonpoint source pollution. In these cases, habitat degradation may be the major problem under investigation. It is possible, however, that sampling the most productive habitat (least stressed) may minimize between-site habitat differences and focus on between-site differences in water quality. This hypothesis reflects the North Carolina experience (see discussion below), but it is largely untested.

EPA Rapid Bioassessment protocols (Plafkin et al. 1989) emphasize collections from the most productive habitat. It is thought that such collections allow between-site evaluations of habitat quality, with fewer complications from between-site differences in habitat diversity. These EPA protocols use an analysis system (largely untested) that attempts to partition adverse biological changes into water quality effects, habitat quality effects, or combined effects. This system incorporates an overall evaluation of habitat structure that produces four habitat quality ratings: "comparable" (to the control), "supporting", "partially supporting" and "nonsupporting".

The most commonly encountered habitat changes are related to sediment inputs from both local erosion (often the cutting of stream banks) and erosion within the stream's catchment. Sediment tends to accumulate in pool, run, and bank areas, and pass through high-current habitats. The scouring effect of this transient sediment is not well understood. The negative effects of scour also vary with flow, being most severe during periods of high flow (Lenat et al. 1981).

Standardized sampling methods have been developed largely for shallow (wadable) streams. Organizations that require the same sampling technique for a wider variety of stream types often use artificial substrates (e.g., Maine). Location of samples in larger rivers requires considerable judgment on the part of the collector. Under high to normal flow conditions, the diversity of organisms of shore/bank areas in deeper streams and rivers may be much greater than the diversity of unstable midstream substrates. In North Carolina's coastal plain rivers, we have found that large (i.e., stable) snags in areas of current have much greater diversity than other microhabitats (Benke et al. 1985).

## QA/QC.

The defensibility of any method depends on some evaluation of the method's variability. For this reason, all aspects of a bioassessment program should be evaluated by a Quality Assurance/Quality Control program. QA/QC procedures are particularly important for organizations that are initiating a bioassessment program. A useful reference for setting up QA/QC programs is the proceedings of a recent workshop on Quality Assurance programs (Hart 1990). This document discusses QA/QC elements and the requirements for testing the validity of ecological surveys.

Development of a detailed Standard Operating Procedures (SOP) manual is important for the assurance of uniform sampling methods, especially if there are changes in collection personnel. Any monitoring group also should incorporate periodic validation tests to show that they can verify water quality conditions in both pristine and polluted areas. A few such tests will go a long way towards building confidence in the ability to detect changes in water quality.

Using a reference site to calibrate different investigator teams is a commonly used QA procedure. A reference site may reflect typical conditions in a catchment, but minus a pollutant of interest. Reference sites also should be located in a largely undisturbed catchment. Such relatively pristine streams may be used to define the conditions for each ecoregion (Hughes and Larsen 1988, Hughes et al. 1994). Periodic sampling of ecoregion reference sites may be used to test for any seasonal changes in the benthic macroinvertebrate community.

Most invertebrate sampling assumes that collections are representative of a larger stream segment. This assumption may be tested by sampling two reaches within a fairly homogeneous stream segment. The use of different collection teams may be tested by "overlap" sites (sites sampled by both teams). Overlap sites are also useful in comparing results from different agencies. Alabama and Mississippi are using overlap sites as part of their ecoregion testing; likewise, North Carolina and South Carolina will be establishing some overlap sites.

Taxonomic QA/QC is difficult, as it may require consultation with authorities outside of the local organization. Part of the difficulty is the potential for conflict in establishing what is a "correct" identification. However, the accuracy of identifications is most commonly assessed by comparing results with regional experts, thereby showing that identifications are consistent with those of other established taxonimists. Within-group exchanges (if a group contains more than one taxonomist) also are essential in maintaining taxonomic consistency.

## Comparability with existing data bases

Contiguous states might share similar ecoregions, i.e., to have areas where they will have similar biological expectations; this would warrant the use of similar biological monitoring approaches. In such situations, existing approaches might be adopted with minimal developmental work. Alabama and Mississippi are examples of state agencies that are attempting to share the costs of sampling ecoregion reference sites. Similarly, Oregon, Washington, and Idaho are cooperating in the development of bioassessment programs, and Delaware, Maryland, and Virginia are evaluating common elements in their coastal plain aquatic ecosystems.

## Making Choices

Setting up a rapid bioassessment program involves some difficult choices on level of taxonomy (e.g., order, family, genus, or species), sampling strategy, and sampling equipment. There are many "correct" choices and it is unlikely that any two water quality agencies will have identical programs. The most appropriate choices may also vary depending on the needs of individual surveys.

## Taxonomy

Most biological monitoring techniques are dependent on the correct identification of the organisms collected. Better taxonomy can be expected to produce more accurate results, with a better ability to detect subtle changes in water quality. A biological monitoring group should always strive to increase its level of taxonomic proficiency, and should interact with regional taxonomic experts. In the long run, much time and money can be saved by hiring qualified and experienced taxonomists; if this is not possible, monitoring agencies should develop adequate training programs. In establishing a new biological monitoring unit, we suggest building in a training phase before starting pollution assessment work. The training phase would be completed when taxonomic QA checks reach some acceptable level. Adequate taxonomic skills also are required to pass other QA/QC checks, especially the validation of known high quality sites.

At the coarsest level of taxonomy, personnel can be trained to identify orders of aquatic insects and then estimate taxa richness per order. This type of taxonomy can be done with a hand lens (as in Cummins and Wilzbach 1985); it is used with the Operational Taxonomic Unit (OTU) approach of Mason (1979) and the Sequential Comparison Index of Cairns and Dickson (1971). With this level of taxonomy, it is particularly important that investigators prove that they can verify unpolluted conditions. Such approaches are most appropriate for citizen monitoring groups (Klein 1983).

Family level taxonomy has been recommended by many investigators, including Kaesler and Herricks (1979), Osborne et al. (1980), Furse et al. (1984), and Hilsenhoff (1988a). EPA's Rapid Bioassessment Protocol II (Plafkin et al. 1989) also was set up using family-level identifications. It is argued that less taxonomic training is required to identify macroinvertebrates at the family level, thus minimizing costs of hiring and/or training personnel. Furthermore, there may be a large savings in sample processing time with family-level identifications; most organisms can be identified in the field to a family level. This "reduction in sample processing time" argument is most valid for personnel with limited expertise; more experienced taxonomists can identify many organisms to the genus/species level almost as fast as to the family level. The latter statement is based on the experience of the authors, as well as conversations with many working taxonomists. However, such taxonomic proficiency may require 3-4 years of experience in collecting and identifying aquatic macroinvertebrates.

The use of family-level identifications may be a compromise between production of rapid results and the accuracy of water quality evaluations. Analysis of family-level data must make certain assumptions about the composition of species assemblages and their ecological sensitivities. For example, an identification of Hydropsychidae will usually assume the presence of relatively tolerant *Hydropsyche* and *Cheumatopsyche* species, although a much more intolerant assemblage of Hydropsychidae might actually be present.

Genus/species level identifications clearly increase the precision of site classifications (Resh and Unzicker 1975, Furse et al. 1984, Hilsenhoff 1982, Rosenberg et al. 1986), and such identifications are essential for the use of pollution "indicator assemblages" (Resh 1979, Simpson and Bode 1980). The use of a biotic index also is vastly improved with more precise identifications (Hilsenhoff 1988a). Family-level identifications may be inappropriate for examinations of changes over time, assessment of the length of a recovery zone, and the determination of special high quality waters. Family-level identifications are primarily useful for one-time assessments of water quality in a specific area, or in the ranking of sites for additional study. If a monitoring agency intends to gradually build a large regional data base (for future between-site and between-date comparisons), then genus/species level identifications are more appropriate. Family-level identifications, therefore, may save time in the short run, but become inefficient for long-term monitoring efforts.

If a biological monitoring group has very limited resources, they may wish to focus taxonomic expertise on certain intolerant (and easily identified) groups. North Carolina has developed an abbreviated "EPT" collection method, characterized by fewer samples (4) per site and collections limited to the Ephemeroptera, Plecoptera, and Trichoptera. In other geographic areas, a different set of intolerant organisms might be selected, as long as these organisms were relatively large, long-lived, easy to identify, and diverse at regional reference sites.

Species-level identification are most troublesome for specimens that must be slide-mounted for definitive identifications, i.e., Chironomidae (midges) and Oligochaeta (worms). However, biological monitoring results will be greatly enhanced by the identification of these groups. Taxa richness of Chironomidae may not be linearly related to water quality (Lenat 1983), but their indicator value offsets such problems. Information on these groups is a crucial step in deducing both the type of pollution as well as level of stress. Chironomid taxonomy can be very difficult, especially if expert help is not readily available. However, the literature on chironomid taxonomy and ecology is steadily improving; two recent examples include Wiederholm (1983) and Hudson et al. (1990) There are many methods that can be used to speed up identification of Chironomidae, especially sorting before slide-mounting specimens, subsampling, and learning to recognize some genera without slide mounting of specimens.

## Sampling method

The selection of a sampling method is probably the most difficult choice in establishing a rapid bioassessment method. Generally speaking, a sampling method is made "rapid" by some restriction on the number of organisms picked and identified. An investigator also may restrict the type of organisms selected (EPT sampling, chironomid pupal exuviae, etc.). Restrictions on the number of organisms to be identified can be acomplished either by using an ordinal scale (rare, common, abundant, etc.), or by subsampling the organisms at each location. Sample splitters are sometimes used, often with the idea of subsampling until reaching some target number of organisms. With any type of subsampling, the assumption is made that all organisms have the same probability of being selected. Both ordinal-scale counting and subsampling de-emphasize a quantitative measurement of abundance per unit area, although they may still include quantitative measurement of abundance or dominance. Abundance per unit area is a highly variable measurement spatially and temporally, even in the absence of any water quality change (Resh and Rosenberg 1989, Lenat 1990).

The concept of rapid bioassessment has embraced a variety of sampling strategies, while adhering to certain basic concepts. At one end of the spectrum, Plafkin et al. (1989) recommended a single-habitat sample (this may be a composite collection), using a subsampling technique to withdraw approximately 100 organisms. At the other end of the spectrum, Lenat (1988) suggests multiple-habitat collections with ordinal-scale counts. The selection of an appropriate sampling strategy will depend on many factors, including the purpose of the survey, the methods used by nearby monitoring groups, the expertise of collecters, preference for quantitative versus qualitative data, and the characteristics of streams in a particular geographic area. Some agencies have attempted to combine strategies by supplementing standardized collections with additional "visual" sampling from large rocks and logs.

Multiple-habitat collections more completely census the invertebrate fauna than single-habitat collections. Method-testing in Plafkin et al. (1989) indicated that 100-count samples (from riffles) collected 35 – 68% (n = 5) of the taxa in multiple-habitat samples. The relative variability of the two methods has not been adequately evaluated; each method will be certain to have its own proponents, based largely on their unquantified experience in collecting invertebrate samples. Additional testing is needed to evaluate the reliability of site classifications using each approach.

Multiple-habitat sampling must be paired with either qualitative (presence/absence) or semiquantitative (ordinal) enumerations. Semiquantitative sorting of samples usually puts taxa into categories (absent, rare, common, or abundant), depending on their relative abundance in the final sample (Lenat 1988). Organisms are picked out of samples "in proportion to abundance", or until no new taxa are encountered. This process may seem too variable to some investigators, because the stopping point is not readily apparent, but it has worked well in practice for North Carolina Division of Environmental Management (DEM) biologists. While such a sorting technique requires (or works better) with experienced people, North Carolina often uses nonspecialists as part of collecting teams without affecting data quality.

Another sampling choice is whether samples should be sorted in the field or in the lab. Lab sorting of samples should result in more efficient removal of organisms, especially the smaller (less conspicuous) taxa. The laboratory also provides a more controlled environment for sample processing and permits more detailed quality assurance procedures. However, field sorting produces less battered specimens and has immediate feedback on the adequacy of samples. The problem of sorting efficiency can be dealt with by "saturation" sampling: many samples and many habitats. There are also some techniques that help in field sorting: field preservation, elutriation, and special fine-mesh samplers (Lenat 1988). North Carolina biologists regularly collect about 25 chironomid taxa per site (maximum around 40 per site) using these techniques.

## Sampling gear

In setting up a biological monitoring program, decisions must be made concerning the type of sampling equipment and the mesh size used to process samples. Most benthos workers (in streams) have used a 400–600 micron mesh size to collect stream invertebrates. In multiple habitat sampling, however, this is often supplemented with a smaller mesh size (200 – 300 micron) aimed at the collection of Chironomidae and other small invertebrates. Generally, the smaller the mesh size, the smaller the area of substrate that may be sampled.

Rapid bioassessment programs require relatively versatile collecting gear; equipment that can be used across a wide range of depths, velocities, and substrate types. This decision tends to favor large kick nets and "sweep" nets (D-frame nets, A-frame nets, and dip nets). This type of equipment collects large composite samples, often covering a range of microhabitats. Such composite samples are intended to reduce the variability inherent in small-scale samples. Small-scale (microhabitat) differences in current, canopy, substrate, etc., should not be allowed to obscure between-site changes in water quality. When samples from specific substrates are desired, Hess and Surber samplers are widely used.

Some agencies supplement collections of larvae with sampling of adults (light traps, sweeps) or chironomid pupal exuviae. Given the necessary taxonomic skills, such collections are useful in assigning species names. In high quality areas, it may be very important to document the presence of rare species, or new species.

Visual collections are very important for a complete census of the macroinvertebrate community. Simple disturbance samplers may entirely miss some of the dominant taxa: tightly attached genera (e.g., *Psychomyia*), stone-cased Trichoptera, cryptic genera (e.g., *Nyctiophylax*), or those limited to special microhabitats. There will always be some taxa that will only be found on the largest logs or largest boulders in the stream. Visual collections may be the most important type of collection in large and slow-flowing rivers. This type of collection is the most difficult; experienced collectors will always get more taxa than novices. Therefore, this is an important area on which to focus training efforts.

## Level of sampling intensity

Some rapid bioassessment programs establish a variety of sampling methods; the particular method selected is determined by the survey objectives. Plafkin et al. (1989) established three levels of sampling intensity for invertebrates: Rapid Bioassessment Protocols I, II, and III. Protocol I (reconnaissance survey) is intended to document the existence of obvious water quality impairment; it also may be used to see if more detailed studies are required. In this protocol, multiple-habitat collections of invertebrates (field identifications) are conducted by an experienced biologist. It is expected that this process would only require 1 to 2 hours per site with a single investigator. Protocol II uses a 100-count organism subsample from two composited kick-net collections, plus a coarse particulate organic material (CPOM) sample. Organisms are identified in the field to a family level. Protocol II data can be used for site rankings, and requires about 1.5 to 2.5 hours per site with one biologist and one technician. Protocol III uses a similar sampling strategy, but organisms are identified in the lab to the genus/species level.

The idea of different levels of sampling intensity has been adopted by a number of state agencies. Both New York and Ohio have quantitative and qualitative sampling options. North Carolina may use up to three levels of sampling intensity, although only two of these methods have been rigorously tested. Genus/species level-identification is used for all these collection methods, and all samples are field-picked. North Carolina's standard semi-quantitative collections (10 samples) requires 4.5 to 6 person-hours at each site, usually using a three-person collecting team. A more rapid survey (3 person-hours per site) can be achieved using a four-sample collection, limited to EPT taxa. In this sytem, the more rapid methods are subsamples of the standard collection method. The system has been set up so that all stations may be compared, even if different collection methods have been used. Such comparisons are based on regression equations that predict 10-sample EPT taxa richness from the four-sample collections (Eaton and Lenat 1991).

## Data Analysis: Metrics

Rapid bioassessment analyses frequently rely on several different ways of examining and summarizing macroinvertebrate data. This use of multiple metrics originates with the Index of Biotic Integrity used with fisheries information. Just as the IBI had to be modified for regional differences, so must invertebrate indices be retested and adapted for different geographic areas. Some analysis metrics also may vary with flow, season, and ecoregion; and they may not be linear over the entire range of water quality. For example, total taxa richness may increase with moderate enrichment, due to increases in facultative and tolerant groups.

Metrics may be evaluated by either a percent change (versus a reference site) or as a category variable (poor, fair, good, etc.). Categorization is more difficult to establish, as it implies comparisons with a large amount of information from ecoregion reference sites. A "percent change" analysis must take into account both differences due to water quality and the inherent variability of invertebrate communities. There may be a 20 – 30% difference between replicates from identical habitats (c.f. Shackleford 1988), so this level of change can be established as "no impact". Slight-moderate impact usually falls into the range of 30 – 50% change, while severe impact is often established by changes greater than 50%. The exact percentages used to define impact need to be tested for each metric, stream size, stream type, ecoregion, and collection method.

Many agencies rely heavily on paired stations (examples: Arkansas and Maine). This kind of approach automatically adjusts for some temporal changes (especially normal seasonal changes), but control sites must be carefully established to compensate for normal spatial changes. Look carefully at metrics that define any change (versus reference) as stress. For example, do not confuse natural longitudinal changes in the stream community (Hynes 1972) with changes in water quality.

It is difficult to statistically evaluate most rapid bioassessment results if no replicates are taken. Method testing can generate information on the normal variance of any metric, and such tests are important in order to establish the validity of rapid bioassessment data. When data are generated for certain subcategories (usually orders of aquatic insects), it is possible to statistically compare sites with a chi-square analysis or a Wilcoxon signed-rank test. Parrish and Wagner's (1983) modification of the chi-square statistic also has potential for testing between-site differences, but has not yet been widely tested.

The discussion of analysis metrics given below utilizes information from many sources, especially Resh and Jackson (1993).

## Structure metrics

Metrics in this category usually can accommodate qualitative or quantitative data; they include taxa richness (number of species, S) analyses, and similarity coefficients. The most commonly used richness metrics are total taxa richness and EPT (Ephemeropta + Plecoptera + Trichoptera) taxa richness. Ohio biologists use the number of mayfly species and the number of caddisfly species as separate metrics. For large studies, the number of "unique" species (limited to single site, with tolerant species excluded) has proved useful (Crawford and Lenat 1989) for comparing sites, although this metric is not suitable for assigning water quality ratings.

EPT taxa richness is the single most reliable metric used by North Carolina biologists. EPT taxa richness was very sensitive to changes in water quality, and it proved to be less variable than total taxa richness in relation to between-year changes in flow (unpublished data from North Carolina's ambient monitoring network).

Simple taxa richness metrics may not always demonstrate impact if drift specimens can easily colonize a stressed site. To deal with this problem, North Carolina uses an "EPT abundance" metric, which is the sum of abundance values (1=rare, 3=common, 10=abundant) for all EPT taxa. This metric responds to a combination of both taxa richness and the relative abundance of the more intolerant groups. Examination of taxa richness data also must take into consideration sources of normal variation, including:

1) Increases (especially for total taxa richness) with mild enrichment.
2) Decreases in nutrient-poor water, especially for single-habitat or subsampled collections. Species accumulation curves may be flatter in such habitats, but intensive, multiple-habitat samples can be used to demonstrate high taxa richness.
3) Seasonal changes, especially taxa richness increases during recruitment periods. Spring peaks are particularly troublesome in the southeastern United States. Limited experiments by North Carolina biologists suggest that seasonal patterns in small streams differ from those in large streams.

4) Effects of stream size, especially decreases in smaller streams. Taxa richness also may decrease in larger rivers, largely as a function of current speed. It is extremely difficult to assign reliable bioclassification to rivers that are transitional between lotic and lentic habitats.

Similarity coefficients are commonly combined with paired-site surveys in order to detect water quality problems. These metrics are useful for identifying compositional changes in the invertebrate community that may not be evident in taxa richness metrics. Resh and Jackson (in press) have reviewed some of the more common similarity coefficients, indicating that suitable equations exist for both qualitative data (e.g., Jaccard Index) or quantitative data (Pinkham-Pearson Index). Maine biologists have had success using a Community Loss Index (Courtemanch and Davies 1987), and this index also has been used for EPA's Rapid Bioassessment Protocols (Plafkin et al. 1989). However, subsequent analyses (Barbour et al, 1992) have shown the Community Loss Index to be highly variable among relatively unimpacted sites. Arkansas (Shackleford 1988) uses both a Common Taxa Index (CTI) and a Common Dominants Index (CDI), and have published classification criteria (versus reference) to define slight, moderate and severe impacts. North Carolina also uses the Arkansas system, modifying the CDI to include all "abundant" taxa.

## Community balance

Community balance metrics need some measure of abundance (or relative abundance) and may not be used with presence/absence data. These metrics attempt to measure the evenness (i.e., redundancy) of the invertebrate community, assuming that a highly redundant community (with few species dominating) is evidence of stress. Most ecologists now prefer to examine taxa richness and evenness separately, rather than combine them into a diversity index (Godfrey 1978, Hughes 1978). Florida and Ohio, however, continue to use the Shannon diversity index.

Many mathematical formulas can be used to compute evenness (Peet 1975), but most rapid bioassessment programs prefer very simple calculations. The simplest number is the percent contribution of the dominant (most abundant) taxon (Plafkin et al. 1989). Other biological monitoring groups look at the percent contribution of some tolerant or intolerant group:

1) EPT abundance/Total abundance
2) EPT abundance/Chironomid abundance (Plafkin et al. 1989, also see Shackleford's (1988) modification of this metric)
3) Hydropsychidae abundance/Total Trichopera, Baetidae abundance/ Total Ephemeroptera (Barbour et al., 1992)
4) Tanytarsini abundance/Total Chironomidae expressed as a percentage (Ohio)
5) Maine uses the abundance (or log abundance) of key groups, including Ephemeroptera, *Hydropysche, Cheumatopsyche*, Polycentropidae, Hydropsychidae, *Brachycentrus*, Perlidae, Tanypodinae, and *Rheotanytarsus*.

These metrics must be used with extreme caution, due to *large* differences that may occur between ecoregions, stream sizes, flow conditions, and seasons. In particular, the abundance of Chironomidae at stressed sites can be highly variable (i.e., dominant during low flow, but sharply reduced by scour at high flow). Barbour et al. (1992) recommends deleting the EPT abundance/Chironomidae abundance metric. Community balance metrics may also be influenced by mesh sizes, especially if they include some estimate of chironomid abundance.

## Tolerance metrics

Tolerance metrics assume the presence of a large data base, with some pollution tolerance rating assigned to each species. At the lowest level of discrimination, organisms are simply categorized as intolerant, facultative, or tolerant. These categories are then used to compute the percent of intolerant species, based either on taxa richness or abundance.

More precise categorization of water pollution tolerance is needed to compute a biotic index. In North America, most biotic indices are based on Chutter's (1972) system as modified by Hilsenhoff (1982). Hilsenhoff used a large Wisconsin data base (2000+ collections) to assign tolerance values (integer values from 0–5; this was later expanded to a 0–10 range). Recently, a similar attempt was made to derive tolerance values from a large North Carolina data base (Lenat, 1993a). While each list of tolerance values should show the same overall patterns, there will be many regional differences. For example, many of the southeastern species would not be on a Wisconsin list, and a population in North Carolina may prove to be more or less tolerant than a Wisconsin population.

The presence of deformities also falls into the category of "tolerance metrics". While such analyses are an integral of fish Index of Biological Integrity (IBI) surveys, they are not frequently used for macroinvertebrates. The exception is some excellent work on morphological deformities in chironomid larvae (Warwick 1988, Warwick and Tisdale 1988). Preliminary observation of this concept in North Carolina indicated that it is an excellent (and quick) way to separate the effects of organic pollution from added toxic stress, especially at sites dominated by *Chironomus* larvae (Lenat 1993b).

## Feeding group metrics

Feeding group metrics are an outgrowth of river continuum theory (Vannote et al. 1980). This theory predicts that energy flow (and therefore feeding types) should change in a predictable fashion from small streams to large streams. Specifically, the proportion of the invertebrate community that is represented by any given functional feeding group should be predictable from some measure of stream size. Any change in the proportions of the various feeding groups (versus a control site) may indicate water quality problems, but it is important not to equate all such changes with a decline in water quality. This type of water quality analysis requires quantitative data.

Advantages of this approach include the ease of sample collection (the CPOM sample of Plafkin et al. 1989) and the possibility of assigning feeding categories in the field (Cummins and Wilzbach 1985, Cummins 1993). The disadvantages of this approach relate to the difficulty of assigning feeding categories (Resh and Jackson, 1993). It is difficult to assign feeding types to groups which are poorly known or poorly identified. For example, Chironomidae are frequently left at a family level, but many include predators, filter-feeders, collector-gatherers, shredders, and grazers. Many taxa are flexible in their feeding strategies and the same species may have significant between-stream and between-season changes in diet (Chapman and Demory 1963, Kawecka 1977). Comprehensive investigations of feeding habits of Wisconsin Plecoptera,

Ephemeroptera and Trichoptera (Shapas and Hilsenhoff 1976) have pointed out the difficulty of correctly predicting feeding strategies solely from literature reviews. These investigators found that even some species in the same genus had completely different diets. Some taxa also may shift feeding strategies as they grow older (Anderson and Cummins 1979), with some Plecoptera and Trichoptera switching from algal/detrital diets to animal diets in later larval stages. At this time, there is little published information to indicate whether these complicating factors are a major or minor problem in the evaluation of stream water quality.

Feeding groups are usually examined through ratios: shredders/total abundance, scrapers/filterer-gatherers, specialists/generalists (Maine). Initial tests of the first two metrics suggest that they work better when expressed as a percentage, rather than as simple ratio (Barbour et al.). Shackleford (1988) also suggests comparing the proportional makeup of paired sites with a similarity index. The shredder ratio should work well in areas where the shredder guild is relatively abundant at the control site. This restriction will limit its use in the warmer and/or drier portions of the United States. Resh and Jackson found that only the scraper ratio detected water quality changes in the "Mediterranean" climate of northern California. As expected, they also noted sharp seasonal changes in feeding group metrics.

Feeding group metrics will show any change in the proportional composition of feeding types as "stress". Investigators should look for any other habitat changes that might affect food resources in the stream, including changes in riparian vegetation, substrate, and canopy cover.

## Special Problems

There are a number of situations that may create special difficulties in the interpretation of rapid bioassessment data. These difficulties, however, are not unique to rapid bioassessment, but will apply to almost any type of chemical and biological sampling. Problems arise when criteria are applied to streams that do not correspond with the data base used to develop these criteria. If your reference sites include a full range of ecoregions, stream sizes, and seasons, these problems are reduced. Few agencies, however, have developed their reference sites to this extent, and these special situations must be examined cautiously.

## Small streams

Invertebrate taxa richness in small streams may be limited by low flow (even drying up), lower habitat diversity, and greater thermal constancy (Vannote and Sweeney 1980). Ecoregions with poorly drained or desert soils are especially susceptible to low flow during droughts. North Carolina investigations (unpublished data) indicate that small streams (less than 3 meters wide) have expected EPT taxa richness that are at least 20% lower than larger streams. If it is not possible to adjust taxa richness values from smaller streams, use of tolerance metrics may be a better method of water quality evaluation.

## Atypical lotic habitats

Atypical lotic habitats must always be carefully evaluated, including areas below dams, habitats with low current velocity, springs, and swamps. Larger rivers may be difficult to evaluate because of reduced current speed and the problems of finding a comparable unstressed reference station.

## Season

Seasonal changes in the macroinvertebrate fauna are a major headache for routine water quality monitoring. Almost any metric may vary seasonally (in the absence of any pollution), and the expected amount of seasonal change may be different for each year (warm versus cold years), ecoregion, and stream size. When possible, periods of rapid changes should be avoided, especially major recruitment periods. If this is not possible, some adjustment must be made to summarize statistics by comparison to reference site data. North Carolina metrics are based on collections during the summer months; these metrics may be used with little adjustment from June to September. Outside of this period, especially in spring, it is often necessary to resample a reference site to assess the degree of seasonal change. In surveys of large catchments, it is advisable to separately assess seasonal changes for streams of different sizes. Ohio and Maine also have established "optimal" months for sampling invertebrates.

North Carolina DEM biologists have experimented with several ways of seasonally adjusting taxa richness values (Trish MacPherson, unpublished data). An equal proportional adjustment at all sites has not worked, as highly stressed sites often had less seasonal change than control sites. Instead, North Carolina biologists are experimenting with "subtraction" methods, where certain species with short life cycles (especially spring species) are deleted before applying taxa richness criteria. Changes in the Plecoptera assemblage are the largest problem in North Carolina, but Ephemeroptera taxa richness also may have sharp seasonal peaks. Seasonal adjustment methods (which differ according to month and ecoregion) must be tested by their success in predicting summer taxa richness at reference sites.

Biotic index values also vary seasonally. Hilsenhoff based his biotic index criteria on spring sampling; he recommends subtracting 0.5 (for a 0–10 scale) for periods "when biotic index values are abnormally high" (i.e., summer) (Hilsenhoff 1988b). North Carolina biotic index criteria (Lenat 1993a) are based on summer sampling. Recommended seasonal adjustments to winter/spring sample, were +0.2 for piedmont and coastal plain areas and +0.5 for mountain areas. (0–10 scale).

# Appropriate Use of
# Rapid Bioassessment Methods

The shortcuts used in rapid bioassessment methods may be associated with either greater data variability or a reduced ability to detect water quality problems. For certain problem situations, one may wish to switch to more intensive methods.

## Upstream-downstream studies

Almost any rapid bioassessment method will identify severe impact in a typical "upstream-downstream" survey. A comparison of paired sites allows the use of many different kinds of metrics. More subtle impacts, however, may not be detected with family-level identifications or 100-count samples.

## Basin-wide studies

Basin-wide studies (with a range of stream sizes and habitats). This type of survey may be difficult to establish with paired sites, although USGS basin-wide studies are set up in this manner (Gurtz 1994). No direct comparisons should be made betweens streams of different sizes, and normal longitudinal changes should not be interpreted as water quality problems. A cautious use of similarity indices, feeding group metrics, etc., is suggested. North Carolina biologists often utilize a mixture of sampling methods in large basin-wide surveys: EPT surveys in tributaries and full-scale qualitative surveys at key mainstream sites. Preliminary reconnaissance sampling may be useful in establishing a final study plan.

## Trend studies

Looking at changes over time ("before-and-after" surveys, trend monitoring sites) requires the best quality data. Only precise taxonomy and complete sampling will be able to detect subtle changes; some rapid bioassessment methods may be inappropriate for ambient monitoring networks. It can be very difficult to separate the effects of between-year changes in flow from actual changes in water quality.

## Conservation biology

There is an increasing use of macroinvertebrate studies to identify special "Outstanding Resource Waters" and other high quality aquatic habitats. North Carolina's Bioassessment Group may expend up to one-third of its total efforts in this type of survey. European authors (Jenkins et al. 1984) refer to such sampling as "conservation biology". Conservation biology requires precise taxonomy (species-level where possible); supplemental collections of adults also will improve the reliability of identifications. Both verification of high diversity and identification of rare species are important components of conservation biology. For this reason neither subsampling nor family-level identifications are appropriate survey techniques.

## Examples

EPA (1991) compiled a summary of bioassessment programs and biological criteria in Arkansas, Florida, North Carolina, Maine, and Ohio. Case studies also were presented from Connecticut, Delaware, Minnesota, Nebraska, New York, Texas, and Vermont. Only two examples are given here, representing two contrasting styles of rapid bioassessment: North Carolina's multiple-habitat samples and the EPA single habitat approach using a 100-organism subsample.

### North Carolina

The North Carolina examples are drawn from a large study of discharger toxicity (Table 11.1). We have selected two surveys indicating a toxic impact on stream fauna and two surveys indicating no significant impact. Sites are rated with taxa richness criteria and a biotic index; these metrics are evaluated against ecoregion expectations. Paired-site metrics also are employed here: a Common Dominants Index, a Common Taxa Index and a Wilcoxon Signed Rank test. All data are in an ordinal scale (absent=0, rare=1, common=—3, abundant=10), which precludes the use of evenness measures or feeding group metrics.

Of the two studies with an apparent impact, only one developed tolerant "indicator" assemblages; only in this study was there a change in biotic index values and a large change in community composition. The similarity measures may be overly sensitive; they sometimes indicated "slight impact" even when there appeared to be no significant change in water quality.

Of the four studies, two were in agreement with the effects predicted by water chemistry and standard toxicity tests. The other two studies, however, produced unexpected results. Discharger B was in compliance with chemical permit limits and was passing *Ceriodaphnia* chronic toxicity tests. Conversely, Discharger D was failing their toxicity tests, but did not appear to be affecting stream biota. This information clearly indicated that rapid bioassessment methods are often better than water chemistry or toxicity tests in detecting water quality problems.

**Table 11.1.** Examples of information from North Carolina's standardized qualitative collection method: taxa richness by group, and summary statistics. Four "upstream/downstream" studies of wastewater treatement plants, two surveys indicating impact vs. two surveys indicating no impact. All data collected September 1990.

| Station | With Impact | | | | No Impact | | | |
|---|---|---|---|---|---|---|---|---|
| | A1 | A2 | B1 | B2 | C1 | C2 | D1 | D2 |
| **Group** | | | | | | | | |
| Ephemeroptera | 6 | 1 | 8 | 1 | 11 | 10 | 7 | 7 |
| Plecoptera | 3 | 0 | 0 | 0 | 3 | 4 | 2 | 3 |
| Trichoptera | 6 | 5 | 5 | 3 | 8 | 7 | 6 | 8 |
| Coleoptera | 2 | 3 | 3 | 3 | 6 | 7 | 6 | 8 |
| Odonata | 3 | 5 | 6 | 7 | 5 | 6 | 8 | 10 |
| Megaloptera | 2 | 2 | 1 | 2 | 1 | 3 | 2 | 2 |
| Diptera: Misc. | 4 | 5 | 5 | 1 | 6 | 4 | 3 | 3 |
| Diptera: Chironomidae | 22 | 17 | 19 | 7 | 25 | 25 | 27 | 22 |
| Oligochaeta | 2 | 2 | 3 | 2 | 1 | 3 | 4 | 5 |
| Crustacea | 1 | 1 | 1 | 1 | 1 | 2 | 2 | 4 |
| Mollusca | 0 | 0 | 1 | 0 | 2 | 1 | 2 | 3 |
| Other | 1 | 0 | 0 | 0 | 0 | 0 | 3 | 4 |
| | | | | | | | | |
| EPT Taxa Richness | 12 | 6 | 13 | 4 | 22 | 21 | 15 | 18 |
| EPT Abundance | 52 | 28 | 59 | 4 | 127 | 100 | 88 | 102 |
| Total Taxa Richness | 50 | 40 | 52 | 27 | 69 | 72 | 72 | 80 |
| Taxa Richness Rating | Fair | Poor | Fair | Poor | G-F | G-F | Fair | G-F |
| Biotic Index (1-5 scale) | 3.49 | 3.53 | 3.24 | 3.88* | 2.95 | 3.06 | 3.10 | 2.93 |
| | Fair | Fair | G-F | Poor | G-F | G-F | G/F | G/F |
| **Indicator Assemblages** | | | | | | | | |
| Organics/Enrichment | - | + | - | - | - | - | + | - |
| Toxics | - | - | - | + + | - | - | - | + + |
| | | | | | | | | |
| **Between-site Tests** | | | | | | | | |
| Common Taxa | 44% (Mod) | | 29% (Severe) | | 76% (None) | | 63% (Slight) | |
| Common Dominants** | 50% (Mod) | | 8% (Severe) | | 76% (Slight) | | 68% (Slight) | |
| | | | | | | | | |
| Wilcoxon Signed Rank | NS | | Significant | | NS | | NS | |
| | | | | | | | | |
| Impact? | Yes | | Yes | | No | | No | |
| Agreement with Chemistry and Toxicity data? | Yes | | No | | Yes | | No | |

*Between-site difference >0.5, significant change in water quality, bioclassifications based on biotic index numbers are tentative. Scale = 0.5
**Dominants defined as any abundant taxa (≥10 specimens/site).
***GF = Good-Fair

## EPA Protocol III example

Example data sets are presented from four of the EPA rapid bioassessment workshops (Table 11.2). Both nonpoint source problems and point source discharges were included in these test cases. Two surveys were selected that indicated slight impact (Kansas and Texas), while two other surveys indicated moderate impact (Massachusetts and District of Columbia). Impact assessment followed the procedures outlined in Plafkin et al. (1989).

**Table 11.2.** Examples of data using EPA's rapid Bioassessment Protocol III, 100-count samples. Information from four rapid bioassessment workshops, 1990[1]. Values are relative abundance (percent of total).

| | Slight Impact | | | | Moderate Impact | | | |
|---|---|---|---|---|---|---|---|---|
| | A1 | A2 | B1 | B2 | C1 | C2 | D1 | D2 |
| **Group** | | | | | | | | |
| Ephemeroptera | 19 | 35 | 35 | 17 | 4 | 1 | 47 | 4 |
| Plecoptera | 0 | 0 | 21 | 11 | 4 | 0 | 26 | 0 |
| Trichoptera | 0 | 11 | 2 | 3 | 4 | 0 | 28 | 1 |
| Coleoptera | 11 | 23 | 10 | 2 | 2 | 0 | 0 | 0 |
| Megaloptera | 0 | 0 | 0 | 0 | 1 | 0 | 2 | 2 |
| Diptera: Misc. | 2 | 14 | 24 | 3 | 13 | 0 | 6 | 5 |
| Diptera: Chironomidae | 8 | 18 | 10 | 48 | 74 | 7 | 8 | 85 |
| Oligochaeta | 10 | 0 | 0 | 3 | 0 | 3 | 0 | 0 |
| Crustacea | 23 | 3 | 0 | 2 | 1 | 63 | 0 | 0 |
| Mollusca | 25 | 2 | 0 | 5 | 5 | 4 | 3 | 3 |
| Other | 2 | 0 | 0 | 1 | 0 | 7 | 2 | 17 |
| **Selected Metrics** | | | | | | | | |
| Total Taxa Richness | 16 | 14 | 12 | 17 | 25 | 11 | 21 | 26 |
| EPT Taxa Richness | 3 | 2 | 5 | 5 | 8 | 1 | 11 | 2 |
| HBI[2] | 6.54 | 5.94 | 6.86 | 8.66 | 4.98 | 7.59 | 4.29 | 6.31 |
| Scrapers/Filterers | 1.13 | 0 | 0.22 | 0.07 | 0.02 | 0 | 0.30 | 0 |
| **Between-site Tests** | 65% (Slight) | | 75% (Slight) | | 42% (Moderate) | | 30% (Moderate) | |

[1]A: Upstream/downstream of WWTP (Kansas), B: Nonpoint source problem (Texas), C: Upstream/downstream of WWTP and paper mill (Massachusetts), D: Nonpoint source problem (Washington, D.C.)
[2]Hilsenhoff Biotic Index, based on tolerance range of 0-10, adjusted to include non-arthropods.

These data suggest that relying solely on traditional metrics (taxa richness, Hilsenhoff Biotic Index) may not be sufficient to deduce the health of aquatic communities. In the first two data sets (slight impact), most metric values did not differ substantially between sites. Information on abundance by taxonomic group, however, was more indicative of between-site differences, producing overall between-site comparisons of 65 – 75%. For sites with moderate impact, the EPT index was the most informative metric, while total taxa richness showed between-site differences for only one of these examples. Overall between-site comparisons for these two examples were in the range of 30 – 42%.

## ACKNOWLEDGMENTS

Information from North Carolina's program reflects the work of many biologists, including Dave Penrose, Trish MacPherson, Larry Eaton, Ferne Winborne, and Neil Medlin. Likewise, the development of the "rapid bioassessment" protocols reflects the work of the late Jim Plafkin, and Kim Porter, Sharon Gross, and Bob Hughes. Major contributors to the RBP concept include Ken Cummins, William Hilsenhoff, Paul Leonard, and James Karr. Critical reviews of this manuscript were provided by James Karr, David Penrose, Vincent Resh, and Dave Courtemanch.

## REFERENCES

Anderson, N. H. and K. W. Cummins. 1979. Influences of diet on the life histories of aquatic insects. *Journal Fisheries Research Board Canada* 36:335-342.

Barbour, M. T., J. L Plafkin, B. P. Bradley, C. G. Graves, and R. W. Wisseman. 1992. Evaluation of EPA's rapid bioassessment benthic metrics: metric redundancy and variability among reference stream sites. *Environmental Toxicology and Chemistry*. 11:437-449.

Benke, A. C., R. L. Henry, III, D. M. Gillespie, and R. J. Hunter. 1985. Importance of snag habitat for animal production in southeastern streams. *Fisheries* 10:8-13.

Cairns, J., Jr. and K. L. Dickson. 1971. A simplified method for the biological assessment of the effects of waste dischargers on aquatic bottom-dwelling organisms. *Journal of the Water Pollution Control Federation* 43:755-772.

Chapman, D. W. and R. L. Demory. 1963. Seasonal changes in the food ingested by aquatic insect larvae and nymphs in two Oregon streams. *Ecology* 44:140-146.

Chutter, F. M. 1972. An empirical biotic index of the quality of water in South African streams and rivers. *Water Research* 6:19-30.

Courtemanch, D. L. and S. P. Davies. 1987. A coefficient of community loss to assess detrimental change in aquatic communities. *Water Research* 21:217-222.

Crawford, J. K. and D. R. Lenat. 1989. Effects of land use on the water quality and biota of three streams in the piedmont province of North Carolina. U.S. Geological Survey, Water Resources Investigations Report 89-4007. 67 p.

Cummins, K. W. and M. A. Wilzbach. 1985. *Field Procedures for Analysis of Functional Feeding Groups of Stream Macroinvertebrates*. Contribution 1611, Appalachian Environmental Laboratory, University of Maryland. 36 p.

Eaton, L. E. and D. R. Lenat. 1991. Comparison of a rapid bioassessment method with North Carolina's macroinvertebrate collection method. *Journal North American Benthological Society* 10:335-338.

Furse, M. T., D. Moss, J. F. Wright, and P. D. Armitage. 1984. The influence of seasonal and taxonomic factors on the ordination and classification of running water sites in Great Britain and on the prediction of their macro-invertebrate communities. *Freshwater Biology* 14:257-280.

Gurtz, M. E. 1994. Design Considerations for Biological Components of the National Water-Quality Assessment (NAWQA) Program. *In*: S. L. Loeb and A. Spacie, eds. *Biological Monitoring of Aquatic Systems*. Lewis Publishers, Boca Raton, FL.

Godfrey, P. J. 1978. Diversity as a measure of benthic macroinvertebrate community response to water pollution. *Hydrobiologia* 57:11-122.

Hart, D. R., ed. 1990. *Proceedings of the Third Annual Ecological Quality Assurance Workshop*. United States Environmental Protection Agency and Environment Canada. April 1990. Burlington, Ontario, Canada. 205 p.

Hilsenhoff, W. L. 1982. *Using a Biotic Index to Evaluate Water Quality in Streams*. Technical Bulletin No. 132, Wisconsin Department of Natural Resources. 22 p.

Hilsenhoff, W. L. 1988a. Rapid field assessment of organic pollution with a family-level biotic index. *Journal North American Benthological Society* 7:65-68.

Hilsenhoff, W. L. 1988b. Seasonal correction factors for the biotic index. *Great Lakes Entomologist* 21:9-13.

Hudson, P. L., D. R. Lenat, B. A. Caldwell, and D. Smith. 1990. Chironomidae of the southeastern United States: checklist of species and notes on biology, distribution and habitat. *U.S. Fish and Wildlife Service, Fish and Wildlife Research* 7:1-46.

Hughes, D. D. 1978. The influence of factors other than pollution on the value of Shannon's diversity index for benthic macroinvertebrates in streams. *Water Research* 12:359-364.

Hughes, R. M. and D. P. Larsen. 1988. Ecoregions: an approach to surface water protection. *Journal Water Pollution Control Federation* 60:486-493.

Hughes, R. M., S. A. Heiskary, W. J. Matthews, and C. O Yoder. 1994. Use of Ecoregions in Biological Monitoring. *In*: S. L. Loeb and A. Spacie, eds. *Biological Monitoring of Aquatic Systems*. Lewis Publishers, Boca Raton, FL.

Hynes, H. B. N. 1972. *The Ecology of Running Water*. University of Toronto Press. 555 p.

Jenkins, R. A., K. R. Wade, and E. Pugh. 1984. Macroinvertebrate-habitat relationships in the River Teifi catchment and the significance to conservation. *Freshwater Biology* 14:23-42.

Kawecka, B. 1977. The food of the dominant species of bottom fauna larvae in the River Raba (Southern Poland). *Acta Hydrobiologia* 19:191-213.

Kaesler, R. L. and E. E. Herricks. 1979. Hierarchical diversity of communities of aquatic insects and fishes. *Water Research Bulletin* 15:1117-1125.

Karr, J. R., K. D. Fausch, P. L. Angermeier, P. R. Yant, and I. J. Schlosser. 1986. Assessing biological integrity in running waters: a method and its rationale. *Illinois Natural History Survey Special Publication* 5:1-28.

Karr, J. R. 1991. Biological integrity: a long-neglected aspect of water resource management. *Ecological Applications* 1:66-85.

Klein, R. 1983. *Stream Quality Assessment*. Maryland Save Our Streams.

Lenat, D. R., D. L. Penrose, and K. W. Eagleson. 1981. Variable effects of sediment addition on stream benthos. *Hydrobiologia* 79:187-194.

Lenat, D. R. 1983. Chironomid taxa richness: natural variation and use in pollution assessment. *Freshwater Invertebrate Biology* 2:192-1987.

Lenat, D. R. 1988. Water quality assessment of streams using a qualitative collection method for benthic macroinvertebrates. *Journal North American Benthological Society* 7:222-233.

Lenat, D. R. 1990. Reducing variability in freshwater macroinvertebrate data. pp. 19-32. *In*: W. S. Davis, ed., *Proceedings of the 1990 Midwest Pollution Control Biologists Meeting*. U.S. EPA Region V, Environmental Sciences Division, Chicago, IL. EPA-905-9-90/005.

Lenat, D. R. 1993a. A biotic index for the southeastern United States: derivation and list of tolerance values, with criteria for assigning water-quality ratings. *Journal North American Benthological Society* 12:279-290.

Lenat, D. R. 1993b. Using mentum deformities of *Chironomus* larvae to evaluate the effects of toxicity and organic loading in streams. *Journal North American Benthological Society* 12:265-269.

Lenat, D. R. In press. Reducing variability in freshwater macroinvertebrate data. *In*: W. S. Davis, ed., *Proceeding of the 1990 Midwest Pollution Control Biologists Meeting* (Chicago, IL). U.S. EPA Region V, Instream Biocriteria and Ecological Assessment Committee, Chicago, IL.

Mason, W. T., Jr. 1979. A rapid procedure for assessment of surface mining impacts to aquatic life. pp. 310-323. *In*: *Coal Conference and Expo V* (Symposium proceedings), October 23-25, 1979, Louisville, KY. McGraw-Hill, NY.

Ohio Environmental Protection Agency. 1989. *Biological Criteria for the Protection of Aquatic Life: Volume I-III*. Division of water quality monitoring and assessment, surface water section, Columbus, OH. 351 p.

Osborne, L. L., R. W. Davies, and K. J. Linton. 1980. Use of heirarchical diversity indices in lotic community analysis. *Journal of Applied Ecology* 17:567-580.

Parrish, F. K. and J. A. Wagner. 1983. An index of community structure sensitive to water pollution. *Journal of Freshwater Ecology* 2:103-107.

Patrick, R. 1994. What are the Requirements for an Effective Biomonitor? *In*: S. L. Loeb and A. Spacie, eds., *Biological Monitoring of Aquatic Systems*. Lewis Publishers, Boca Raton, FL.

Paulson, S. G. and R. A. Linthurst. 1994. Biological Monitoring in the Environmental Monitoring and Assessment Program. *In*: S. L. Loeb and A. Spacie, eds., *Biological Monitoring of Aquatic Systems*. Lewis Publishers, Boca Raton, FL.

Peet, R. K. 1975. Relative diversity indices. *Ecology* 56:496-498.

Plafkin, J. L., M. T. Barbour, K. D. Porter, S. K. Gross, and R. M. Hughes. 1989. *Rapid Bioassessment Protocols for Use in Streams and Rivers*. U.S. EPA, Office of Water, EPA/444/4-89-001.

Resh, V. H. 1979. Biomonitoring, species diversity indices, and taxonomy. pp. 241-253. *In*: J. F. Gassle et al., eds., *Ecological Diversity in Theory and Practice*. Statistical Ecology Series Volume 6, International Co-operative Publishing House, Burtonsville, MD.

Resh, V. H. and J. K. Jackson. 1993. Rapid assessment approaches to biomonitoring using benthic macroinvertebrates. pp. 195-233. *In*: Rosenburg, D. M. and V. H. Resh (editors). *Freshwater Biomonitoring and Benthic Macroinvertebrates*. Chapman and Hall, NY. 488 p.

Resh, V. H. and J. D. Unzicker. 1975. Water quality monitoring and aquatic organisms: the importance of species identification. *Journal Water Pollution Control Federation* 47:9-19.

Resh, V. H. and D. M. Rosenberg. 1989. Spatial and temporal variability and the study of aquatic insects. *Canadian Entomologist* 121:941-963.

Rosenberg, D. M., H. V. Danks, and D. M. Lehmkuhl. 1986. Importance of insects in environmental impact assessment. *Environmental Management* 10:773-783.

Shackleford, B. 1988. *Rapid Bioassessments of Lotic Macroinvertebrate Communities: Biocriteria Development*. Arkansas Department of Pollution Control and Ecology, 45 p.

Shapas, T. J. and W. L. Hilsenhoff. 1976. Feeding habits of Wisconsin's predominant lotic Plecoptera, Ephemeroptera and Trichoptera. *Great Lakes Entomologist* 9:176-188.

Simpson, K. W. and R. W. Bode. 1980. Common larvae of Chironomidae (Diptera) from New York state streams and rivers. *New York State Museum Bulletin* 439:1-105.

U.S. Environmental Protection Agency. 1991. Biological criteria—state development and implementation efforts. EPA-440/5-91-003, 37 pp.

Vannote, R. L., G. W. Minshall, K. W. Cummins, J. R. Sedell, and C. E. Cushing. 1980. The river continuum concept. *Canadian Journal of Fisheries and Aquatic Sciences* 37:370-377.

Vannote, R. L. and B. W. Sweeney. 1980. Geographic analysis of thermal equilibria: a conceptual model for evaluating the effect of natural and modified regimes on aquatic insect communities. *American Naturalist* 115:667-695.

Warwick, W. F. 1988. Morphological deformities in Chironomidae (Diptera) larvae as biological indicators of toxic stress. pp. 281-320. *In*: M. S. Evans, ed., *Toxic Contaminants and Ecosystem Health, a Great Lakes Focus*. John Wiley, NY.

Warwick, W. F. and N. A. Tisdale. 1988. Morphological deformities in *Chironomus*, *Cryptochironomus*, and *Procladius* larvae (Diptera: Chironomidae) from two differentially stressed sites in Tobin Lake, Saskatchewan. *Canadian Journal of Fisheries and Aquatic Sciences* 45:1123-1144.

Wiederholm, T., ed. 1983. Chironomidae of the holarctic region. Keys and diagnoses. Part 1:Larvae. *Entomologia Scandinavica Supplement* 19:1-457.

# CHAPTER 12

# Landscape Position, Scaling, and the Spatial and Temporal Variability of Ecological Parameters: Considerations for Biological Monitoring

**Timothy K. Kratz, John J. Magnuson, Thomas M. Frost, Barbara J. Benson** and **Stephen R. Carpenter**, Center for Limnology, University of Wisconsin, Madison, WI

## INTRODUCTION

Designing and implementing an effective biological monitoring program requires making appropriate choices of what and where to measure. The parameters selected for measurement must obviously be sensitive to the environmental stress or stresses anticipated to occur. However, a change in a parameter in a stressed system must also be distinguishable from the natural variability exhibited by that parameter in an unstressed system. To the extent that sensitive parameters also tend to be more variable, choosing which parameters to monitor will involve a balance between these factors (Frost et al., 1992).

Another important consideration in the design and implementation of a biological monitoring program is the ability to extrapolate results to a larger number of locations than can be measured directly. One factor directly influencing the ease to which results can be scaled up to broader regions is the degree to which systems in different locations behave similarly through time. This property, which we term temporal coherence (Magnuson et al. 1990), is a useful concept because the more coherent different locations are, the easier it is to generalize about specific regional responses to environmental stress.

Therefore, an ideal parameter is one that is sensitive to the anticipated stress, has low variability (natural or unpredictable), and is temporally coherent among different locations in the landscape. Is it possible to choose the parameters and locations we measure to increase the likelihood of achieving these characteristics? In this paper we assess the sensitivity, variability, and coherence of selected limnological parameters. We divide the paper into two parts. In the first, *Choosing Parameters and Places*, we explore how sensitivity and variability are influenced by the level of taxonomic resolution of the data, and we show how the position of a lake in the landscape can affect the annual variability exhibited by limnological parameters. In the second part, *Scaling Up*, we evaluate the temporal coherence shown by a broad set of parameters. We use data from two long-term projects: the Northern Temperate Lakes Long-Term Ecological Research Project (Magnuson et al. 1984, Kratz et al. 1986, Magnuson and Bowser 1990) and the Little Rock Lake Acidification Project (Brezonik et al. 1986, Watras and Frost 1989). The results summarized here have been published individually (Frost et al. 1992, Magnuson et al. 1990, Kratz et al. 1991). We have taken this opportunity to provide a review and synthesis of these individual pieces.


# CHOOSING PARAMETERS AND PLACES

## Choosing Parameters: Effects of Aggregation on Variability and Sensitivity.

The choice of biological parameters to measure in a monitoring and assessment program can range widely along a gradient of aggregation, from the responses of individual organisms to the integrated behavior of entire ecosystems (Little and Finger 1990, Fausch et al. 1990, Odum 1985, Karr 1981, Frost et al. 1992). There are reasons to expect that a parameter's sensitivity to stress and its natural variability will be a function of the parameter's level of aggregation. As Schindler (1987) points out, compensatory processes, in which a decline in one taxa occurs simultaneously with an increase in another, decrease the sensitivity and variability of highly aggregated parameters such as primary production, relative to the responses of individual populations. However, if an entire class of processes, for example, ability of organisms to fix nitrogen (Howarth et al. 1988, Schindler et al. 1991), were affected by a stress and compensatory interactions did not occur, then highly aggregated parameters could exhibit a high degree of sensitivity (Odum 1985).

As much as possible, parameters should be chosen in order to maximize sensitivity to stress relative to unpredictable natural variability. Therefore, it is important to know the relationship between the level of aggregation, sensitivity to stress, and unpredictable natural variability when choosing among potential indicator parameters for inclusion in a biological monitoring program.

In this section we give an example of how sensitivity and variability vary across a gradient of aggregation. We use data from a whole-lake acidification experiment in which one half of Little Rock Lake, Wisconsin, was acidified over a five-year period from an original pH of 6.1 to a pH of 5.2 (Brezonik et al. 1986, Watras and Frost 1989). The other half of the lake was maintained as a reference. In preliminary analyses we examined the sensitivity and variability of zooplankton using data that ranged in resolution from the biomass of individual taxa to the summed behavior of the biomass of the entire community (Frost et al. 1992). We used data from the reference half of the lake to calculate natural variability and comparisons of reference half and treatment half to estimate sensitivity to acidification.

As an example of the way aggregation affects sensitivity and variability, we present results from one particular agggregation series, that of a rotifer genus *Polyarthra*. We chose four levels of aggregation: species (four for *Polyarthra* in Little Rock Lake), genus, total rotifers, and total zooplankton. Data are expressed in units of biomass (mg dry mass/L) and have been transformed as [ln(biomass + 1)]. To compute natural variability we used data from the reference half of the lake. Data from 188 separate sampling periods between 1984 and 1989 were used. To compute sensitivity, we used data from a baseline period prior to any acid condition (20 samplings periods from January 1984 through April 1985) and for a two-year period when the treatment half was maintained at pH 5.2 (38 sampling periods from May 1987 through April 1989). Details of the sampling methods are given in Frost and Montz (1988) and Frost et al. (1992). We calculated the difference between the treatment and reference halves for each sampling date and then compared the average difference for the baseline period with the average difference for the pH 5.2 period. The magnitude of this difference is our measure of sensitivity.

The natural variability as a function of aggregation level is shown in Figure 12.1. The solid line indicates the observed variability. The dashed line indicates the variability that would have occurred if no compensation had taken place, that is, if each species, genus, or major group was truly independent of other species, genera, or major groups. These values were calculated by summing the variances of the members of the next lowest level in the aggregation series. For example, if species are independent, then the variance of the genus is equal to the sum of the variances of the individual species (Snedecor and Cochran 1980). The observed variability is relatively low at each level in the aggregation series, and because it is below the level that would be obtained if independence held, compensation occurs throughout the aggregation series.

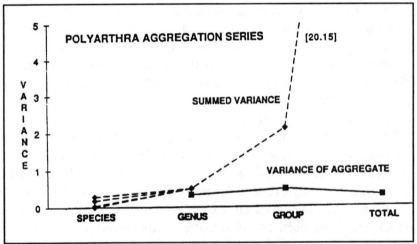

**Figure 12.1.**    Variance of zooplankton biomass versus aggregation level for the rotifer genus *Polyarthra*. Values are presented for the actual calculated variance of the aggregated parameter and for the no-compensation case in which variance is determined as the sum of all individual species variances. The maximum value of the summed variance for total zooplankton, 20.15, is not plotted so that detail is visible at the lower aggregation levels. The lowermost symbol at the species level of aggregation represents the similar values for two *Polyarthra* species. (Taken from Frost et al. 1992.)

The sensitivity to acidification of pH 5.2 as a function of aggregation is shown in Figure 12.2. Species are highly variable in their sensitivity to acidification. However, genus is the most sensitive level of the aggregation series.

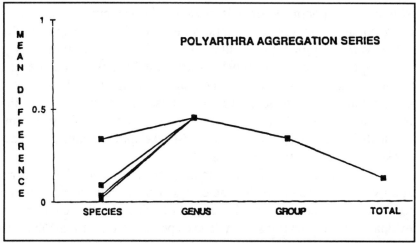

**Figure 12.2.** Shifts in the mean, inter-basin difference (Biomass in the Reference Basin – Biomass in the Treatment Basin) between the baseline period and pH 5.2 acidification period versus aggregation level for the rotifer genus *Polyarthra*. (Taken from Frost et al. 1992.)

Synthesizing the information on variability and sensitivity, it appears that in this example, the best level of aggregation for indicating acidification effects appears to be the genus level. The genus level has nearly the same natural variability as other levels, but exhibits the most sensitivity.

Our analyses are still preliminary, but we have found that aggregation series based on other taxa differ in variability and sensitivity patterns. Genus will not always be the best level to use and the result shown here for *Polyarthra* should not be taken as a general rule. However, it is clear that the effects of aggregation level on sensitivity and variability should be considered when parameters for biological monitoring are being selected.

## Choosing Places: Effects of
## Landscape Position on Variability and Sensitivity

The sensitivity and variability of parameters can be affected not only by the choice of the parameter itself, but also by the location at which the parameter is measured. In some cases the importance of location on the sensitivity of a system to stress is obvious. An example is the effect of acid precipitation on surface waters, where sensitive systems are those with low acid neutralizing capacities (ANC). ANC is determined largely by landscape influences on geochemistry and hydrology (Eilers et al. 1983) such that even within the same general landscape, lakes in certain locations will be more sensitive than those in others, due to these hydrological effects. Clearly, in such a case the design of a program to detect acidification of surface waters must take into account the position of a lake or stream in the landscape.

Often, however, there is little or no *a priori* evidence that a certain location will be more or less variable or sensitive than others. For these situations it would be helpful to have a set of principles that could be used to evaluate which locations in a landscape are likely to exhibit more temporal variability than others. As a first step toward developing these principles we compared the annual variability of a broad set of ecological parameters at different locations within four contrasting North American landscapes (Kratz et al. 1991). The landscapes included the Northern Highland Lake District in northern Wisconsin (Magnuson et al. 1984, Kratz et al. 1986), the Hubbard Brook Experimental Forest in New Hampshire (Likens et al. 1977, Bormann and Likens 1979), the North Inlet estuary in South Carolina (Dame et al. 1986, Blood et al. in press), and the Jornada Desert in southern New Mexico (Whitford et al. 1987, Wondzell et al. 1987). Ecological data spanning at least five years were available for 3 – 7 locations at each of these four sites. The locations at each of the four sites were distinct parts of the landscape. For example, at the Northern Highland Lake District in Wisconsin the locations were different lakes, and at the Jornada Desert the locations were different vegetation/landform units along an elevational gradient from mountain to playa. At each of the sites the locations could be ordered along an important, but not necessarily large, elevational gradient. For each of the four landscapes we computed the coefficient of interannual variability for each parameter at each of the 3 – 7 locations. Then for each parameter we ranked the 3 – 7 locations in order of increasing coefficient of variation.

Finally, for each of the 3 - 7 locations we computed the average rank. Detailed descriptions of the data sets, landscapes, and computational methods are given in Kratz et al. (1991).

We tested whether the annual variability exhibited at a location in the landscape was related to the location's position in the elevational gradient at each of the four sites. Results from each of the four sites are shown in Figure 12.3. We found that although the specific mechanisms leading to patterns of annual variability observed in each landscape were quite different, three important generalizations could be drawn. First, within each of the four landscapes, locations differed significantly in the annual variability exhibited by ecological parameters. Second, for at least a

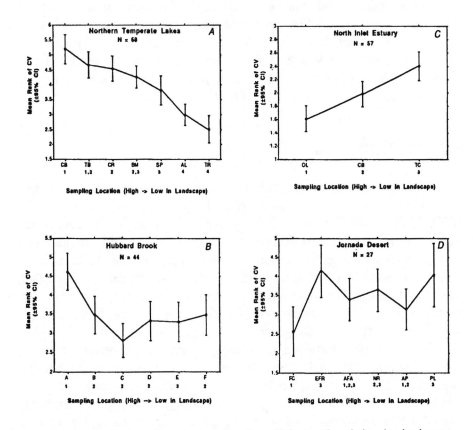

**Figure 12.3.** The overall average ranks of coefficients of variation by landscape position for the four LTER sites. Larger numbered ranks indicate larger cv's. Locations having same number below id are not significantly different (p > 0.05, Fischer's Least Significant Difference). (Taken from Kratz et al. 1991.)

subset of the parameters, this variability was related to the location's relative position in the elevational gradient. Finally, water movement across the landscape was an important underlying factor determining variability patterns in all four landscapes.

As these results demonstrate, the position of sampling location within the landscape can have an important influence on the annual variability exhibited by ecological parameters. At this point it is too early to make sweeping generalizations, but the results from the lake district allow the formulation of some interesting hypotheses. For example, if sensitivity to stress is correlated with natural variability, then intermediate locations in the landscape may provide the power to detect effects of stress. Lakes high in the landscape may be so sensitive and variable that it will be difficult to detect real change from the background of high natural variability. Similarly, lakes low in the landscape may exhibit such little variability because they are simply not as sensitive to fluctuating conditions, and it may take a major stress to cause these systems to change. It may take years to test these hypotheses but the generalizations that they could provide would be important.

## SCALING UP

With the increasing recognition that many environmental problems are regional, continental, or global in scale, there has been interest in the difficult problem of how to scale up from single site studies to larger areas (Turner et al. 1989). There are a number of strategies to deal with this problem, ranging from sampling the entire population of sites of interest, to sampling a randomly selected subset of the sites, to sampling sites picked in a nonrandom way in order to meet certain criteria. No single strategy will be appropriate for all circumstances; however, it is usually not possible to sample the entire population of sites.

In cases where some subset of sites is sampled, the extent to which sites act similarly through time is important. For example, if the fact that one lake has unusually high chlorophyll in a certain year means that it is likely that other lakes in the region also had high chlorophyll that year, then fewer sites would need to be sampled in order to assess regional chlorophyll patterns than if each lake acted independently. We use the term temporal coherence (Magnuson et al. 1990) to describe the tendency of systems to behave similarly through time.

In this section we summarize results reported in Magnuson et al. (1990) on the temporal coherence exhibited by a suite of 37 limnological parameters measured on seven neighboring lakes at the North Temperate Lakes Long-Term Ecological Research site in northern Wisconsin. We tested whether lakes more similar in exposure to the atmosphere were more temporally coherent than lakes that differ more in exposure, and whether temporal coherence in lakes was stronger for variables directly influenced by climate than for those less directly influenced by climate. We chose variables ranging from those we expected to respond directly to climatic influences, to those where direct climatic influences would be minor. Variables were grouped into six ranks of decreasing responsiveness to climatic influences: physical variables, including water level and temperature; atmospherically dominated chemical variables, including $SO_4$ and pH; groundwater dominated chemical variables, including K and Ca; biologically dominated chemical variables, including $SiO_2$ and total dissolved phosphorus; biologically dominated physical variables, including Secchi depth and light extinction coefficient; and biological variables, including chlorophyll concentration and zooplankton abundance.

We calculated two indices of temporal coherence for each variable and for each lake pair. The first index was the mean correlation. For a particular lake pair we calculated the arithmetic average of correlation coefficients determined for each of the 37 variables. For a particular variable we calculated the average correlation coefficient for the 21 lake pairs. The second index was the percentage of strong correlations, which was the percentage of the 37 or 21 correlation coefficients larger than +0.67. For details on the variables used, their measurement, and indices of temporal coherence see Magnuson et al. (1990).

We found that lake pairs that were similar in exposure to climate (as measured by similarity in area/mean depth) tended to be more temporally coherent than lakes differing in exposure to climate (Figure 12.4). This was the case for both measures of temporal coherence. We also found that variables that we expected to be more responsive to direct influences of climate, such as water level, were more temporally coherent than those less directly influenced by climate (Figure 12.5).

These results raise two important points. First, lakes and variables differed substantially in among-year temporal coherence. This suggests that the choice of lakes and variables will directly influence the ability to scale up results from studies of a few locations to a broader spatial region. For variables that are not coherent it may be necessary to develop a

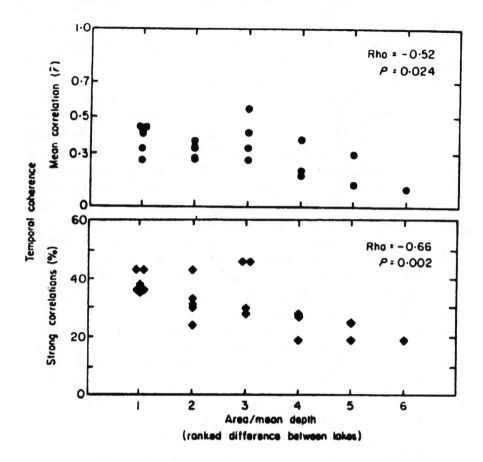

**Figure 12.4.** Temporal coherence of lake pairs measured as mean correlation (r) and strong correlations (%); lake pairs are ordered in decreasing similarity of exposure to climatic factors measured as a ranked difference between lakes in ratios of "area/mean depth" (Taken from Magnuson et al. 1990.)

mechanistic understanding of important controlling factors at a landscape level (Burke et al. 1990). Second, it may be possible to make *a priori* predictions about which types of lakes or variables are likely to be temporally coherent given a particular environmental "stress". In the example we used, the "stress" was normal climatic variability, but the same reasoning could be applied to other anticipated stresses. By assessing the expected degree of temporal coherence it may be possible to make wise selections of sampling lakes and variables.

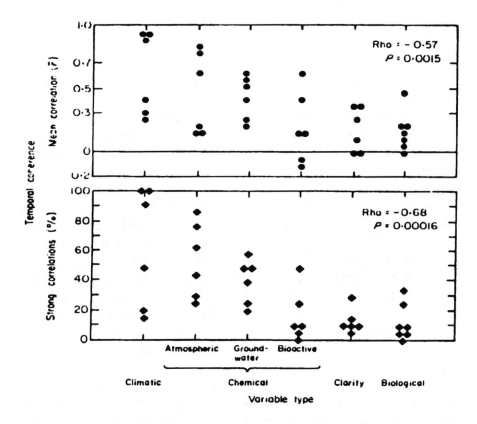

Figure 12.5. Temporal coherence of limnological variables measured as mean
correlation (r) and strong correlations (%) aggregated by categories of
responsiveness to climatic factors in order of hypothesized decreasing
responsiveness. (Taken from Magnuson et al. 1990.)

It should also be possible to construct criteria for stratifying the
selection of study lakes based on defendable and easily measured
characteristics of lakes. For example, if potential effects of climatic
change are of interest, then stratifying lakes by their area/mean depth
would be appropriate. Other stratifications could be determined for other
anticipated problems.

## CONCLUSIONS

At the beginning of this paper we stated that an ideal parameter for inclusion in a monitoring program is one that is sensitive to the anticipated stress, has low variability (natural or unpredictable), and is temporally coherent among different locations in the landscape. Using examples from a northern Wisconsin lake district, we outlined techniques that can determine the effect of level of taxonomic resolution of the measurement, the particular lake that is sampled, and the type of parameter that is sampled on sensitivity, variability, and temporal coherence. The results of these studies suggest that sensitivity, variability, and temporal coherence are affected by the choice of parameters and places to study.

The specific results presented here should not be overgeneralized. Details of the results are likely to be dependent on the type of lakes present in Northern Highland Lake District and the time scale of interest. Details of the results would likely be different for streams in the lake district or for lakes in other regions. Similarly, if the time scale of interest was other than annual, it is likely that specific results would differ. However, the techniques and approaches presented here should be applicable in a general way.

Finally, we have presented a clear example of the way basic research can be used in a very applied manner. The major goals of the studies that collected the data used here were not to address the question of design and implementation of monitoring projects. Yet, the data are quite appropriate for that purpose. We suspect that many other data sets exist that could be used in the same way we used these data. These sources of information should not be overlooked by those designing and implementing biological monitoring projects.

## ACKNOWLEDGMENTS

This work was supported by the NSF through grant BSR 85 14330 to the Northern Temperate Lakes Long Term Ecological Research Project, by cooperative agreements with the USEPA, and by a grant from the Andrew W. Mellon Foundation. We thank the myriad of individuals who helped with the field collection and management of the original data. Logistic support was provided by the University of Wisconsin Trout Lake Station.

# REFERENCES

Blood, E., W. Swank, and T. Williams. 1989. Precipitation, throughfall, and stemflow chemistry in a coastal loblolly pine forest. pp. 61-78. *In:* Sharitz, R. R. and J. W. Gibbons, eds., *Freshwater Wetlands and Wildlife.* Department of Energy Symposium Series No. 61. U.S. DOE Office of Scientific and Technical Information, Oak Ridge, TN. 1265 p.

Bormann, F. H., and G. E. Likens. 1979. *Pattern and Process in Forested Ecosystem.* Springer-Verlag, NY. 253 p.

Brezonik, P. L., L. A. Baker, J. R. Eaton, T. M. Frost, P. Garrison, T. K. Kratz, J. J. Magnuson, J. Perry, W. J. Rose, B. K. Shepard, W. A. Swenson, C. J. Watras, and K. E. Webster. 1986. Experimental Acidification of Little Rock Lake, Wisconsin. *Water, Air, and Soil Pollution* 31:115-121.

Burke, I. C., D. S. Schimel, C. M. Yonker, W. J. Parton, L. A. Joyce, and W. K. Lauenroth. 1990. Regional modeling of grassland biogeochemistry using GIS. *Landscape Ecology* 4:45-54.

Dame, R., T. Chrzanowsky, D. Bildstein, B. Kjerfve, H. McKellar, D. Nelson, J. Spurrier, S. Stancyk, H. Stevenson, J. Vernberg, and R. Zingmark. 1986. The outwelling hypothesis and North Inlet. *South Carolina, Mar. Ecol. Prog. Ser.* 32:71-80.

Eilers, J. M., G. E Glass, K. E. Webster, and J. A. Rogalla. 1983. Hydrologica control of lake susceptibility to acidification. *Can. J. Fish. Aquat. Sci.* 40:1896-1904.

Fausch, K. D., J. Lyons, J. R. Karr, and P. L. Angermeier. 1990. Fish communities as indicators of environmental degradation. *American Fisheries Society Symposium* 8:123-144.

Frost, T. M. and P. K. Montz. 1988. Early zooplankton response to experimental acidification in Little Rock Lake, Wisconsin, U.S.A. *Verh. Internat. Verein. Limnol.* 23:2279-2285.

Frost, T. M., S. R. Carpenter, and T. K. Kratz. 1992. Choosing ecological indicators: effects of taxonomic aggregation on sensitivity to stress and natural variability. pp. 215-217. *In:* D. H. McKenzie, D. E. Hyatt, and J. McDonald, eds., *Ecological Indicators.* Elsevier Applied Science Publishers, Essex, England. 1567 p.

Howarth, R. W., R. Marino, and J. J. Cole. 1988. Nitrogen fixation in freshwater, estuarine, and marine ecosystems: 2. Biogeochemical controls. *Limnol. Oceanogr.* 33:688-701.

Karr, J. R. 1981. Assessment of biotic integrity using fish communities. *Fisheries* (Bethesda) 6:21-27.

Kratz, T. K., J. J. Magnuson, C. J. Bowser, and T. M. Frost. 1986. Rationale for data collection and interpretation in the Northern Lakes Long-Term Ecological Research Program. pp. 22-33. *In*: B. G. Isom, ed., *Rationale for Sampling and Interpretation of Ecological Data in the Assessment of Freshwater Systems*. ASTM STP 894, American Society for Testing and Materials, Philadelphia. 193 p.

Kratz, T. K., B. J. Benson, E. Blood, G. L. Cunningham, and R. A. Dahlgren. 1991. The influence of landscape position on temporal variability in four North American ecosystems. *American Naturalist*. 138:355-378.

Likens, G. E., F. H. Bormann, R. S. Pierce, J. S. Eaton, and N. M. Johnson. 1977. *Biogeochemistry of a Forested Ecosystem*. Springer-Verlag, NY. 146 p.

Little, E. E. and S. E. Finger. 1990. Swimming behavior as an indicator of sublethal toxicity in fish. *Environmental Toxicology and Chemistry* 9:13-19.

Magnuson, J. J. and C. J. Bowser. 1990. A network for long-term ecological research in the United States. *Freshwater Biology* 23:137-143.

Magnuson, J. J., C. J. Bowser, and T. K. Kratz. 1984. Long-term ecological research on north temperate lakes (LTER). *Verh. Internat. Verein. Limnol.* 22:533-535.

Magnuson, J. J., B. J. Benson, and T. K. Kratz. 1990. Temporal coherence in the limnology of a suite of lakes in Wisconsin, U.S.A. *Freshwater Biology* 23:145-159.

Odum, E. P. 1985. Trends expected in stressed ecosystems. *Bioscience* 35:419-422.

Schindler, D. W. 1987. Detecting ecosystem responses to anthropogenic stress. *Can. J. Fish. Aquat. Sci.* 44 (Supp. 1):6-25.

Schindler, D. W., T. M. Frost, K. H. Mills, P. L. Brezonik, P. S. S. Chang, I. J. Davies, P. J. Garrison, J. M. Gunn, W. Keller, D. F. Malley, J. A. Shearer, W. A. Swenson, M. A. Turner, C. J. Watras, K. E. Webster, and N. D. Yan. 1991. Comparisons between experimentally–and atmospherically–acidified lakes during stress and recovery. Proceedings of the Royal Society of Edinburgh. 97B:193-226.

Snedecor, G. W. and W. G. Cochran. 1980. *Statistical Methods*. The Iowa State University Press, Ames, Iowa, U.S.A. 593 p.

Turner, K. G., V. H. Dale, and R. H. Gardner. 1989. Predicting across scales: theory development and testing. *Landscape Ecology* 3:245-252.

Watras, C. J. and T. M. Frost. 1989. Little Rock Lake (Wisconsin): perspectives on an experimental ecosystem approach to seepage lake acidification. *Arch. Env. Cont. and Toxicology* 18:157-165.

Whitford, W. G., J. F. Reynolds and G. L. Cunningham. 1987. How desertification affects nitrogen limitations of primary production in Chihuahuan desert watersheds. pp. 143-152. *In*: E. F. Aldon, C. E. Gonzales-Vicente, and W. H. Moir, eds., *Strategies for Classification and Management of Native Vegetation for Food Production in Arid Zones*. USDA Forest Service Gen. Tech. Rept. RM-150.

Wondzell, S. M., G. L. Cunningham, and D. Bachelet. 1987. A hierarchical classification of land forms: some implications for understanding local and regional vegetation dynamics. pp. 15-23. *In*: E. F. Aldon, C. E. Gonzales-Vicente, and W. H. Moir, eds., *Strategies for Classification and Management of Native Vegetation for Food Production in Arid Zones*. USDA Forest Service Gen. Tech. Rept. RM-150.

# CHAPTER 13

# Paleolimnological Approaches to Biological Monitoring

**Donald F. Charles[1], John P. Smol[2] and Daniel R. Engstrom[3]**

[1]Indiana University, c/o U.S. EPA Environmental Research laboratory, 200 SW 35th St., Corvallis, Oregon 97333, USA. Current Address: Patrick Center for Environmental Research, The Academy of Natural Sciences of Philadelphia, 1900 Benjamin Franklin Parkway, Philadelphia, PA 19103-1195.

[2]Paleoecological Environmental Assessment and Research Lab (PEARL) Department of Biology, Queen's University, Kingston, Ontario K7L 3N6, Canada.

[3]Limnological Research Center, University of Minnesota, Minneapolis, Minnesota 55455, USA.

## INTRODUCTION

In response to increasing concern about the effects of human activities on aquatic ecosystems, a growing number of biological monitoring programs are being established to track the environmental condition of lakes and ponds for the purpose of making management and policy decisions. These programs are usually designed to determine if change is occurring, and if so, how much and how fast, and to attempt to identify the causes. The programs are designed to detect changes early, monitor recovery from disturbance, determine compliance with regulations, and establish baseline or reference conditions. Continuous monitoring programs provide information on average or typical conditions that one-time measurements taken during surveys do not. Biota are being included more frequently in monitoring programs because they provide historical information and because concern about protecting the biota is often the reason for the monitoring programs in the first place.

Although high-quality data can be obtained in biological monitoring programs, their usefulness can be severely limited for decision making purposes. These limitations can both endanger the existence of the program itself, and result in delay in taking actions that may be necessary to protect the systems being monitored. For example, many years of sampling may be necessary to collect adequate data to detect a trend, especially in variable systems, or, it may be impossible to determine if a trend, once detected, is due to anthropogenic causes or is simply within the range of natural variability. These and other limitations can be eliminated or reduced by incorporating paleolimnological approaches in biological monitoring programs and using the very long-term data that can be obtained with these techniques.

This chapter describes the specific information that can be provided by paleolimnological approaches, explains how paleolimnological studies can be performed, and informs readers about the most up-to-date techniques. The number of new paleolimnological studies is accelerating and techniques have advanced rapidly in the last 5 to 10 years, particularly in the areas of dating, identification of characteristics that can be measured, taxonomy, ecological data, multivariate and statistical data analysis, data management, and quality assurance. We focus on the use of biological remains in sediments for monitoring lakes and reservoirs.

We direct this chapter particularly toward aquatic ecologists and administrators who are interested in or responsible for designing monitoring programs and who want to know more about how to incorporate paleolimnological approaches. We decided to present the information in the form of responses to the questions most frequently asked by ecologists and administrators. We think it is an effective way to provide the reader with the guidance perceived to be most relevant. Our objective in addressing these questions is to provide the simplest possible answer, without lengthy qualifications and detailed explanations, and to provide references that readers can consult for further information.

We address three main topics: (1) the benefits of paleolimnological approaches for biological monitoring, (2) methods and techniques available and the information they provide, and (3) examples of how paleolimnological approaches have been used to address some environmental issues.

# IN WHAT GENERAL WAYS CAN PALEOLIMNOLOGICAL APPROACHES CONTRIBUTE TO BIOMONITORING PROGRAMS?

## Why are Long-Term Data Important for Biomonitoring and Why Should Paleolimnological Studies Be Undertaken to Provide Them?

Historical data are important in assessing surface water trends because they provide key information on the systems being monitored, including (1) the nature of environmental conditions existing before natural or anthropogenic disturbances, (2) the nature and magnitude of natural variability, including rare and infrequent events, and (3) significant trends that began before the initiation of a monitoring program. There are four primary ways to obtain information on historical conditions:

1) Historical measurements, which are usually rare and are often made using methods different from those employed in modern monitoring programs
2) Computer model simulations, which are often quite uncertain because input data is inadequate or because important mechanisms are not included in the models
3) Space-for-time substitutions, for example, comparing the system being monitored with an undisturbed reference system, which requires the uncertain assumption that the two systems are reasonably similar
4) Paleoecological analysis of the archives contained in the sediment record—paleolimnology (Frey 1969)

Of these, paleolimnology is one of the best and often the only way to obtain historical data. In addition to providing background information on systems, sediments can provide a record of disturbances that have occurred in the lakes in the form of natural processes and anthropogenic perturbations. The response of a lake to these disturbances provides insight into how the monitored system functions, and it also provides a better understanding of the significance of trends observed in the modern monitoring program. For example, paleolimnological approaches can be used to determine whether a lake has been as productive as it is now, or whether plant biomass and any associated problems have increased.

## How Can Information Provided by Paleolimnological Techniques Contribute to the Understanding and Interpretation of Monitoring Data?

Paleolimnological approaches can provide information relevant to monitoring on a wide range of spatial and temporal scales. Sediment records can be obtained from a single lake or from many lakes representative of a region. Temporal trends can be resolved over single years (and in some cases even seasonally) or over tens of thousands of years. Because sediments receive materials from the atmosphere, from the watershed, and from the lake, a skilled analysis of a sedimentary profile can provide information on all three sources (Figure 13.1). Different combinations of sediment characteristics can represent the watershed alone (e.g., trophic state indicated by algae, erosion indicated by inorganic sediment input) or the airshed around the lake (e.g., vegetation type represented by pollen). Studies can be designed to examine the temporal trends in various ecological characteristics and determine when and how the characteristics changed (e.g., lake level), usually by analyzing a single core, or studies can be designed to calculate rates of accumulation of materials to the lake (e.g., atmospheric deposition of metals), which may require multiple cores.

A wide variety of environmental issues can be addressed, including acidic precipitation, climatic change, eutrophication, toxic substances, soil erosion, habitat quality, biotic integrity, fishability, and human recreational use (see later Sections for examples). These issues can be assessed at different levels of intensity and thoroughness, using an array of approaches and techniques. Paleolimnological approaches in biological monitoring programs can help in many ways, such as to

- Determine past trends in the absence of other consistent data sources
- Determine whether measured trends fall outside natural variability
- Determine the extent to which trends may be anthropogenic or natural in origin

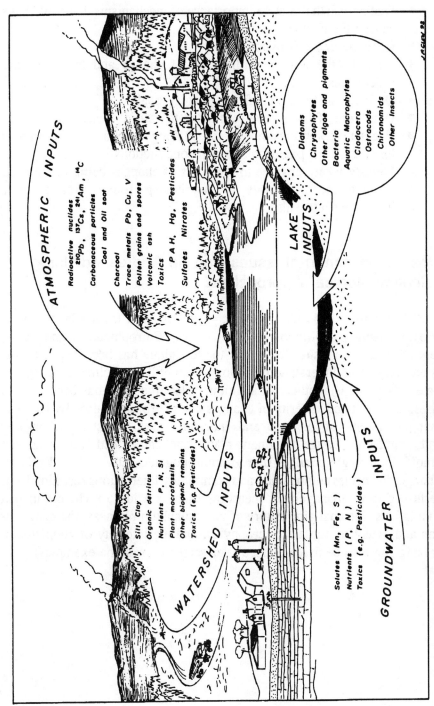

**ATMOSPHERIC INPUTS**

Radioactive nuclides
$^{210}Pb$, $^{137}Cs$, $^{241}Am$, $^{14}C$
Carbonaceous particles
Coal and Oil soot
Charcoal
Trace metals  Pb, Cu, V
Pollen grains and spores
Volcanic ash
Toxics
P.A.H., Hg, Pesticides
Sulfates  Nitrates

**WATERSHED INPUTS**

Silt, Clay
Organic detritus
Nutrients P, N, Si
Plant macrofossils
Other biogenic remains
Toxics (e.g. Pesticides)

**LAKE INPUTS**

Diatoms
Chrysophytes
Other algae and pigments
Bacteria
Aquatic Macrophytes
Cladocera
Ostracods
Chironomids
Other Insects

**GROUNDWATER INPUTS**

Solutes (Mn, Fe, S)
Nutrients (P, N)
Toxics (e.g. Pesticides)

**Figure 13.1.** Common components of lake sediment and their sources.

- Distinguish the relative importance of mechanisms that could have caused the change
- Generate and test hypotheses
- Describe the long-term direction of change and predict future trajectories
- Define realistic targets for regulatory goals
- Determine if lakes have responded to control measures
- Evaluate computer models by making comparisons of model hindcasts with paleoecological reconstructions
- Identify lakes sensitive to the types of changes being monitored
- Provide the possibility of monitoring some lakes at very infrequent intervals

## What Environmental Issues Can Be Addressed Using Paleolimnological Approaches?

Most natural and anthropogenic perturbations affecting lakes can be assessed using paleolimnologic approaches. Eutrophication and acidic deposition are the issues for which the technique has been applied most extensively. The understanding of long-term effects of climate change on water resources is another. In addition, paleolimnology has been used to assess the onset and severity of changes due to erosion caused by land use changes, inputs of organic and inorganic toxic substances, organic pollution from sewage treatment plants and cattle feed lots, road salt runoff, hydrologic changes due to anthropogenic activities, and others. Changes in species composition and community interactions resulting from perturbation or management measures such as a loss or change in fish populations, the changing nature of phytoplankton biomass, the composition and extent of aquatic macrophytes, and the intensity of zooplankton grazing can also be assessed (see later sections for some examples).

# WHAT METHODS AND TECHNIQUES ARE AVAILABLE?

## How Can Sediment Cores be Taken?

**What coring devices are available?** A wide variety of well-tested coring equipment is now available for collecting sediment samples of appropriate quality and age to be used in biological monitoring programs. In most North American lakes, significant culturally related disturbances are limited to the last 200 years or so, and therefore, short sediment cores (usually less than 1 m and often less than 0.5 m in length) are sufficient to characterize background (i.e., pre-cultural) or reference conditions and to trace culturally related changes in aquatic systems. Although the basic design of most corers remains the same, considerable refinements to paleolimnological equipment are constantly being sought and implemented. Many recent technological advances have been stimulated by the heightened interest in obtaining high-quality (undisturbed) cores of recent lake sediments that can be sectioned into closely spaced sedimentary intervals (e.g., 0.25 cm), which provide temporal resolution of less than five years. Some of the approaches discussed under the next question (e.g., tape-peel adhesions) can even provide subannual resolution.

Aaby and Digerfeldt (1986) have reviewed various sampling techniques for lake and bog sediments; more recently Smol and Glew (1992) have summarized and illustrated the major types of sediment samplers. Only a few salient points that are most appropriate to monitoring programs need to be mentioned here.

Piston corers are widely used for collecting sediment cores up to several meters in length. Most piston corers follow the design originally popularized by Livingstone (1955) and later by Cushing and Wright (1965), whereby the core tubes are driven and controlled by push rods from the lake surface. Brown (1956) modified the design so that undisturbed piston cores of recent lake sediments could be obtained (0.5–1.0 m), preserving the often flocculent sediment stratigraphy near the mud-water interface, the region of most interest for biological monitoring studies (i.e., the recent past). Recently, Wright (1991) provided a practical guide to piston coring. The main advantage of piston corers is that they minimize the shortening of sediment cores that can occur with gravity devices (Hongve and Erlandsen 1979).

Gravity or open-drive samplers are now commonly used to collect cores from the top 50 cm of sediment. The development of simple and reliable gravity coring devices and associated core-handling equipment has seen continued improvement over the last decade, and several models are now available that, when used properly, can provide sedimentary records of high quality and high resolution. In its simplest design, the gravity corer is operated by line or rope from the lake surface and consists of an open tube with a mechanism that places a plug in the top of the tube after the corer has penetrated the bottom sediments. The gravity corer is lowered slowly into the sediment using the corer's weight as a driving force. The plug-placing device is then triggered by a messenger weight that travels down the recovery line. The plug keeps sediment from falling out of the tube as it is raised from the bottom and out of the water. Gravity corers are lighter and easier to use than piston corers, but cannot be easily used for taking large diameter cores (greater than 10–15 cm).

A third technique for sediment sampling is known as the crust-freezing method (Renberg, 1981), or freeze sampling, as it is often called. The freeze sampler consists of a hollow metal chamber (probe) that is filled with dry ice and alcohol, thus lowering the temperature to about —70°C. The chilled probe is lowered into the lake sediment, where it is allowed to remain stationary for about 20 minutes while a rind of sediment freezes to its outer surface. A chief advantage of this type of sampler is its effectiveness in collecting undisturbed stratigraphic sequences from a lake's most recent sediments, which are usually very watery and are sometimes difficult to collect using other devices. This advantage is particularly important in collecting cores from lakes with laminated sediments.

In addition to these three types of corers, lightweight corers have been designed to sample just the sediment surface (e.g., 0–1 cm) (Hongve 1972, Wright 1990, Glew 1989, 1991). These devices are used primarily for surveys of a large number of lakes designed to provide information on the distribution and ecological characteristics of biota found in the sediments. Other corers include Russian peat samplers and box corers.

**When should coring be done?** Lakes can be cored any day of the year (weather permitting), although some researchers working in colder climates prefer to sample lakes through ice-cover, as this provides a very stable platform from which to work. When working from a boat, secure anchoring is crucial.

**How can cores be subsampled?** Once collected, sediment cores must be subsampled. It is very important that the plan for sectioning the core into intervals be well thought out so that the amount of time represented by each sediment interval is appropriate to address questions that the monitoring program was designed to answer (i.e., the thicker the sediment subsamples, the coarser the temporal resolution one can attain; see Section IIc: How can paleolimnological trends be dated?). Ideally, sediment cores less than 0.5 m long, collected using either piston or gravity corers, should be sectioned at the lake shore (i.e., as soon as possible after collection), because significant disturbance in sediment stratigraphy may occur if cores are transported before sectioning. Extrusion of the watery surface sediments can be efficiently accomplished using a vertical extrusion system, such as the one designed by Glew (1988). This device allows for the accurate sectioning of cores in precise increments of any size, for example 0.25 cm intervals, which in many lake systems represents a temporal resolution of under three years. If desired, contiguous samples can be analyzed to obtain a continuous trend of past environmental change. For other types of fine-resolution analyses, sediments can be impregnated with a variety of waxy or plastic resins and then thin-sectioned (e.g., Tippett 1964, Lotter 1989). Frozen sediment can be sectioned with a saw, or, if close-interval analyses are required, a tape-peel method can be used. In this process, adhesive tape strips are used to transfer a thin layer of sediment material from the cleaned surface of a frozen core directly onto a microscope slide, so that the original stratigraphy is preserved (Simola 1979, Davidson 1988). Many of these fine-interval techniques allow detection of environmental changes occurring at the subannual level, the same time scale as many monitoring programs employing direct observation.

## For What Range of Habitats are Paleolimnological Approaches Applicable?

Sediment cores suitable for paleolimnological analysis are usually obtained from quiescent depositional basins where fine-grained sediments accumulate conformably over time. Aquatic habitats characterized by erosion and scouring, slumping or turbidity flows, or episodic drying are generally unsuitable for coring, because these processes severely disrupt sediment stratigraphy. In relatively large lakes, for example, wave and current action tends to erode fine-grained materials from littoral areas into deep-water habitats offshore (Håkanson and Jansson 1983, but see

Anderson 1990a); the removal is often episodic and the littoral record is therefore often incomplete. Although most sediment cores are collected from the deeper parts of lakes and reservoirs, the sediments from these profundal regions integrate environmental signals from pelagic, littoral, epilimnetic, hypolimnetic, and benthic habitats.

The beds of streams and rivers, almost by definition, do not preserve a sedimentary record; exceptions include cut-and-fill deposits along low-gradient streams, which are used occasionally as environmental archives for palynological and archaeological investigations (Chumbly et al. 1990, Baker et al. 1991), and other riverine sedimentary deposits (Klink 1989, Petts et al. 1989). Natural and artificial impoundments, however, represent an important sink for sediments in lotic environments, and stratigraphic sections of these deposits can be used to assess water quality changes in the inflowing rivers (Brush and Davis 1984, Trefry et al. 1985). Depositional areas are also present in estuaries and reservoirs, as well as wetlands (Stevenson and Flower 1991).

## Where Should Cores Be Taken?

In paleolimnological studies, it is traditional to infer past lake environmental trends from a single core taken from the deepest point in the basin. This choice of coring sites is reasonable, because sediment accumulation rates are usually highest in deep-water environments. Processes of erosion and transport, which predominate in shallow water habitats, focus sediments to the deeper parts of the basin, so that cores taken from profundal regions generally exhibit the greatest temporal resolution (M. B. Davis et al. 1984, Hilton et al. 1986). However, in some wind-stressed lakes with low sedimentation rates, deposition is asymmetric with respect to lake depth, and the deepest core site is not always the best (Anderson 1990c). Deep-water sediments are also the least prone to erosion or desiccation from a drop in water level and are most likely to provide an uninterrupted sedimentary sequence. In addition, the preservation of labile (easily degraded) compounds (e.g., photosynthetic pigments) and sedimentary structures such as varves (annual laminations) is favored in deep-water environments (O'Sullivan 1983, Hurley and Armstrong 1991). The primary disadvantage to the deep-water core is that it is often unrepresentative of sedimentary conditions for the lake as a whole (Evans and Rigler 1983, Engstrom and Swain 1987).

Basin slope is another constraint on the location of good core sites. Because of episodic slumping of accumulated sediment, steep slopes (or areas near the base of steep slopes) should be avoided for coring (cf. Ludlam 1981). Locations where rivers enter lakes or reservoirs may or may not make good coring sites, depending on the hydrodynamics of the system and the questions at hand. In situations where river-borne sediment loads are high, sediment deposition at the proximal end of an impoundment can be very rapid (i.e., too rapid for $^{210}$Pb dating) and dominated by coarse-grained sediments (Håkanson and Jansson 1983). If current velocities are high, deltaic processes of scour and fill may predominate. On the other hand, a core taken closer to a stream discharge is more likely to contain fossil remains of stream organisms that might be used as direct biological indicators of stream conditions.

## How Many Cores are Needed?

**Trend Studies.** A single sediment core can reveal a great deal about lake history if the processes of sediment redistribution and fossil taphonomy are taken into account. For example, fossil diatom assemblages in different cores from the same lake usually express the same environmental trends irrespective of core location (Battarbee 1978b, Dixit and Evans 1986, Kreis 1986, 1989, Anderson 1989, Anderson 1990c). The relative abundance of the various fossils may differ from site to site (e.g., Bradbury and Winter 1976, Frey 1988), as may the clarity of the signal, but the direction of change should be qualitatively the same. Even quantitative reconstructions of environmental variables, such as diatom-inferred pH, may be relatively insensitive to core location if the inference technique is robust (i.e., based on a diverse fossil assemblage) (Charles et al. 1991, Anderson 1990c). The same argument holds for geochemical profiles. Changes in nutrient loading, erosion, or atmospheric deposition that alter the chemical composition of the sediments should register the same stratigraphic trends at different core sites, although the actual composition of the sediments may differ greatly among sites (Williams et al. 1976, Dillon and Evans 1982). Although single cores have usually proven satisfactory, multiple cores should be taken if significant variability of sediment characteristics is expected.

**Budget Studies.** The interpretation of accumulation rates is another matter. Single cores may at times be misleading because changes in accumulation may reflect shifts in the pattern of sediment deposition within the basin, as opposed to changes in material input to the lake (Battarbee 1978a, Davis and Ford 1982, Engstrom and Swain 1987, Anderson 1990b). The possibility that sediment focusing underlies an increase (or decrease) in sedimentation at a single core site is not always recognized in the paleoecological literature. Furthermore, because sediment deposition and composition are spatially variable, accumulation rates at a single core site cannot be automatically extrapolated to the entire lake bottom (Lehman 1975, Dearing 1983, Hilton and Gibbs 1984, Downing and Rath 1988). If whole-basin fluxes are desired, multiple cores representing different depositional environments are required (Dearing 1986).

The actual number of cores needed to estimate whole-lake accumulation depends on the size and morphometry of the basin and the nature of the environmental signal under investigation. Sediment deposition is spatially more variable in large deep lakes with irregular morphometry than in small basins of uniform depth, and more cores are required from the former to attain the precision that a few cores would provide in the latter. Impoundments and lakes with major river inputs typically exhibit strong depositional gradients relative to their inflows (Håkanson and Jansson 1983) and require a higher density of cores to calculate whole-basin sediment loading. Finally, only a few cores might be necessary to confirm that accumulation changes in one core are qualitatively representative of the entire lake, whereas a quantitative estimate of lake-wide sediment flux could require perhaps a dozen cores or more (Evans and Rigler 1980).

## How Can Paleolimnological Trends Be Dated?

**Why date sediments?** An accurate sediment chronology is an essential ingredient in any paleolimnological study. Sediment cores are usually dated to (1) establish the timing of past environmental change or (2) determine the flux of materials to a lake or reservoir. In the first case, sediment chronology allows us to determine the date at which stratigraphic changes occurred and to compare them with known historical events. Accurate sediment chronology can be particularly useful for investigating cause-and-effect relationships. In the second case, the desired metric is the rate at which materials accumulate at the core site, in itself an important paleoenvironmental signal.

**What dating methods are available?** Dating techniques fall into two general classes: those that rely on stratigraphic markers to identify a particular date or event and those that assign dates to all or many intervals in a core. A stratigraphic marker is an identifiable and discrete signature in sediments, produced by a known historical event; it is usually restricted to one or a few core intervals. Widely employed examples of this type of dating include:

- Pollen analysis, to identify the onset of European agriculture in the northeastern United States and adjacent Canada–the so-called settlement horizon or rise of *Ambrosia* (ragweed) (Bruland et al. 1975, Brugam 1978, Charles et al. 1990)
- Radioisotopic analysis, for peak concentrations of $^{137}$Cs and $^{241}$Am produced by atomic weapons testing in the mid-1960s (Pennington et al. 1973, Appleby et al. 1990, Ritchie and McHenry 1990)
- Geochemical analysis of volcanic ash layers associated with known dated eruptions (Sarna-Wojcicki et al. 1983, Einarsson 1986)

Examples of more localized markers include those produced by discharge of mining and sawmill wastes, atmospheric soot emissions, silt bands from road-building, and copper sulfate treatment of algal blooms (Allott 1978, Coard et al. 1983, Renberg and Wik 1984, Smol and Dickman 1981, Engstrom et al. 1985). Methods that provide a continuous sequence of dates from a single core utilize either the radioactive decay of naturally occurring isotopes such as $^{14}$C or $^{210}$Pb, or the visible sequence of annual laminations (varves) preserved in the deep-water sediments of some lakes (O'Sullivan 1983). Dating approaches are explained in more detail in following subsections.

**What time-span and resolution are required?** The choice of dating methods depends on the time span of interest, the temporal resolution required by the investigation, and the sedimentary material at hand. Most biological monitoring studies of human impacts on aquatic systems need to resolve recent events on the order of 2–10 years from sedimentary records spanning the last 100–200 years, and perhaps up to a millennium,

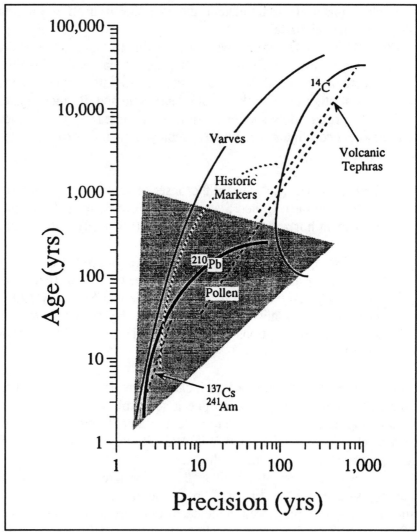

**Figure 13.2.** Time-scale and precision for dating methods commonly used in paleo-limnology. The shaded triangle approximates the time window of most interest in biological monitoring.

where information on long-term variability is required. This generalized time window overlaps with a number of dating methods (Figure 13.2) that are most effectively used in concert to arrive at a secure chronology. Because each dating technique has its own limitations and uncertainties, a combination of approaches provides the greatest reliability through corroboration among independent chronologies (Engstrom et al. 1985) (Figure 13.3).

## How Do Dating Methods Work and What Are Their Limitations?

**Stratigraphic Markers.** Marker horizons are usually located by analysis of contiguous levels in a core until the signal of interest is narrowly constrained; the known date is applied to that level, and if

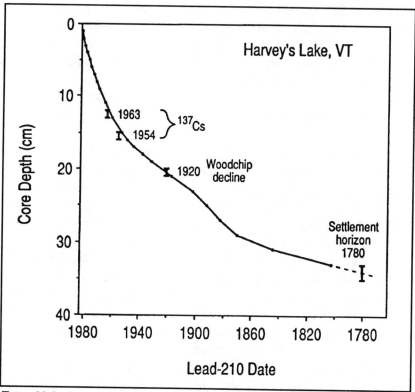

**Figure 13.3.** A composite of dating results from Harvey's Lake, Vermont, showing how several independent dating methods provide a more secure chronology. The woodchip decline represents the historically documented cessation of sawmill discharge into the lake, and the settlement horizon represents the initiation of land clearance and European agriculture as revealed by pollen analysis (from Engstrom et al. 1985).

desired, sediment accumulation is estimated by interpolation to another dated stratum (typically the present-day core surface). General limitations of this approach include (1) the geographic distribution of the marker, (2) the intensity and duration of the event that created it, (3) its stratigraphic definition in the core, and (4) the number of dates required by the study.

Most temporally discrete events that provide stratigraphic markers are local in extent (e.g., sawmill waste discharges) rather than regional (e.g., land clearance and agriculture). Only pollen and human-made radionuclides ($^{137}$Cs and $^{241}$Am) are at all widespread. Even where European settlement was rapid and well-documented, the resulting change in the pollen signal is still gradual because of long-distance dispersal of disturbance indicators (e.g., *Ambrosia*) from regions of earlier settlement. Pollen markers typically contain a dating uncertainty of at least a decade. When exotic weeds are involved, a temporal resolution on the order of $\pm$ 5 years is possible if the history of introduction is well documented (Jacobson and Engstrom 1989).

Even temporally discrete events can leave gradual stratigraphic signals because of sedimentary processes such as mixing, resuspension, and diffusion (Robbins 1982, Moeller et al. 1984, Anderson 1990c). If sedimentation rates are low relative to the intensity of these processes, a marker with a resolution in years may be smeared across sediments representing decades. For example, the well-defined input signal for $^{137}$Cs is commonly distorted beyond recognition by pore-water diffusion in unproductive low acid neutralizing capacity (ANC) lakes (R.B Davis et al. 1984, Heit and Miller 1987). This problem can be surmounted in some cases by recourse to the less mobile $^{241}$Am (Appleby et al. 1990, 1991); however, few laboratories currently possess the low-background, gamma-detection systems necessary for measuring environmental levels of this radionuclide. Cesium diffusion is less of a problem where sedimentation rates and clay content are high (e.g., oxbow lakes and reservoirs of large rivers) (Wise 1980).

Finally, stratigraphic markers typically provide only a single date, despite a fair expenditure of effort to locate the dated stratum. Chronology must be generalized to other core levels by interpolation (or extrapolation) with considerable uncertainty, particularly if sediment accumulation rates have changed over time. Stratigraphic markers are most effectively used to corroborate other dating methods such as $^{210}$Pb when a more secure chronology is required.

**Varves.** Annual laminations (varves) are produced by seasonal changes in the composition of sedimenting materials; where preserved, they provide the most detailed and accurate chronology of any dating technique (O'Sullivan 1983, Saarnisto 1986). Varve chronologies are often used as standards for validating other dating methods (Appleby et al. 1979, Lotter 1991). The procedure requires the annual nature of the laminations to be verified, typically by microscopic examination of the light/dark couplets representing each year, and the varves to be carefully counted, usually with a dissecting microscope along a cleaned surface of the core (Renberg 1981). Surface cores of unconsolidated watery sediments must be collected by freeze-sampling to preserve varve structure. The primary limitation of varve dating is the relative rarity of annually laminated sediments in lakes worldwide. In theory, most lakes in seasonal environments would produce varves were it not for mixing of surface sediments by benthic organisms. However, varves are usually found only in the profundal regions of relatively deep lakes, where the benthos are excluded by near-permanent anoxia, although varves can occur in shallower systems as well. Varved sediments are usually identified by color changes, but varves in some lakes can be detected only by such techniques as X-ray analysis.

**Radiocarbon.** Carbon-14 dating of lacustrine sediments has been a mainstay of paleoecology since 1949. Unfortunately, the optimum time scale for this technique barely overlaps the time scale required for most biological monitoring studies (Figure 13.2). Because of the long half-life of $^{14}C$ (5,730 years) and the long-term changes in its production in the upper atmosphere, radiocarbon dates from relatively young materials (<300 years old) have poor precision (Krishnaswamy and Lal 1978, Olson 1986). Recent advances in analytical methods, particularly accelerator mass spectrometry (AMS), which provides dates from terrestrial macrofossils such as needles, seeds, and twigs, can improve dating precision and eliminate other uncertainties associated with bulk sediment analysis. Nonetheless, radiocarbon dating is generally too coarse for biological monitoring studies, except where dates for long-term reference conditions are required.

**Lead-210.** The most widely applied dating tool for recent lake sediment is $^{210}Pb$ (e.g., Krishnaswamy and Lal 1978, Robbins et al. 1978, Appleby and Oldfield 1983). A naturally occurring radioisotope in the $^{238}U$-decay series, $^{210}Pb$ enters lakes primarily through precipitation and dry deposition, following the decay of atmospheric $^{222}Rn$ (an inert gas). Cores are typically dated by analyzing a series of stratigraphic levels from the mud surface to a depth where unsupported $^{210}$Pb is no longer measurable (roughly 5–8 half-lives, or about 150 years). From the resulting profile of $^{210}$Pb activity, dates and sediment accumulation rates are calculated according to one of several mathematical models that make assumptions regarding the delivery of $^{210}$Pb and sediment to the lake bottom (Figure 13.4). For most lakes with moderate sediment accumulation rates, the model of choice (the c.r.s., or constant rate of supply model) assumes a constant input of $^{210}$Pb to the core site and allows sediment input to vary. General reviews of dating models and their assumptions are given in Appleby and Oldfield (1983) and Oldfield and Appleby (1984).

An important limitation of $^{210}$Pb dating is that sediment accumulation must be conformable—there should be no major hiatus in deposition that would interrupt the flux of $^{210}$Pb or sediment to the core site. This means that cores from environments where sediment accumulation is episodic (e.g., littoral areas or steep slopes) are not reliably dated by this method. Bioturbation of surface sediments is another problem that may compromise $^{210}$Pb dating results, especially in large lakes with low sedimentation rates (Robbins et al., 1977). However, sediment accumulation in most small and moderate-sized basins is fast enough to overwhelm the effect of mixing. Mixing factors can be incorporated into the various dating models if the rate of mixing or the mixed depth is known (Robbins 1982, Oldfield and Appleby 1984). Finally, $^{210}$Pb is unreliable in lakes and reservoirs with extremely high sedimentation rates. High inputs of clastic sediments (sands, silts, clays) can dilute concentration of unsupported $^{210}$Pb to the point where it is difficult to detect.

## Can a Large Number of Sites Be Dated Economically?

The primary limitations in using $^{210}$Pb for biological monitoring purposes are cost and availability. At present, few laboratories are equipped to handle the large number of samples that could be generated by even a modest survey of lakes. It is possible, however, to economize on the number of samples required for each core and to thereby increase geographic coverage under a fixed budget.

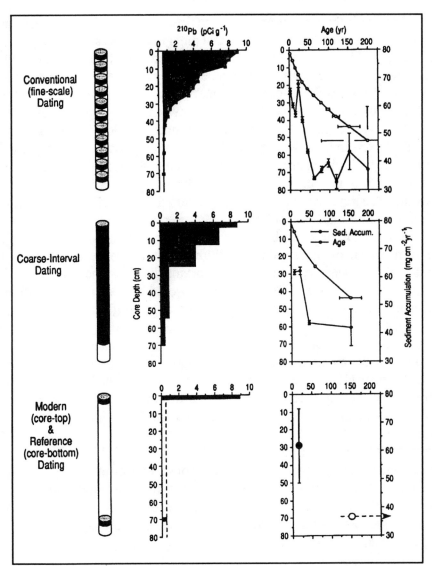

**Figure 13.4.** Lead-210 dating applied to a sediment core to obtain different levels of temporal resolution. Cores are conventionally dated by analyzing $^{210}Pb$ activity at relatively close intervals (top row), but sediments can be integrated into a smaller number of contiguous intervals to obtain a coarsely dated profile (middle row). Dates and sedimentation rates are calculated according to the constant rate of supply dating model (Appleby and Oldfield 1983). Modern sedimentation rates can be estimated from a single $^{210}Pb$ measurement from the top of the core (Binford and Brenner 1986), and a minimum age of about 150 years can be established for deeper sediments containing no unsupported $^{210}Pb$ (i.e., where total $^{210}Pb = ^{226}Ra$ (lower row).

If a coarsely dated profile will suffice, it is possible to analyze only a few contiguous depth intervals that integrate large sections of a core (Figure 4). The total inventory of unsupported $^{210}$Pb in the profile, which is required by the c.r.s. dating model, is calculated by summing the average $^{210}$Pb activity of each section. When core sections are integrated, there is no need to interpolate activity between widely spaced samples (a risky enterprise), and dates and sediment accumulation can be calculated by the c.r.s. model with no additional assumptions. The only major difficulty with this approach is that core intervals, selected *a priori*, may be either too fine or too coarse to adequately describe the full $^{210}$Pb profile. A careful choice of sampling intervals based on best estimates of sedimentation rates is required.

For some purposes, biological monitoring studies may require only a comparison of present-day conditions with reference conditions—the path of the trajectory being less important than the end points. In this case, a single date near the core surface would provide modern sediment accumulation rates; a single basal date would confirm that the deeper sediments were of predisturbance (reference) age (Figure 13.4). A core top date can be approximated by using regional estimates of atmospheric $^{210}$Pb flux in a modification of the c.r.s. dating model proposed by Binford and Brenner (1986). The regional flux may be estimated from well-dated cores from nearby sites, and if sediment focusing is not too extreme, the sediment accumulation rates so obtained should be within $\pm50\%$ of the true value.

At the other end of the scale, a minimum date of about 150 years can be established for deeper sediments that contain no unsupported $^{210}$Pb. The major uncertainty of a minimum date is that the sediments might possibly be far older than expected; that is, they could represent climatic (and limnological) conditions quite different from those prevailing at the time of settlement.

# What Biological Indicators are Contained in Sediment Records, How Can They Be Measured, and What Lake and Watershed Variables Can They Tell us About?

Lake sediments contain a surprisingly large suite of indicators that can be used in a biological monitoring program. Almost all organisms usually considered as prime indicators for these programs are preserved in lake sediments, either as morphological remains or biogeochemical markers. Two recent volumes (Gray 1988, Warner 1990) collate a number of review articles summarizing many of these indicator groups; thus, only a brief overview of the major biological indicators is presented here.

Algae have long been considered excellent indicators of limnological conditions so it is not surprising that they are also valuable paleo-indicators (Smol 1987). Fortunately, algal fossils are among the fossil groups best represented in lake sediments. Their analysis and interpretation have reached a high degree of sophistication in recent years.

Diatoms (Bacillariophyceae) frequently form the mainstay of paleolimnological studies. Their taxonomically distinct, siliceous cell walls, called frustules and composed of two valves, are usually abundant, diverse, and well-preserved in sedimentary profiles. Because diatom species are ecologically diverse and their ecological optima and tolerances can be quantified (see following section), they have been used to infer past changes in acidification (Charles et al. 1989) and recovery (Dixit et al. 1991), trophic variables (Engstrom et al. 1985, Whitmore 1989), salinity (Fritz 1990), dissolved organic carbon (Kingston and Birks 1990), metals (Dixit et al. 1991) land use, and other anthropogenically related activities (Tuchman et al. 1984), as well as changes in lake levels and climatic variables (Smol et al. 1991). Using diatoms, it is even possible to infer trends in other biotic communities, such as changes in fishery resources (Kingston et al. 1992) or past macrophyte development (Osborne and Moss 1977).

Chrysophycean algae are also well represented by siliceous fossils, which include endogenously formed resting stages called stomatocysts or statospores (see Sandgren 1991), which are formed by all taxa, and species-specific siliceous scales, spines, and bristles, which are formed by important genera such as *Synura* or *Mallomonas*. In general, chrysophytes and diatoms indicate similar limnological variables, except that the former are almost exclusively planktonic.

Other algal groups are preserved in lake sediments as nonsiliceous fossils, such as the organic-walled resting cysts of dinoflagellates, vegetative remains of many green algae (such as *Pediastrum* and desmids), and reproductive structures of charophytes. In addition, biochemical fossils, such as the photosynthetic pigments and their derivatives, can be used to trace past changes in algal and bacterial populations that do not leave reliable morphological fossils. For example, Leavitt et al. (1989) used annually resolved sedimentary records and fossil pigment analyses to trace past changes in phytoplankton populations and related these changes to other trophic variables in two manipulated lakes. Fossil pigment analyses may be especially important in studies dealing with eutrophication because blue-green algae and other noxious algae can be traced in sedimentary profiles by specific carotenoids (Sanger 1988).

Bacterial groups can also be studied using paleolimnological approaches (see review by Renberg and Nilsson 1992). For example, Nilsson and Renberg (1990) used viable endospores of specific bacteria to document the agricultral history of Swedish lakes. Many photosynthetic bacteria can be studied using fossil pigments (Brown 1968). These analyses are useful adjuncts to many limnological studies, such as those dealing with past trophic status (Brown et al. 1984).

A large number of animal fossils, many of which are amenable to long-term biological monitoring programs, can also be used in paleo-limnological studies (Walker et al. 1991). For example, protozoan parts preserve as fossils (e.g., Warner 1990, Douglas and Smol 1987) and can be used as indicators of peatland development and water level fluctuations. Crustacean fossils, such as the chitinous parts of the exoskeletons of Cladocera, are commonly studied (Hann 1991) and have been used to infer past changes in trophic status, acidification, and other forms of pollution. In well-buffered alkaline sediments, the calcite bivalve shells of Ostracoda are often abundant and species composition has been used to infer past changes in lakewater oxygen, salinity, and climate-related variables (DeDeckker et al. 1988). Moreover, techniques have recently been developed whereby ostracod shell chemistry is being used to infer other long-term environmental trends, especially those related to past climatic changes (Chivas et al. 1986).

Aquatic insect groups, such as Chironomidae (midge) larvae are often considered important biological monitors. These larvae are well represented in the fossil record by their sclerotized, chitinous head capsules, which can generally be identified to the generic level, and sometimes to the species level. Fossil midge distributions have been used to assess a variety of environmentally related issues (Walker 1987) including lake eutrophication (Warwick 1980), acidification (Uutala 1986), and climatic change (Walker et al. 1991). The analysis of *Chaoborus* mandibles can provide additional information on past limnological trends, and, may also provide indications of the distribution of past fish stocks. Many other aquatic insect groups, such as caddisfly (Williams 1989) and black fly (Currie and Walker 1992) larvae, can also be used.

Analysis of the remains of biota can be used to help understand species interactions and community relationships, which can be important if the goal of a biological monitoring program is to track changes due to management activities, species invasions, or other perturbations affecting an existing food web. Invertebrate fossils can provide sophisticated interpretations of predator-prey interactions. Because selective predation by vertebrate planktivores influences zooplankton species composition, the relative level of predation can be inferred from an analysis of the size distribution of fossils from pelagic zooplankton. For example, in many north temperate lakes, *Bosmina longirostris* dominates when and where fish planktivory is high, whereas *Daphnia pulex* are indicative of conditions where piscivory is more intense. Intermediate-sized *Daphnia* (e.g., *D. rosea*) are common during moderate planktivory (Leavitt et al. 1989).

Predation by invertebrate planktivores (*Chaoborus*, copepods) can be estimated by studying changes in the abundance of fossils of the predators themselves or by examining alterations in prey morphology to favor predator-resistant forms, such as long mucros in *Bosmina* (Kerfoot 1981). Uutala (1990) and Johnson et al. (1990) have shown that by examining the species-specific, sclerotized, chitinous mandibles preserved in lake sediments, we can identify species changes in large, nonmigratory chaoborids (e.g., *C. americanus*), which indicate fish and fishless conditions.

Lake sediments also entomb a wide selection of indicators from terrestrial systems, although a discussion of this would be outside the purview of this contribution. Of most relevance to this review is palynology, the study of pollen grains and plant spores. A considerable literature exists on this subject, including many books and several journals. By interpreting sedimentary profiles of pollen and spores, palynologists can reconstruct past terrestrial vegetation, and to a lesser extent aquatic vegetation, from which other characteristics of the environment can be inferred (e.g., forest and soil development, climatic change, and agricultural activity).

## What Qualitative and Quantitative Procedures Can Be Used to Interpret Biological Data?

Primary data on biological fossils in lake sediments consist of the number of remains found per unit of sediment ($cm^3$, or gram dry weight). These concentrations are used to calculate percent abundance of taxa for each sediment interval, or to calculate accumulation rates (numbers/$cm^2$/year) by multiplying the concentration of remains by the sedimentation rate, the latter being determined from dating procedures. These metrics, in turn, can be analyzed in three basic ways in order to address questions about past limnological conditions and to make quantitative inferences (see Charles and Smol (1993), for more detailed discussion). First, profiles of relative abundance of individual taxa can be interpreted qualitatively (Figure 13.5), past conditions or trends being determined on the basis of known ecological characteristics of the taxa (e.g., tolerates low dissolved oxygen, found only in acid lakes). Information obtained with this approach is often adequate for assessing the general nature and direction of large environmental changes.

Second, taxa can be combined into ecological groups (e.g., acidophilic, halophobic, oligotrophic, and planktonic; Lowe 1974), and changes in percent abundance of these groups can be analyzed qualitatively and quantitatively (Figure 13.6). Ecological groups can be constructed arbitrarily, using general knowledge of taxa within a region, or quantitatively, using techniques such as cluster analysis (e.g., Davis et al. 1990, Dixit et al. 1990). Many types of indices have been proposed based on ratios of the abundance of organisms in different ecological categories. They have, for example, been derived for pH/acidity (Meriläinen 1967, Renberg and Hellberg 1982), organic pollution (Watanabe et al. 1988), and habitat type (e.g., plankton/benthic ratio). Categories have also been used in multiple regression analyses to quantify relationships between groups and various ecological factors (e.g., Charles 1985, Charles and Smol 1988, Whitmore 1989). In both approaches, the indices or ratios are designed to provide quantitative or semi-quantitative information on limnological characteristics.

The third approach is multivariate analysis of assemblage composition (e.g., Brugam 1983, Gasse et al. 1983, Lipsey 1988, Metcalfe 1988, Servant-Vildary and Roux 1990). This approach provides the most comprehensive understanding of ecological characteristics of taxa and

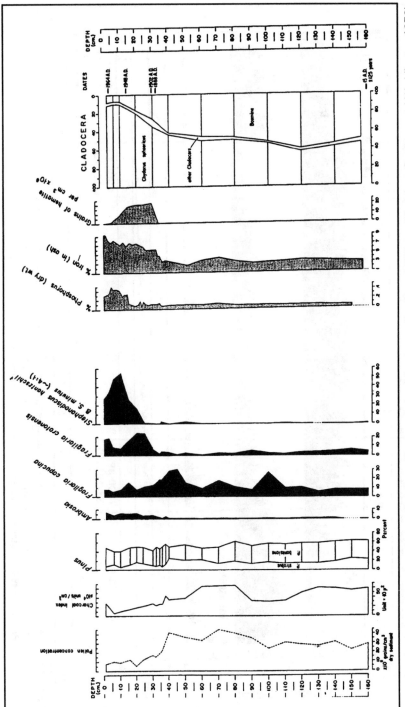

**Figure 13.5.** Selected stratigraphic profiles from Shagawa Lake, Minnesota, (Bradbury and Waddington 1973). (Bradbury 1975; modified from Bradbury and Waddington 1973). The increases in *Fragilaria crotonensis*, *Stephanodiscus*, and *Chydorus* indicate a shift from mesotrophic-eutrophic to eutrophic conditions.

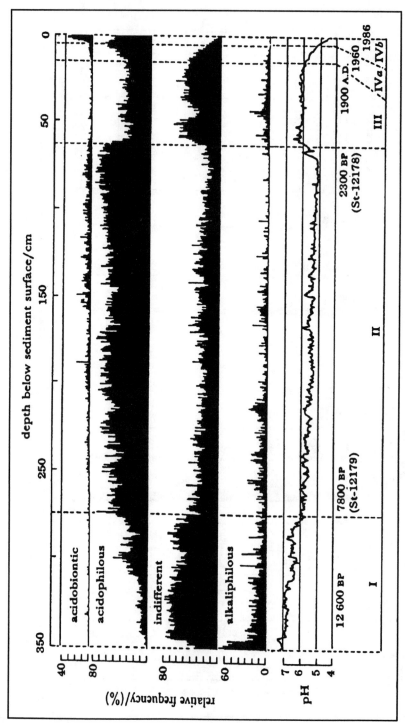

**Figure. 13.6.** Relative frequency of occurrence of diatoms grouped into four Hustedt pH categories, diatom inferred pH values (weighted averaging), calibrated radiocarbon dates, $^{210}$Pb dates and pH periods in the history of Lilla Öresjön, southwest Sweden (from Renberg 1990). Diatoms were analyzed at 0.5 cm intervals, 700 samples total.

the most accurate techniques for inferring lake characteristics. Many multivariate approaches have been used to analyze environmental gradients (Birks 1985), but the best technique at present is Canonical Correspondence Analysis (CCA), developed by ter Braak (ter Braak 1986, ter Braak and Prentice 1988). This technique is an extension of the DECORANA program of Hill (1979). Both programs, plus Reciprocal Averaging (also known as Correspondence Analysis), Principal Components Analysis, and Redundancy Analysis, are available in the program CANOCO (CANOnical Community Ordination), by C.J.F. ter Braak (available from the Agricultural Mathematics Group, TNO Institute for Applied Computer Science, Box 100, NL-6700 AC Wageningen, The Netherlands, and from Microcomputer Power, 111 Clover Lane, Ithaca, NY 14850, U.S.A.)

Canonical Correspondence Analysis has been used to investigate relationships among several characteristics, including pH and related factors (Stevenson et al. 1989, Birks et al. 1990a,b, Sweets et al. 1990a), metals (Dixit et al. 1991), trophic state (Hall and Smol 1992, Christie and Smol 1993), salinity (Fritz 1990), and temperature (Walker et. al. 1991). Input data for this program are counts of taxa for several sites and measurements of the environmental characteristics of those sites. The program ordinates both taxa and sites along one or more axes and calculates relationships among the taxa, sites, and environmental characteristics. The ordination process places the taxa or sites most different from each other at opposite ends of an axis, and puts all others in between, their specific location being based on their similarity to the other taxa or lakes. Taxa are considered similar if they occur in lakes with similar water chemistry. Sites (lakes) are considered similar if their biotic assemblages have similar composition. Relationships can be shown on biplots, which consist of the taxa, sites, or both, displayed as a function of their ordination values on two perpendicular axes, and arrows representing the environmental variables contributing to the axes (Figure 13.7). The direction of the arrow indicates increasing values. The length of the arrow indicates the strength of the correlation between the environmental characteristics and the distributions of the taxa or sites (Figure 13.7). The distance between sites or taxa and the arrow is a measure of the closeness of the relationship with that variable. The closer together they are, the stronger the relationship. However, the distance to arrows on a biplot can be misleading because the arrows are multi-dimensional. An important characteristic of CCA is that it provides an

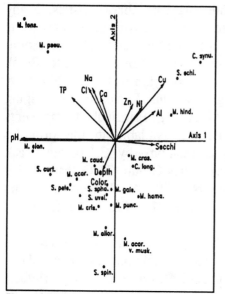

**Figure 13.7a.**

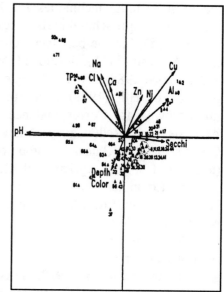

**Figure 13.7b.**

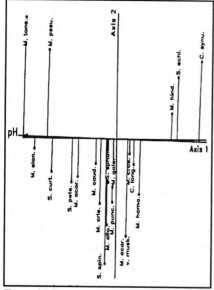

**Figure 13.7c.**

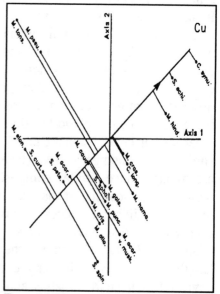

**Figure 13.7d.**

**Figure 13.7.** Canonical correspondence analysis ordination biplots for 72 lakes near Sudbury, Ontario. (a) environmental variables (arrows) and chrysophyte species distribution (closed circles), (b) environmental variables (arrows) and coordinates of 72 study lakes (triangles), (c) positioning of chrysophyte species along the pH arrow indicates the approximate ranking of the weighted averages of taxa with respect to pH, and (d) copper (from Dixit et al. 1989).

understanding of the relative strengths of association between taxa and environmental variables. This knowledge is useful for biological monitoring because it can be used to help identify groups of taxa that might be particulary good indicators of change in the variables of interest, and to determine what and how many factors have a strong influence on species composition and whether problems in interpreting changes in assemblage composition could arise.

One of the most powerful recent advancements in paleolimnology is the development of techniques for quantitatively inferring specific lake characteristics from biological remains in lake sediments. This development is primarily a result of efforts to use diatom assemblages to reconstruct acidification trends in lakes. The basic approach is to quantitatively relate asssemblage composition with environmental characteristics by using a calibration or training data set consisting of data on surface sediment assemblages and lake measurements for many lakes, usually 30 or more (Birks et al. 1990a). Extant populations of plankton algae from many lakes have also been used for calibration sets (Siver and Hamer 1990). The first and most basic technique developed was a simple regression of indices, as described earlier, with a single lake characteristic, such as pH (Meriläinen 1967). A more flexible and objective approach is to avoid the derivation of indices and develop inference techniques by doing multiple regression of the percentage of diatoms of other biota in the ecological categories with the lake characteristic to be inferred (e.g., trophic categories such as mesotrophic, eutrophic, with total phosphorus; Whitmore 1989). The most robust technique, in our view, is weighted averaging (WA) regression and calibration (ter Braak and Looman 1986, ter Braak and Barendregt 1986, ter Braak and van Dam 1989, Birks et al. 1990a,b). First, data from a calibration set is used to calculate the abundance weighted mean (AWM) of each taxon (Birks et al. 1990a):

$$\hat{u}_k = \sum_{i=1}^{n} y_{ik} x_i \Big/ \sum_{i=1}^{n} y_{ik}$$

where

$\hat{u}_k$ = abundance weighted mean, or optimum of taxon $k$ for a particular environmental variable,

$y_{ik}$ = percentage abundance of taxon $k$ in sample (lake) $i$,

$x_i$ = value of environmental variable $x$ in sample $i$, and
$n$ = number of lakes in the calibration data set.

Once this is done, past environmental conditions can be inferred from sediment core assemblages using the following equation:

$$\hat{x}_i = \sum_{k=1}^{m} y_{ik}\, \hat{a}_k \; / \; \sum_{k=1}^{m} y_{ik}$$

where

$\hat{x}_i$ = the inferred value of $x$ for sample $i$, and
$m$ = number of taxa in the core sediment sample assemblage.

The computer program WACALIB (Line and Birks 1990) is designed to do the necessary calculations. This program performs several additional functions necessary or desirable in the analysis, including calculation of error estimates. Using the techniques described here, primarily WA and CCA, it is possible to infer lakewater pH with a standard error (SE) value of 0.25 to 0.35 pH units (Charles et al. 1990), ln total alkalinity (μeq/L) with an SE value of 0.31 (Cumming et al. 1991), ln monomeric aluminum (μg/L) with an SE value of 0.43 (Cumming et al. 1992, Birks et al. 1990a), dissolved organic carbon with an SE value of 80–150 μmol/L (Birks et al. 1990a, Kingston and Birks 1990), phosphorus and chlorophyll trophic state indices (Whitmore 1989), and salinity (with an SE value of 0.48 ln salinity, ppt; Fritz 1990).

Interpretation of the biological record in sediments can be supplemented and enhanced by other research and monitoring approaches, such as use of sediment traps, artificial substrates, sampling of planktonic and benthic habitats, and bioassay and mesocosm studies.

## What Can Be Learned From Chemical Analysis of Sediment Cores?

Although geochemistry is not strictly within the purview of biological monitoring, the chemical analysis of sediment cores can greatly enhance the picture of limnological change that is reconstructed from fossil organisms. A sizeable array of geochemical tools is available, including:

- The analysis of trace metals (Mg, Sr, Ba) and stable isotopes ($^{18}O$, $^{13}C$) in biogenic and inorganic carbonates to reconstruct the paleosalinity and paleohydrology of closed-basin lakes (Chivas et al. 1986, Siegenthaler and Eicher 1986, Engstrom and Nelson 1991)
- The use of fossil photosynthetic pigments to infer past changes in algal composition and trophic interactions in the lacustrine food web (Swain 1985, Sanger 1988, Leavitt et al. 1989)
- The analysis of biogenic components (phosphorus, nitrogen, carbon, silica) to reconstruct the history of nutrient loading and eutrophication (Williams et al. 1976, Schelske et al. 1986)
- The measurement of redox-sensitive elements (Fe, Mn, S) to infer changes in the hypolimnetic oxygen regime (Mackereth 1966, Engstrom et al. 1985)
- The analysis of metal pollutants (e.g., Hg, Zn, Pb, Cu, Cd) and organic contaminants (PCBs, DDT, PAHs) in lake sediments as a record of industrial emissions (Eisenreich et al. 1989, Norton et al. 1990)
- The use of nondetrital pools of S, Ca, Mg, and Al to reconstruct biogeochemical changes in elemental cycling associated with lake acidification (Nriagu and Soon 1985, Charles et al. 1987, White et al. 1989)
- The analysis of major elements derived from catchment soils (Na, K, Mg, Al, Si, Ti, Fe, etc.) to infer changes in erosional intensity and soil development (Engstrom and Hansen 1985, Whitehead et al. 1989)

## What Geochemical Methods are Most Suitable for Biological Monitoring?

**Loss-on-ignition.** In this simple procedure, organic matter and carbonate content are calculated from loss of mass following high-temperature ignition (Dean 1974). Sedimentary organic matter, which is roughly 50% carbon by weight, is a combination of autochthonous and allochthonous production. An increase in this sedimentary component may signify eutrophication or vegetational and edaphic changes in the catchment (Pennington 1978, Davis et al. 1985). Lacustrine carbonates are primarily endogenic or biogenic, although a sizeable fraction may be detrital (eroded from soils) in some systems. The carbonate content of sediments in high ANC lakes frequently increases with eutrophication as a result of increased algal uptake of $CO_2$ during summer stratification (Dean 1981, Brenner and Binford 1988, Anderson 1989). The remaining inorganic fraction is composed primarily of detrital silicate minerals, biogenic silica (from diatoms and chrysophytes), and amorphous iron oxides. Because silicates are the dominant phase in most lake sediments, the inorganic fraction is often a good proxy for erosional intensity (Brugam 1978, Engstrom et al. 1985).

**Major element chemistry.** The gross sediment stratigraphy provided by loss-on-ignition can be further refined by analyzing major elemental composition to yield additional information about the source (provenance) and environmental significance of the inorganic components. In a general review of this subject, Engstrom and Wright (1984) discuss extraction procedures, analytical methods, and environmental interpretations for a number of major elements, some of which could be applied economically in a lake sediment survey.

Certain elements (K, Na, Mg, Ti) are almost entirely associated with detrital silicates in lake sediments, and their stratigraphy provides an unambiguous proxy for erosional intensity, especially if combined with sedimentation rates from [210]Pb dating (Burden et al. 1986, Engstrom et al. 1991). Other elements, such as phosphorus and base-extractable silica (biogenic $SiO_2$), can be used to reconstruct the history of nutrient loading and trophic change. Although phosphorus (P) interpretations can be confounded by other limnological factors affecting P retention in sediments (hypolimnetic oxygen regime, Fe content of the sediments), P stratigraphy is still a fairly reliable indicator of major anthropogenic changes in nutrient loading (Deevey et al. 1979, Moss 1980). Biogenic silica, which can be distinguished from silicate-$SiO_2$ by alkaline extraction techniques, provides a rapid and inexpensive method for quantifying past changes in diatom/chrysophyte production (Schelske et al. 1985, 1987, Conley 1988).

**Metal pollutants.** Numerous studies have shown lake sediments to be excellent archives for historic changes in the industrial discharge of potentially toxic metals such as Hg, Zn, Pb, Cu, Ni, Cd, and Co (Goldberg et al. 1981, Rippey et al. 1982, Trefry et al. 1985, Norton et al. 1990). Most heavy metals have short residence times in the water column and are quantitatively retained in the sediments, making stratigraphic interpretations relatively straightforward (but see Carignan and Tessier 1985, White et al. 1989). Sediment records have provided compelling evidence for the role of atmospheric deposition in contaminating aquatic environments far removed from direct (point source) industrial influence (Evans 1986, Rada et al. 1989). These studies have been used to document the history of emissions, the effect of abatement, and the geochemical cycling of trace metals in the aquatic environment (Carignan and Nriagu 1985, Renberg 1986, Johnson 1987).

## What Are the Sources of Variability in Paleolimnological Analyses? How Can Error Be Assessed and Quantified?

There are several potential sources of variability in paleolimnological investigations. These may or may not be important for a particular study because different levels of uncertainty are acceptable or unacceptable, depending on the questions being addressed. Fortunately, it is usually possible to identify situations in which variability might cause significant problems in meeting project objectives, and to quantify the error associated with measurements so it can be accounted for when interpreting results. The following paragraphs identify major sources of variability and describe standard approaches for quantifying error.

**Within-lake variability.** Lake sediment depth and composition vary spatially (Häkanson and Jansson 1983). Multiple cores taken at widely spaced locations can provide information on this variability and an indication of how well any one core represents the sediment record of the lake as a whole (Kreis 1986, 1989). Multiple core studies have been used to evaluate distribution in sediments of diatoms (Meriläinen 1971, Bradbury and Winter 1976, Battarbee 1978b, Anderson 1986, 1990a,b,c, Charles et al. 1991), Cladocera (Mueller 1964, Goulden 1969), pollen (e.g. Davis et al. 1969, Edwards 1983), and other biological groups. The number of replicate cores that should be taken and the type and number of analyses to be made depends on the complexity of the lake's morphometry and sedimentary deposits and on the level of uncertainty that is acceptable (see also Section II.D.: How many cores are needed?). To assess representativeness, complete analysis of all replicate cores is not necessary. They can be limited to:

- Dating only
- Analysis of only selected intervals
- Analysis of easily measured characteristics that have irregular and easily matchable profiles, such as magnetic susceptibility (Edwards and Thompson 1984, Anderson 1986)
- Analysis of total lead, pollen (e.g., Edwards 1983), or carbonaceous particles (Renberg and Wik 1984)

Cores can be compared most effectively by analyzing intervals having corresponding dates and using computer programs that compare sequences among cores (e.g., Thompson and Clark 1989).

**Within-site variability.** Analysis of three or more cores, taken from within a few to several meters of each other, can be used to assess the variability within a particular site and the variability associated with the coring processes (Charles et al. 1991, Kreis 1986, 1989). Within-site variability is usually small compared to among-site variability, and it is not usually quantified in smaller studies unless there is a specific need for the information, such as concern that sediments have been disturbed by anthropogenic activities.

**Mixing, bioturbation.** Sediment mixing by any of several mechanisms can weaken or obscure the signal of an event recorded in a core. Changes might seem less abrupt than they actually were, and they might appear to begin sooner or end later than they actually did. However, the point of maximum change still represents the time that an event occurred or the time when change was greatest. In many cases, this is the most important information required. Mixing can be caused by water movement from wind stress, localized underwater currents and turbidity flows, and activities of animals living on or near the sediment surface (Häkanson and Jansson 1983). There is usually evidence of at least limited mixing of the top few centimeters in most cores, but in most cases it does not significantly disrupt sediment profiles. Because lakes with anoxic hypolimnia have few animals living on or in the bottom, sediments are relatively undisturbed, are sometimes laminated, and contain a record with high temporal resolution. The extent of mixing in a core is best evaluated by examining the dating profile (e.g., $^{210}$Pb) or the rates of change of particulate sediment characteristics (e.g., diatom assemblages) near the core surface. Slopes of profiles that are steep near the surface indicate little mixing. If the $^{210}$Pb profile is nearly vertical for the first few centimeters, mixing may be a problem, although this type of profile can also result from recent increases in sedimentation rate, which can dilute $^{210}$Pb concentrations near the surface.

**Subsampling, sample preparation, and counting.** Error resulting from subsampling and sample handling can be assessed by replicating each stage of the analysis process and using standard statistical procedures to express error and to test for significant differences among sample treatments (Kreis 1986, 1989). Most studies have found that error associated with these procedures is small, usually amounting to only a few percent (Kreis 1986, Charles et al. 1991).

**Sediment chronology and dating.** Several factors can affect determination of sediment core chronology and they are discussed in the sections on dating. The best way to determine the nature and magnitude of uncertainties of radiometric dating is to examine the profile of the radionuclide. In addition, methods are available to calculate standard errors of $^{210}$Pb dates (Binford 1990).

**Taxonomy.** The greatest source of error in using biological remains to reconstruct past lake conditions is incorrect and/or inconsistent taxonomy. If taxa are identified incorrectly, they may be associated with the wrong ecological characteristics, and inferences calculated from assemblage composition may indicate erroneous results (trends). Taxonomic identifications and quality control must be emphasized. Those doing identifications and counting should:

- Have complete and up-to-date literature references
- Consult museum collections
- Send samples to experts for confirmation of the identifications of unknown or difficult taxa
- Prepare and periodically consult reference material from the sites involved in a study
- Create a collection of photographs of taxa

Because considerable expertise is required for most biological analyses, samples are usually sent to experienced scientists qualified to do the analyses, or sufficient time and funding is provided to develop the expertise for a particular study. However, in some cases, where only minimal information about trends is needed, or where biota in sediments are well known and adequate taxonomic documentation is already available, analyses could be done by less qualified individuals, as long as they have adequate guidance and the results are reviewed by knowledge-able experts. In certain well-characterized systems, it may be possible to obtain the required information concerning conditions and trends by counting only some of the most informative and easily identifiable taxa.

**Ecological characteristics of taxa.** The extent to which ecological changes can be inferred from shifts in the abundance of individual taxa is limited by how well the ecological characteristics of the taxa are known and defined. Problems with interpretations can occur, for example, if a taxon is found in abundance in a core and little data exist in the literature or elsewhere describing its ecological characteristics. The best ways to insure sufficient ecological data are to develop an adequate calibration data set that contains many lakes similar to those studied and to do a thorough search of the literature. If resources are available, additional ecological information on taxa can be obtained by sampling a variety of habitats during different times of the year (DiNicola 1986, Jones and Flower 1986), collecting samples in sediment traps (Flower 1991), and doing laboratory or mesocosm studies (Gensemer 1990).

A concern often expressed is that ecological characteristics of taxa may change significantly over time, such that interpretations of past conditions based on ecological data obtained from studies of modern assemblages may be biased and incorrect. There are at least three reasons why this is not a problem for groups of organisms commonly used in paleolimnological studies. First, similar taxa occur together today that occured together millions of years ago in the same types of environments. Second, many taxa have geographic distributions spanning one or more continents, yet exhibit similar ecological characteristics throughout their range. Third, the smallest organisms (e.g., diatoms, chrysophytes, chironomids) are easily and widely dispersed to new habitats, and this dispersion, in combination with the existence of many taxa in each lake, results in relatively low selection pressure forcing genetic adaptation. All of these lines of evidence support the idea that rates of evolution (genetic change) are relatively low. This reasoning is not meant to suggest that these organisms do not evolve, or that a few taxa may not change fairly rapidly, only that the total amount of change in most communities is small enough that it does not preclude their use in inferring ecological change over the past few hundred years.

**Inference techniques.** The variability of values derived from inference techniques is a function of the quality and representativeness of the calibration data set and the nature of the specific inference techniques. Factors important in the calibration data set include: (1) the quality, appropriateness, and frequency of measurements of limnological characteristics; (2) the accuracy and level of identification of taxa; and (3) the number of lakes and how well they represent the lakes to be cored and analyzed stratigraphically (see also the section concerning biological indicators). In our view, weighted averaging regression and calibration with appropriate deshrinking steps is the best inference technique presently available for a biological monitoring program (Birks et al. 1990a,b). The error associated with inferred values can be calculated and expressed quantitatively in several ways, the best of which is boot-strapping (Birks et al. 1990a,b). Another expression of uncertainty is the standard error derived from the correlation of the measured values of lakes in the calibration data set with the inferred values for the same lakes.

## Have Data Management Strategies and Programs Been Developed for Paleolimnological Studies? What is Available?

Because monitoring with biological indicators involves generation, storage, and analysis of considerable amounts of data of several kinds, it is essential to have a computer-based data management system. The use of computers for analyzing assemblage data has increased rapidly in the past 10 years, and though more sophisticated software is continually being developed, there are no standard data management systems for paleolimnological biological monitoring. The best approach is to use one of the few existing documented management systems, such as those developed for the Paleoecological Investigation of Recent Lake Acidification (PIRLA) Project (Charles et al. 1989, Ahmad and Charles 1988) or the Surface Waters Acidification Programme (SWAP) (Munro et al. 1990). The next best approach, especially for smaller projects or for those in which data bases will be developed for a variety of monitoring characteristics, is to use a standard software package. The package should be flexible, expandable, and transportable to other computers. It should be able to output files into different formats and to accommodate different types of data, such as lake location and morphometry, water chemistry, core descriptions, taxa lists and ecological data, and taxa counts. Development of a standard set of retrievals for obtaining data in different combinations and formats is useful for efficient analysis of results. A good data management system can also be very helpful in maintaining quality control. For example, data entry programs can be written so that the name of a taxon is retrieved and displayed following each entry of a numerical taxon code to help insure that the correct number has been entered.

# HOW CAN PALEOLIMNOLOGICAL APPROACHES BE APPLIED IN MONITORING PROGRAMS?

## How Have Paleolimnological Approaches Been Applied? What Are Some Examples?

Paleolimnology has been used to address all the questions on monitoring that were raised in the introduction and in the first two topics. The following paragraphs describe examples of specific approaches and their applications to major environmental issues.

**Acidification.** Many studies have been performed to investigate the effects of acidic deposition on lakes (see Charles et al. 1990). The largest of these are the PIRLA-I and PIRLA-II projects in eastern North America (Charles and Whitehead 1986, Charles and Smol 1990) and the paleolimnology component of SWAP in the UK and Scandanavia (Battarbee and Renberg 1990). These and similar projects developed calibration data sets for their regions, cored many lakes, and reconstructed pH and alkalinity. Soot, metals, and other chemical characteristics were analyzed to determine temporal trends in the input of acidic deposition and associated materials. Primary questions addressed in these and related studies were: Have lakes acidified? If so, how much, how fast, and when? What were the relative roles of acidic deposition, changes in land use, and natural process? What were conditions like before the onset of acidic deposition? Results demonstrated that indeed many lakes had recently become more acidic and that acidic deposition was a primary cause.

In the PIRLA-II project, a statistically selected set of Adirondack lakes was cored, and diatom and chrysophyte assemblages in the tops and bottoms of the cores were analyzed in order to infer changes since pre-1850 in pH, acid neutralizing capacity (total alkalinity), aluminum, and dissolved organic carbon (Cumming et al. 1992, Sullivan 1990, Figure 13.8). Population projections were made of the extent of acidification of the 3,000 Adirondack lakes, making use of the statistical basis of the selection. A geographic information system was used to map regions where most change had occurred. In another component of the PIRLA-II project, cores from sensitive lakes (ANC near 0) were sectioned at closely spaced intervals in order to determine the response of lakes to the decrease in sulfur deposition that has occurred in the northeastern United States since 1970 (Sullivan 1990). Results showed that some lakes were recovering, some were becoming more acidic, and some were not changing. These results are being compared with recent direct measurements on some of the lakes in the U.S. EPA's Long-Term Monitoring (LTM) Project. The recent measurements provide a check on the diatom-inferred trends and the longer term paleolimnological data provide a background record of change to help with interpretation of the LTM data. The Adirondack paleolimnological data have also been used to evaluate a chemical model devised to help select lakes that are most sensitive to changes in acidic deposition, which makes these good lakes to monitor (Young 1991).

In the United Kingdom, periodic sampling and analysis of diatom assemblages in surface sediment samples is an integral part of a program to monitor changes in the acidity of lakes over a large area (Battarbee, personal communication). Also, Flower et al. (1990) have analyzed recent sediment diatom assemblages in the UK using CCA and have shown acidification and some recovery from the effects of acidic deposition (Figure 13.9). Paleolimnological data have also been used to help validate dynamic watershed geochemical models (Sullivan et al. 1992, Jenkins et al. 1990).

**Eutrophication.** Similar to the acidification examples cited in the previous section, paleolimnological approaches are very effective tools in the study of eutrophication. Almost all of the indicators we have discussed can and have been used in studies tracing past changes in lake trophic status, but once again diatoms have often provided the most detailed and reliable information. Several indices have been proposed over the years to infer past trophic variables from diatom assemblage composition (e.g., Stockner and Benson 1967, Agbeti and Dickman 1989, Whitmore 1989), however, as with the other examples we cite, we believe that the strongest and most robust inference models are based on calibration sets developed using CCA and weighted averaging. Several such calibration sets are now available for diatoms (e.g., Christie and Smol 1993, Hall and Smol 1992).

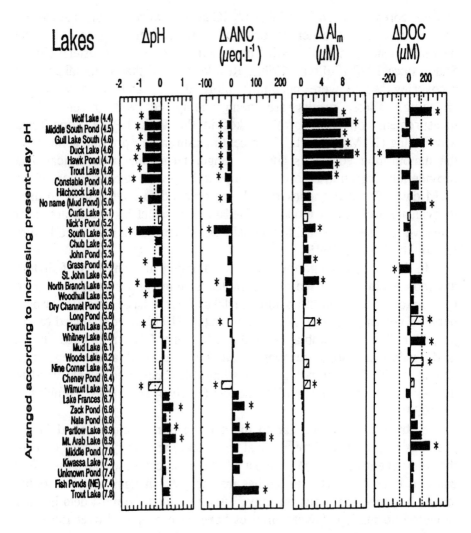

**Figure 13.8.** Diatom-based estimates of historical change (pre-industrial to the present) in pH, acid neutralizing capacity (ANC), total monomeric aluminum (Al$_m$), and dissolved organic carbon (DOC) from 37 Adirondack study lakes. The estimates are presented as the differences in inferred water chemistry between the top (0–1 cm) and bottom (> 20 cm, usually > 25 cm) sediment core intervals. The lakes are arranged according to increasing present-day measured pH. Lakes indicated by hatched bars were limed within 5 years before coring and/or assessment of water chemistry. The dotted lines represent ± bootstrap root mean squared error (RMSE$_{boot}$) of prediction for pH and DOC. Estimates of change in ANC and Al$_m$ are presented as differences between back-transformed estimates since these models were developed on log$_e$ data; thus simple lines for ± RMSE$_{boot}$ cannot be presented. Asterisks denote changes that are > RMSE$_{boot}$ for the various inference models (from Cumming et al. 1992).

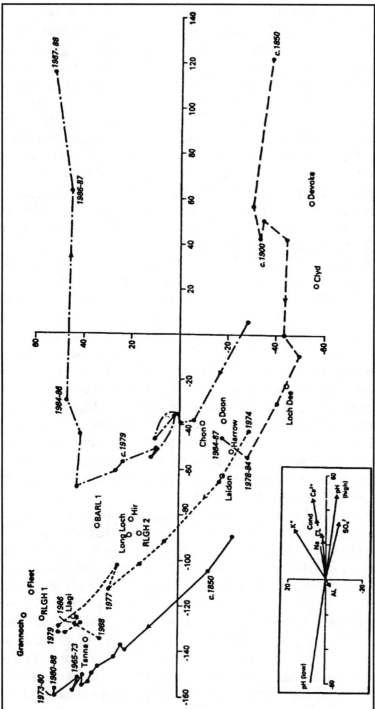

**Figure 13.9.** Canonical correspondence analysis (CCA) of diatom assemblages in the Round Loch of Glenhead, Loch Grannoch, Loch Fleet, and Llyn Hir cores. Time-tracks of floristic change are constructed by using CCA axis 1 and 2 sample scores and are constrained by modern U.K. chemistry (see inset) and diatom samples (O), which approximately define axis 1 as the pH axis. Loch Fleet was limed in the early 1980s. Note, RLGH axis 1 scores are offset by −30 to improve diagram clarity. (- - -) Loch Grannoch; (- -) Llyn Hir; (- -) Loch Fleet; (—) Round Loch of Glenhead (from Flower et al. 1990).

Paleolimnological studies can monitor the timing, rate, and extent of eutrophication as a result of cultural disturbances (e.g., Bradbury 1975) or natural processes (e.g., Boucherle et al. 1986). In some regions, such as the United Kingdom, cultural disturbances can date back several thousands of years, but these too can be documented using paleolimnological techniques (e.g., Fritz 1989). These approaches can also be used to study a lake's recovery from eutrophication after remedial action. For example, Schmidt (1991) used close-interval sediment analyses of diatoms to study the re-oligotrophication of Mondensee, following installation of a sewage treatment plant seven years before he collected the core.

In addition to phosphorus and chlorophyll, the role of changing silica concentrations and its effect on algal composition can be studied by analyzing sedimentary diatoms (e.g., Stoermer et al. 1990). This has been a particularly important issue in the Great Lakes.

**Climate change.** Some of the most widespread effects on aquatic ecosystems expected in the coming decades are those due to climatic change. These effects will add to those already occurring, will undoubtedly cause considerable stress to some systems, and will probably be very difficult to distinguish from other effects. Climatic change can have direct and indirect effects on aquatic systems. Water temperatures, stratification and mixing patterns, and duration of ice cover will be affected, as well as amounts of water flowing through a system and the seasonal pattern of the flow. Indirect effects include changes in water chemistry and biotic structure and food-web interactions. All of these changes influence the nature of materials deposited in sediments, thereby creating a record of ecosystem conditions that can be used to help monitor ongoing changes and provide a long-term context (Winter and Wright 1977, Berglund 1986, Smol et al. 1991).

Paleolimnological studies of climatic change have provided considerable insight into long-term climate trends, natural variability, and their causes. Fritz (1990) has developed diatom-based transfer functions to infer salinity changes of lakes in the northern Great Plains and tested them successfully by comparing them with a long-term record of salinity measurements of Devil's Lake, North Dakota. This is a very promising method that is being applied elsewhere in the U.S.A. and Canada. This approach is also being developed more fully as part of research on long-term climate change in Africa (e.g., Gasse et al. 1987) and has also been applied to saline lakes in Bolivia (Servant-Vildary and Roux 1990). Bradbury (1988) has used diatom assemblages in varved sediments of Elk

Lake, Minnesota to investigate year-to-year variations in lake characteristics caused by differing weather patterns.

Several approaches have been used to study lake level fluctuation (Winter and Wright 1977, Berglund 1986). Some of the basic approaches include examining dating profiles along transects of lakes from shallow to deeper depths to determine when and for how long areas of lake bottoms were exposed, changes in composition of sediment texture along similar transects to determine changes in distance of core sites to shore (coarser sediments should be found nearer to shore), ratios of the percentage of planktonic to littoral species of various organism groups, and composition and accumulation rates of seeds and other remains of aquatic macrophytes.

Changes in water temperature have been inferred from composition of chironomid assemblages (Walker et al. 1991), and relationships between diatom assemblages and water temperatures have been established (e.g., Snoeijs 1990). Chrysophyte bristle shape has been related to temperature change (Siver and Hamer 1990); morphological indicators of temperature will undoubtedly be found for other groups of organisms and will likely prove to be valuable paleoindicators of climatic change. Many of these techniques are still being developed, but appear very promising.

**Fisheries.** A major concern of aquatic managers is the status of fish stocks. Present-day fisheries can be assessed using standard approaches; however, if a lake is found not to support fish, an important monitoring issue is whether or not the lake ever supported fish populations. Paleolimnological approaches can once again be used to supply these missing data.

Uutala (1990) and Johnson et al. (1990) have shown that larval *Chaoborus* remains can be used as indicators of past fisheries status. To summarize the approach, five *Chaoborus* species occur in permanent lakes and ponds of the northeastern United States and eastern Canada. The identifiable fossil remains are the mandibles, which are separable into four taxa (Uutala 1990). The key indicator is *C. americanus*, which is highly susceptible to fish predation and is common only in fishless lakes. By studying the distribution of *Chaoborus* mandibles in a dated sedimentary profile, past fisheries status can be deduced. These studies are especially powerful when done in conjunction with fossil algal indicators, which can be used to infer past lakewater chemistry (Kingston et al. 1991).

# What Are Some of the Practical Considerations When Designing Paleolimnological Components of Biological Monitoring Programs?

A primary consideration in designing monitoring programs is cost. Given the importance of historical data, and that it often can be obtained in no other way, paleolimnological biological monitoring is relatively inexpensive. A single study carried out over one to three years can provide trend data covering a very long period of time. The information on reference conditions and natural variability obtained with this approach would be impossible, or much more difficult and expensive, to obtain through a standard monitoring program. Being able to place currently measured data into proper historic perspective can add considerable value to monitoring data for making management and policy decisions. Paleolimnological studies are well worth the cost.

Personnel needs and skill levels are also a consideration. Coring, especially if done with the more complicated devices, should be directed by experienced personnel, though techniques can be learned quickly. Radiometric dating of cores is usually done at laboratories that have the necessary equipment and expertise; several are available. Analysis of biological remains also requires special expertise, and there are many scientists and technicians in colleges and universities, consulting firms, and other facilities who can do this work. Preparing and analyzing a single group of biota from a core sample typically takes two to five hours. Analysis usually requires specialized apparatus for preparing samples, a microscope or dissecting scope for counting, computers for data entry and analysis, and access to ecological data bases for the taxa being analyzed. Because most paleolimnological laboratories do not posess expertise in all areas, paleolimnological programs are usually performed by teams of investigators located at more than one location, each doing a different type of analysis.

## CONCLUSIONS

Paleolimnological investigations can significantly enhance the usefulness and applicability of monitoring data by extending the temporal record of ecosystem conditions for a considerable time into the past, and providing a context for evaluating more recent measurements.

Paleolimnological studies can provide valuable information that often can be obtained in no other way. What were conditions prior to the occurrence of natural or anthropogenic disturbance? What is the natural variability of the system, including occurrence of rare, catastrophic events? What long-term trends have been ongoing prior to the beginning of modern measurements? What sites or taxa have changed most in the past and might be particularly sensitive to future change?

Paleolimnological techniques and methods have developed rapidly in the past 10 years and can be used to provide more accurate information on more ecosystem characteristics, and with greater temporal resolution, than ever before. Costs for these studies are reasonable considering the information they provide.

Analysis of the sediment record can be used to address most environmental issues that affect lakes, such as eutrophication, acid deposition, toxics, erosion, road salt, and changes in community structure and function. In addition, paleolimnological data contribute to the understanding of how systems function and can be particularly important in determining the causes of changes and for assessing regulatory and management options. For these reasons and more, paleolimnological studies should be seriously considered as important components of monitoring programs.

## ACKNOWLEDGMENTS

Susan Christie provided many helpful editorial suggestions. She and Sue Brenard helped prepare final versions of the manuscript. John Glew prepared Figure 13.1. We thank the paleolimnologists in John Smol's laboratory and Peter Leavitt for helpful comments. The information in this document has been funded in part by the U.S. Environmental Protection Agency through Cooperative Agreement No. CR817489 with Indiana University. Preparation of this manuscript also was funded in part by a grant to JPS from the Natural Sciences and Engineering Research Council of Canada.

This manuscript has been subjected to the U.S. Environmental Protection Agency's peer and administrative review and approved for publication. Mention of trade names or commercial products does not constitute endorsement or recommendation for use.

# REFERENCES

Aaby, B. and G. Digerfeldt. 1986. Sampling techniques for lakes and bogs. pp. 181-194. *In*: B. E. Berglund, ed., *Handbook of Holocene Palaeoecology and Palaeohydrology*. John Wiley & Sons, Ltd., NY. 869 p.

Agbeti, M. D. and M. D. Dickman. 1989. Use of lake fossil diatom assemblages to determine historical changes in lake trophic status. *Can. J. Fish. Aquat. Sci.* 46:1013-1021.

Ahmad, H. M. and D. F. Charles. 1988. *PIRLA Data Base Management System User's Manual*. Report Number 12. 2nd ed. PIRLA Unpublished Report Series. 256 p.

Allott, N. 1978. *Recent Paleolimnology of Twin Lake Near St. Paul, Minnesota, Based on a Transect of Cores*. M. S. Thesis. University of Minnesota, St. Paul. 102 p.

Anderson, N. J. 1986. Diatom biostratigraphy and comparative core correlation within a small lake basin. *Hydrobiologia* 143:105-112.

Anderson, N. J. 1989. A whole-basin diatom accumulation rate for a small eutrophic lake in northern Ireland and its palaeoecological implications. *J. Ecology* 77:926-946.

Anderson, N. J. 1990a. Variability of sediment diatom assemblages in an upland wind-stressed lake (Loch Fleet, Galloway, S. W. Scotland). *J. Paleolimnol.* 4:43-59.

Anderson, N. J. 1990b. Spatial pattern of recent sediment and diatom accumulation in a small, monomictic, eutrophic lake. *J. Paleolimnol.* 3:143-160.

Anderson, N. J. 1990c. Variability of diatom concentrations and accumulation rates in sediments of a small lake basin. *Limnol. Oceanogr.* 35:497-508.

Appleby, P. G., N. Richardson, and P. J. Nolan. 1991. [241]Am dating of lake sediments. *Hydrobiologia* 214:35-42.

Appleby, P. G., G. F. Oldfield, R. Thompson, P. Huttunen, and K. Tolonen. 1979. [210]Pb dating of annually laminated lake sediments from Finland. *Nature* 280:53-55.

Appleby, P. E. and F. Oldfield. 1983. The assessment of [210]Pb data from sites with varying sediment accumulation rates. *Hydrobiologia* 103:29-35.

Appleby, P. G., N. Richardson, P. J. Nolan, and F. Oldfield. 1990. Radiometric dating of the United Kingdom SWAP sites. *Phil. Trans. R. Soc. Lond. B* 327:233-238.

Baker, R. G., D. P. Schwert, E. A. Bettis, III, T. J. Kemmis, D. G. Horton, and H. A. Semken. 1991. Mid-Wisconsinian stratigraphy and paleoenvironments at the St. Charles sites in south-central Iowa. *Geol. Soc. Am. Bull.* 103:210-220.

Battarbee, R. W. 1978a. Observations on the recent history of Lough Neagh and its drainage basin. *Phil. Trans. R. Soc. B* 281:303-345.

Battarbee, R. W. 1978b. Biostratigraphical evidence for variations in the recent pattern of sediment accumulation in Lough Neagh, N. Ireland. *Verh. Internat. Verein. Limnol.* 20:624-629.

Battarbee, R. W. and I. Renberg. 1990. The Surface Water Acidification Project (SWAP) Palaeolimnology Programme. *Phil. Trans. R. Soc. Lond.* B. 327:227-232.

Berglund, B. E., ed. 1986. *Handbook of Holocene Palaeoecology and Palaeohydrology.* John Wiley, NY. 869 p.

Binford, M. W. 1990. Calculation and uncertainty analysis of $^{210}$Pb dates for PIRLA project lake sediment cores. *J. Paleolimnol.* 3:253-267.

Binford, M. W. and M. Brenner. 1986. Dilution of $^{210}$Pb by organic sedimentation in lakes of different trophic states, and application to studies of sediment-water interactions. *Limnol. Oceanogr.* 31:584-595.

Birks, H. J. B. 1985. Recent and future mathematical developments in quantitative palaeoecology. *Palaeogeogr. Palaeoclim. Palaeoecol.* 50:107-147.

Birks, H. J. B., S. Juggins, and J. M. Line. 1990a. Lake surface-water chemistry reconstructions from paleolimnological data. pp. 301-313 *In*: B. J. Mason, ed., *The Surface Waters Acidification Programme.* Cambridge University Press, Cambridge, United Kingdom. 522 p.

Birks, H. J. B., J. M. Line, S. Juggins, A. C. Stevenson, and C. J. F. Ter Braak. 1990b. Diatoms and pH reconstruction. *Phil. Trans. R. Soc. Lond.* B 327:263-278.

Boucherle, M. M., J. P. Smol, T. C. Oliver, S. R. Brown, and R. N. McNeely. 1986. Limnologic consequences of the decline in hemlock 4800 years ago in three southern Ontario lakes. *Hydrobiologia* 143:217-225.

Bradbury, J. P. 1975. *Diatom Stratigraphy and Human Settlement in Minnesota.* The Geological Society of America, Boulder, CO. 74 p.

Bradbury, J. P. 1988. A climatic-limnologic model of diatom succession for paleolimnological interpretation of varved sediments at Elk Lake, Minnesota. *J. Paleolimnol.* 1:115-131.

Bradbury, J. P. and J. C. B. Waddington. 1973. The impact of European settlement on Shagawa Lake, northeastern Minnesota, U.S.A. pp. 289-307. *In*: H. J. B. Birks and R. G. West, eds., *Quaternary Plant Ecology.* Blackwells, Oxford. 326 p.

Bradbury, J. P. and C. Winter. 1976. Areal distribution and stratigraphy of diatoms in the sediments of Lake Sallie, Minnesota. *Ecology* 57:1005-1014.

Brenner, M. and M. W. Binford. 1988. A sedimentary record of human disturbance from Lake Miragoane, Haiti. *J. Paleolimnol.* 1:85-97.

Brown, S. R. 1956. A piston sampler for surface sediments of lake deposits. *Ecology* 37:611-613.

Brown, S. R. 1968. Bacterial carotenoids from freshwater sediments. *Limnol. Oceanogr.* 13:133-141.

Brown, S. R., H. J. McIntosh, and J. P. Smol. 1984. Recent paleolimnology of a meromictic lake: fossil pigments of photosynthetic bacteria. *Verh. Internat. Verein. Limnol.* 22:1357-1360.

Brugam, R. B. 1978. Human disturbance and the historical development of Linsley Pond. *Ecology* 59:19-36.

Brugam, R. B. 1983. The relationship between fossil diatom assemblages and limnological conditions. *Hydrobiologia* 98:223-235.

Bruland, K. W., M. Koide, C. Bowser, L. J. Maher, and E. D. Goldberg. 1975. [210]Lead and pollen geochronologies on Lake Superior sediments. *Quat. Res.* 5:89-88.

Brush, G. S. and F. W. Davis. 1984. Stratigraphic evidence of human disturbance in an estuary. *Quat. Res.* 22:21-108.

Burden, E. T., G. Norris, and J. H. McAndrews. 1986. Geochemical indicators in lake sediment of upland erosion caused by Indian and European farming, Awenda Provincial Park, Ontario. *Can. J. Earth Sci.* 23:55-56.

Carignan, R. and J. O. Nriagu. 1985. Trace metal deposition and mobility in the sediments of two lakes near Sudbury, Ontario. *Geochim. Cosmochim. Acta* 49:1753-1764.

Carignan, R. and A. Tessier. 1985. Zinc deposition in acid lakes: the role of diffusion. *Science* 228:1524-1526.

Charles, D. F. 1985. Relationships between surface sediment diatom assemblages and lakewater characteristics in Adirondack lakes. *Ecology* 66:994-1011.

Charles, D. F. and J. P. Smol. 1988. New methods for using diatoms and chrysophytes to infer past pH of low-alkalinity lakes. *Limnol. Oceanogr.* 33:1451-1462.

Charles, D. F. and J. P. Smol. 1990. The PIRLA II project: regional assessment of lake acidification trends. *Verh. Internat. Verein. Limnol.* 24:474-480.

Charles, D. F. and J. P. Smol. 1993. Long-term chemical changes in lakes: quantitative inferences using biotic remains in the sediment record. *In*: L. Baker, ed., *Environmental Chemistry of Lakes and Reservoirs*. Advances in Chemistry series, Amer. Chem. Soc.

Charles, D. F. and D. R. Whitehead. 1986. The PIRLA project: Paleoecological Investigation of Recent Lake Acidification. *Hydrobiologia* 143:13-20.

Charles, D. F., D. R. Whitehead, D. R. Engstrom, B. D. Fry, R. A. Hites, S. A. Norton, J. S. Owen, L. A. Roll, S. C. Schindler, J. P. Smol, A. J. Uutala, J. R. White, and R. J. Wise. 1987. Paleolimnological evidence for recent acidification of Big Moose Lake, Adirondack Mountains, NY (U.S.A.). *Biogeochemistry* 3:267-296.

Charles, D. F., J. P. Smol, A. J. Uutala, P. R. Sweets, and D. R. Whitehead. 1989. The PIRLA DataBase Management System. INQUA - Commission for the Study of the Holocene, Working Group on Data Handling Methods. *Newsletter* 2:3-6.

Charles, D. F., M. W. Binford, B. D. Fry, E. T. Furlong, R. A. Hites, M. J. Mitchell, S. A. Norton, M. J. Patterson, J. P. Smol, A. J. Uutala, J. R. White, D. R. Whitehead, and R. J. Wise. 1990. Paleo-Tecological investigation of recent lake acidification in the Adirondack Mountains, NY *J. Paleolimnol.* 3:195-241.

Charles, D. F., S. S. Dixit, B. F. Cumming, and J. P. Smol. 1991. Variability in diatom and chrysophyte assemblages and inferred pH: paleolimnological studies of Big Moose Lake, New York, (U.S.A). *J. Paleolimnol.* 5:267-284.

Chivas, A. R., P. DeDeckker, and J. M. G. Shelley. 1986. Magnesium and strontium in non-marine ostracod shells as indicators of palaeosalinity and palaeotemperature. *Hydrobiologia* 143:135-142.

Christie, C. E. and J. P. Smol. 1993. Diatom assemblages as indicators of lake trophic status in southeastern Ontario lakes. *J. Phycol.* 29:575-586.

Chumbly, C. A., R. G. Baker, and E. A. Bettis, III. 1990. Midwestern Holocene paleoenvironments revealed by floodplain deposits in northeastern Iowa. *Science* 249:272-273.

Coard, M. A., S. M. Cousen, A. H. Cuttler, H. J. Dean, J. A. Dearing, T. I. Eglinton, A. M. Greaves, K. P. Lacey, P. E. O'Sullivan, D. A. Pickering, M. M. Rhead, J. K. Rodwell, and H. Simola. 1983. Paleolimnological studies of annually laminated sediments in Loe Pool, Cornwall, U. K. *Hydrobiologia* 103:185-191.

Conley, D. J. 1988. Biogenic silica as an estimate of siliceous microfossil abundance in Great Lakes sediments. *Biogeochemistry* 6:161-179.

Cumming, B. F. 1991. Paleoecological assessment of changes in Adirondack (New York, U.S.A.) lake-water chemistry historical (pre-1950) and recent (post-1970) trends. PhD. Thesis, Queen's University, Kingston, Ontario.

Cumming, B. F., J. P. Smol, J. C. Kingston, D. F. Charles, H. J. B. Birks, K. E. Camburn, S. S. Dixit, A. J. Uutala, and A. R. Selle. 1992. How much acidification has occurred in Adirondack region lakes (New York, U.S.A) since pre-industrial times? *Can. J. Fish. Aquat. Sci.* 49:128-141.

Currie, D. C. and I. R. Walker. 1992. Recognition and palaeohydrological significance of fossil black fly larvae, with a key to the Neararctic genera (Diptera: Simuliidae). *J. Paleolimnol.* 7:37-54.

Cushing, E. J. and H. E. Wright, Jr. 1965. Hand operated piston corers for lake sediments. *Ecology* 46:380-384.

Davidson, G. A. 1988. A modified tape-peel technique for preparing permanent qualitative microfossil slides. *J. Paleolimnol.* 1:229-234.

Davis, M. B. and M. S. (J. ) Ford. 1982. Sediment focusing in Mirror Lake, New Hampshire. *Limnol. Oceanogr.* 27:137-150.

Davis, M. B., R. E. Moeller, and J. Ford. 1984. Sediment focusing and pollen influx. pp. 261-293. *In*: E. Y. Haworth and J. W. G. Lund, eds., *Lake Sediments and Environmental History*. University of Leicester Press, Leicester, United Kingdom. 411 p.

Davis, M. B., M. S. Ford, and R. E. Moeller. 1985. Paleolimnology. pp. 345-429. *In*: G. E. Likens, ed., *An Ecosystem Approach to Aquatic Ecology, Mirror Lake and its Environment*. Springer-Verlag, NY. 516 p.

Davis, R. B., L. A. Brewster, and J. W. Sutherland. 1969. Variation in pollen spectra within lakes. *Pollen et Spores* 11:557-571.

Davis, R. B., C. T. Hess, S. A. Norton, D. W. Hanson, K. D. Hoagland, and D. S. Anderson. 1984. $^{137}$Cs and $^{210}$Pb dating of sediments from soft water lakes in New England (U.S.A.) and Scandinavia, a failure of $^{137}$Cs dating. *Chem. Geol.* 44:151-185.

Davis, R. B., D. S. Anderson, M. C. Whiting, J. P. Smol, and S. S. Dixit. 1990. Alkalinity and pH of three lakes in northern New England, U.S.A., over the past 300 years. *Phil. Trans. R. Soc. Lond.* 327:413-421.

Dean, W. E., Jr. 1974. Determination of carbonate and organic matter in calcareous sediments and sedimentary rocks by loss on ignition: comparison with other models. *J. Sediment. Petrol.* 44:242-248.

Dean, W. E., Jr. 1981. Carbonate minerals and organic matter in sediments of modern north temperate hard-water lakes. *SEPM Spec. Pub.* 31:213-231.

Dearing, J. A. 1983. Changing patterns of sediment accumulation in a small lake in Scania, Southern Sweden. *Hydrobiologia* 103:59-64.

Dearing, J. A. 1986. Core correlation and total sediment influx. pp. 247-270. *In*: B. E. Berglund, ed., *Handbook of Holocene Palaeoecology and Palaeohydrology*. John Wiley & Sons, NY. 869 p.

DeDeckker, P., J. P. Colin, and J. P. Peypouquet. 1988. *Ostracoda in the earth sciences*. Elsevier, Amsterdam.

Deevey, E. S., D. S. Rice, P. M. Rice, H. H. Vaughan, M. Brenner, and M. S. Flannery. 1979. Mayan urbanism: impact on a tropical karst environment. *Science* 206:298-306.

DiNicola, D. M. 1986. The representation of living diatom communities in deep-water sedimentary diatom assemblges in two Maine (U.S.A.) lakes. pp. 73-85. *In*: J. P. Smol, R. W. Battarbee, R. B. Davis, and J. Meriläinen, eds., *Diatoms and Lake Acidity*. Dr. W. Junk, Dordrecht, The Netherlands. 307 p.

Dillon, P. J. and R. D. Evans. 1982. Whole-lake lead burdens in sediments of lakes in southern Ontario, Canada. *Hydrobiologia* 91:121-130.

Dixit, S. S. and R. D. Evans. 1986. Spatial variability in sedimentary algal microfossils and its bearing on diatom-inferred pH reconstructions. *Can. J. Fish. Aquat. Sci.* 43:1836-1845.

Dixit, S. S., A. S. Dixit, and J. P. Smol. 1989. Relationship between Chrysophyte assemblages and environmental variables in seventy-two Sudbury lakes as examined by Canonical Correspondence Analysis (CCA). *Can. J. Fish. Aquat. Sci.* 46:1667-1676.

Dixit, S. S., J. P. Smol, D. S. Anderson, and R. B. Davis. 1990. Utility of scaled chrysophytes for inferring lakewater pH in northern New England lakes. *J. Paleolimnol.* 3:269-286.

Dixit, S. S., A. S. Dixit, and J. P. Smol. 1991. Multivariable environmental inferences based on diatom assemblages from Sudbury (Canada) lakes. *Freshwater Biol.* 26:251-266.

Douglas, M. S. V. and J. P. Smol. 1987. Siliceous protozoan plates in lake sediments. *Hydrobiologia* 154:13-23.

Downing, J. A. and L. C. Rath. 1988. Spatial patchiness in the lacustrine sedimentary environment. *Limnol. Oceanogr.* 33:447-457.

Edwards, K. J. 1983. Quaternary palynology: multiple profile studies and pollen variability. *Prog. Physical Geog.* 7:587-609.

Edwards, K. J. and R. S. Thompson. 1984. Magnetic, palynological, and radiocarbon correlation and dating comparisons in long cores from a northern Irish lake. *Catena* 11:83-89.

Einarsson, T. 1986. Tephrochronology. pp. 329-342. *In:* B. E. Berglund, ed., *Handbook of Holocene Palaeoecology and Palaeohydrology.* John Wiley & Sons, NY. 869 p.

Eisenreich, S. J., J. A. Capel, J. A. Robbins, and R. Bourbonniere. 1989. Accumulation and diagenesis of chlorinated hydrocarbons in lacustrine sediments. *Environ. Sci. Technol.* 23:1116-1126.

Engstrom, D. R. and B. C. S. Hansen. 1985. Postglacial vegetational change and soil development in southeastern Labrador as inferred from pollen and chemical stratigraphy. *Can. J. Bot.* 63:543-561.

Engstrom, D. R. and S. Nelson. 1991. Paleosalinity from trace metals in fossil ostracodes compared with observational records at Devil's Lake, N. Dakota. *Palaeogeogr., Palaeoclimatol., Palaeoecol.* 83:295-312.

Engstrom, D. R. and E. B. Swain. 1987. The chemistry of lake sediments in time and space. *Freshwater Biology* 15:261-288.

Engstrom, D. R. and H. E. Wright, Jr. 1984. Chemical stratigraphy of lake sediments as a record of environmental change. pp. 11-67. *In:* E. Y. Haworth and J. W. G. Lund, eds. *Lake Sediments and Environmental History.* Leicester University Press, Leicester, United Kingdom. 411 p.

Engstrom, D. R., E. B. Swain, and J. C. Kingston. 1985. A palaeolimnological record of human disturbance from Harvey's Lake, Vermont: geochemistry, pigments and diatoms. *Freshwat. Biol.* 15:261-288.

Engstrom, D. R., C. Whitlock, S. C. Fritz, and H. E. Wright, Jr. 1991. Recent environmental changes inferred from the sediments of small lakes in Yellowstone's northern range. *J. Paleolimnol.* 5:139-174.

Evans, R. D. 1986. Sources of mercury contamination in the sediments of small headwater lakes in south-central Ontario, Canada. *Arch. Environ. Contam. Toxicol.* 15:505-512.

Evans, R. D. and F. H. Rigler. 1980. Measurement of whole lake sediment accumulation and phosphorus retention using lead-210 dating. *Can. J. Fish. Aquat. Sci.* 37:817-822.

Evans, R. D. and F. H. Rigler. 1983. A test of lead-210 dating for the measurement of whole-lake soft sediment accumulation. *Can. J. Fish. Aquat. Sci.* 40:506-515.

Flower, R. J. 1991. Field calibration and performance of sediment traps in a eutrophic holomictic lake. *J. Paleolimnol.* 5:175-188.

Flower, R. J., N. G. Cameron, N. Rose, S. C. Fritz, R. Harriman, and A. C. Stevenson. 1990. Post-1970 water-chemistry changes and palaeolimnology of several acidified upland lakes in the U. K. *Phil. Trans. R. Soc. Lond. B* 327:427-433.

Frey, D. G. 1969. The rationale of paleolimnology. *Mitt. Internat. Verein. Limnol.* 17:7-18.

Frey, D. G. 1988. Littoral and offshore communities of diatoms, cladocerans, and dipterous larvae, and their interpretation in paleolimnology. *J. Paleolimnol.* 1:179-191.

Fritz, S. C. 1989. Lake development and limnological response to prehistoric and historic land-use in Diss, Norfolk, U.K. *J. Ecology* 77:182-202.

Fritz, S. 1990. Twentieth-century salinity and water-level fluctuations in Devils Lake, North Dakota: test of a diatom-based transfer function. *Limnol. Oceanogr.* 35:1771-1781.

Gasse, F., J. F. Talling, and P. Kilham. 1983. Diatom assemblages in East Africa: classification, distribution and ecology. *Rev. Hydrobiol. Trop.* 16:3-34.

Gasse, F., J. C. Fontes, J. C. Plaziat, P. Carbonel, I. Kaczmarska, P. De Deckker, I. Soulí-Marsche, Y. Callot, and P. A. Dupeuble. 1987. Biological remains, geochemistry and stable isotopes for the reconstruction of environmental and hydrological changes in the Holocene lakes from north Sahara. *Paleogeogr., Palaeoclimatol., Palaeoecol.* 60:1-46.

Gensemer, R. W. 1990. Role of aluminum and growth rate on changes in cell size and silica content of silica-limited populations of *Asterionella ralfsii* var. *americana* (Bacillariophyceae). *J. Phycol.* 26:250-258.

Glew, J. R. 1988. A portable extruding device for close interval sectioning of unconsolidated core samples. *J. Paleolimnol.* 1:235-239.

Glew, J. R. 1989. A new trigger mechanism for sediment samplers. *J. Paleolimnol.* 2:241-243.

Glew, J. R. 1991. Miniature gravity corer for recovering short sediment cores. *J. Paleolimnol.* 5:285-287.

Goldberg, E. D., V. F. Hodge, J. J. Griffin, M. Koide, and D. N. Edgington. 1981. Impact of fossil fuel combustion on the sediments of Lake Michigan. *Environ. Sci. Technol.* 15:466-471.

Goulden, C. E. 1969. Interpretative studies of cladoceran microfossils in lake sediments. *Mitt. Int. Ver. Theor. Angew. Limnol.* 17:43-55.

Gray, J. 1988. *Paleolimnology: Aspects of Freshwater Paleoecology and Biogeography.* Elsevier, Amsterdam. 678 p.

Håkanson, L. and M. Jansson. 1983. *Principles of Lake Sedimentology.* Springer-Verlag, New York. 316 p.

Hall, R. I. and J. P. Smol. 1992. A weighted-averaging regression and calibration model for inferring total phosphorus concentration from diatoms in British Columbia (Canada) lakes. *Freshwater Biol.* 27:417-434.

Hann, B. J. 1991. Cladocera. B. G. Warner, ed., *Methods in Quaternary Ecology.* Geological Association of Canada, St. Johns. 170 p.

Heit, M., and K. M. Miller. 1987. Cesium-137 sediment depth profiles and inventories in Adirondack lake sediments. *Biogeochemistry* 3:243-265.

Hill, M. O. 1979. *DECORANA: a FORTRAN program for detrended correspondence analysis and reciprocal averaging.* Section of Ecology & Systematics, Cornell University, Ithaca, NY. 52 p.

Hilton, J. and M. M. Gibbs. 1984. The horizontal distribution of major elements and organic matter in the sediment of Esthwaite Water, England. *Chem. Geol.* 47:57-83.

Hilton, J., J. P. Lishman, and P. V. Allen. 1986. The dominant processes of sediment distribution and focusing in a small, eutrophic, monomictic lake. *Limnol. Oceanogr.* 31:125-133.

Hongve, D. 1972. En bunnhenter som er lett a lage. *Fauna* 25:281-283.

Hongve, D. and A. Erlandsen. 1979. Shortening of surface sediment cores during sampling. *Hydrobiologia* 65:283-287.

Hurley, J. P. and D. E. Armstrong. 1991. Pigment preservation in lake sediments: a comparison of sedimentary environments in Trout Lake, Wisconsin. *Can. J. Fish. Aquat. Sci.* 48:472-486.

Jacobson, H. A. and D. R. Engstrom. 1989. Resolving the chronology of recent lake sediments: an example from Devils Lake, North Dakota. *J. Paleolimnol.* 2:81-97.

Jenkins, A., P. G. Whitehead, B. J. Cosby, and H. J. B. Birks. 1990. Modeling long-term acidification: a comparison with diatom reconstructions and the implications for reversibility. *Phil. Trans. Roy. Soc. B* 327:435-440.

Johnson, M. G. 1987. Trace element loadings to sediments of fourteen Ontario lakes and correlations with concentrations in fish. *Can. J. Fish. Aquat. Sci.* 44:3-13.

Johnson, M. G., J. R. M. Kelso, O. C. McNeil, and W. B. Morton. 1990. Fossil midge associations and the historical status of fish in acidified lakes. *J. Paleolimnol.* 3:113-127.

Jones, V. J. and R. J. Flower. 1986. Spatial and temporal variability in periphytic diatom communities: palaeoecological significance in an acidified lake. pp. 87-94. *In*: J. P. Smol, R. W. Battarbee, R. B. Davis, and J. Meriläinen, eds. *Diatoms and Lake Acidity*. Dr. W. Junk, Dordrecht, The Netherlands. 307 p.

Kerfoot, W. C. 1981. Long-term replacement cycles in cladoceran communities: a history of predation. *Ecology*. 62:216-233.

Kingston, J. C. and H. J. B. Birks. 1990. Dissolved organic carbon reconstruction from diatom assemblages in PIRLA project lakes, North America. *Phil. Trans. R. Soc. Lond.* B 327:279-288.

Kingston, J. C., R. B. Cook, R. G. Kreis, Jr., K. E. Camburn, S. A. Norton, P. R. Sweets, M. W. Binford, M. J. Mitchell, S. C. Schindler, L. C. K. Shane, and G. A. King. 1990. Paleoecological investigation of recent lake acidification in the northern Great Lakes states. *J. Paleolimnol.* 4:153-201.

Kingston, J. C., H. J. B. Birks, A. J. Uutala, B. F. Cumming, and J. P. Smol. 1992. Assessing trends in fishery resources and lake water aluminum from paleolimnological analyses of siliceous algae. *Can. J. Fish. Aquat. Sci.* 49:116-127.

Klink, A. 1989. The lower Rhine: palaeoecological analysis. pp. 183-201. *In*: G. E. Petts, ed., *Historical Change of Large Alluvial Rivers: Western Europe*. John Wiley, NY. 355 p.

Kreis, R. G., Jr. 1989. Variability Study - Interim results. Section 4. pp. 4-1 to 4-48. *In*: D. F. Charles and D. R. Whitehead, eds., *Paleoecological Investigation of Recent Lake Acidification (PIRLA): 1983-1985*. EN-6526. Project 2174-10. Interim Report. Electric Power Research Institute, Palo Alto, CA. 253 p.

Kreis, R. G., Jr. 1986. Variability study. Section 17. pp. 17-1 to 17-19. *In*: D. F. Charles and D. R. Whitehead, eds., *Paleoecological Investigation of Recent Lake Acidification (PIRLA): Methods and Project Description*. EPRI EA-4906, Project 2174-10. Interim Report. Electric Power Research Institute, Palo Alto, CA. 228 p.

Krishnaswamy, S. and D. Lal. 1978. Radionuclide limnochronology. pp. 153-177. *In*: A. Lerman, ed., *Lakes, Chemistry, Geology, and Physics*. Springer-Verlag, NY. 363 p.

Leavitt, P. R., S. R. Carpenter, and J. F. Kitchell. 1989. Whole-lake experiments: the annual record of fossil pigments and zooplankton. *Limnol. Oceanogr.* 34:700-717.

Lehman, J. T. 1975. Reconstructing the rate of accumulation of lake sediment: the effect of sediment focusing. *Quat. Res.* 5:541-550.

Line, J. M. and H. J. B. Birks. 1990. WACALIB version 2.1 - a computer program to reconstruct environmental variables from fossil assemblages by weighted averaging. *J. Paleolimnol.* 3:170-173.

Lipsey, L. L., Jr. 1988. Preliminary results of a classification of fifty-one selected northeastern Wisconsin Lakes (U.S.A.) using indicator diatom species. *Hydrobiologia* 166:205-216.

Livingstone, D. 1955. A lightweight piston sampler for lake deposits. *Ecology* 36:137-139.

Lotter, A. F. 1989. Evidence of annual layering in Holocene sediments of Soppensee, Switzerland. *Aquat. Sci.* 51:19-30.

Lotter, A. F. 1991. Absolute dating of the late-glacial period in Switzerland using annually laminated sediments. *Quat. Res.* 35:321-330.

Lowe, R. L. 1974. *Environmental Requirements and Pollution Tolerance of Freshwater Diatoms.* EPA/670/4-74/005. U.S. Environmental Protection Agency, Washington, D.C. 334 p.

Ludlam, S. D. 1981. Sedimentation rates in Fayetteville Green Lake, New York. *Sedimentology* 28:85-96.

Mackereth, F. J. H. 1966. Some chemical observations on post-glacial lake sediments. *Phil. Trans. R. Soc. Lond. B* 250:165-213.

Meriläinen, J. 1967. The diatom flora and the hydrogen-ion concentration of the water. *Ann. Bot. Fenn.* 4:51-58.

Meriläinen, J. 1971. The recent sedimentation of diatom frustules in four meromictic lakes. *Ann. Bot. Fenn.* 8:160-176.

Metcalfe, S. E. 1988. Modern diatom assemblages in central Mexico: the role of water chemistry and other environmental factors as indicated by TWINSPAN and DECORANA. *Freshwat. Biol.* 19:217-233.

Moeller, R. E., F. Oldfield, and P. G. Appleby. 1984. Biological sediment mixing and its stratigraphic implication in Mirror Lake (New Hampshire, U.S.A.). *Int. Ver. Theor. Angew. Limnol. Verh.* 22:267-572.

Moss, B. 1980. Further studies on the palaeolimnology and changes in the phosphorus budget of Barton Broad, Norfolk. *Freshwat. Biol.* 10:261-279.

Mueller, W. P. 1964. The distribution of cladoceran remains in surficial sediments from three northern Indiana lakes. *Investigations of Indiana Lakes and Streams* VI:1-64.

Munro, M. A. R., A. M. Kreiser, R. W. Battarbee, S. Juggins, A. C. Stevenson, D. S. Anderson, N. J. Anderson, F. Berge, H. J. B. Birks, R. B. Davis, R. J. Flower, S. C. Fritz, E. Y. Haworth, V. J. Jones, J. C. Kingston, and I. Renberg. 1990. Diatom quality control and data handling. *Phil. Trans. R. Soc. Lond. B* 327:257-261.

Nilsson, M. and I. Renberg. 1990. Viable endospores of *Thermoactinomyces vulgaris* in lake sediments as indicators of agricultural history. *Applied and Environmental Microbiology* 56:2025-2028.

Norton, S. A., P. J. Dillon, R. D. Evans, G. Mierle, and J. S. Kahl. 1990. The history of atmospheric deposition of Cd, Hg, and Pb in North America: evidence from lake and peat bog sediments. pp. 73-102. *In*: S. E. Lindberg, A. L. Page, and S. A. Norton, eds., *Acidic Precipitation: Sources, Deposition, and Canopy Interactions, Vol. 3*. Springer-Verlag, NY. 293 p.

Nriagu, J. O. and Y. K. Soon. 1985. Distribution and isotopic composition of sulfur in lake sediments of northern Ontario. *Geochim. Cosmochim. Acta* 49:823-834.

Oldfield, F. and P. G. Appleby. 1984. Empirical testing of $^{210}$Pb-dating models for lake sediments. pp. 93-124. *In*: E. Y. Haworth and J. W. G. Lund, eds., *Lake Sediments and Environmental History*. University of Minnesota Press, Minneapolis, MN. 411 p.

Olson, I. U. 1986. Radiometric dating. pp. 273-312. *In*: B. E. Berglund, ed., *Handbook of Holocene Palaeoecology and Palaeohydrology*. John Wiley & Sons, NY. 869 p.

Osborne, P. L. and B. Moss. 1977. Paleolimnology and trends in the phosphorus and iron budget of old man-made lake Barton Board-Norfolk. *Freshwat. Biol.* 7:213-233.

O'Sullivan, P. E. 1983. Annually laminated lake sediments and the study of Quaternary environmental changes — a review. *Quat. Sci. Rev.* 1:245-313.

Pennington, W. (Mrs. T. G. Tutin). 1978. Response of some British lakes to past changes in land use on their catchments. *Verh. Internat. Verein. Limnol.* 20:636-641.

Pennington, W., R. S. Cambray, and E. M. Fisher. 1973. Observations on lake sediments using fallout Cs-137 as a tracer. *Nature* 242:324-326.

Petts, G. E., H. Möller, and A. L. Roux. 1989. *Historical Change of Large Alluvial Rivers: Western Europe*. John Wiley & Sons, Chichester. 355 p.

Rada, R. G., J. G. Wiener, M. R. Winfrey, and D. E. Powell. 1989. Recent increases in atmospheric deposition of mercury to north-central Wisconsin lakes inferred from sediment analysis. *Arch. Environ. Contam. Toxicol.* 18:175-181.

Renberg, I. 1981. Improved methods for sampling, photographing, and varve-counting of varved lake sediments. *Boreas* 10:255-258.

Renberg, I. 1986. Concentration and annual accumulation values of heavy metals in lake sediments: their significance in studies of the history of heavy metal pollution. *Hydrobiologia* 143:379-385.

Renberg, I. 1990. A 12,600 year perspective of the acidification of Lilla Öresjön, southwest Sweden. *Phil. Trans. R. Soc. Lond.* B 327:357-361.

Renberg, I. and T. Hellberg. 1982. The pH history of lakes in southwestern Sweden, as calculated from the subfossil diatom flora of the sediments. *Ambio* 11:30-33.

Renberg, I. and M. Nilsson. 1992. Dormant bacteria in lake sediments as palaeoecological indicators. *J. Paleolimnol.* 7:127-135.

Renberg, I. and M. Wik. 1984. Dating recent lake sediments by soot particle counting. *Verh. Internat. Verein. Limnol.* 22:712-718.

Rippey, B., R. J. Murphy, and S. W. Kyle. 1982. Anthropogenically-derived changes in the sedimentary flux of Mg, Cr, Ni, Cu, An, Hg, Pb, and P in Lough Neagh, Northern Ireland. *Environ. Sci. Technol.* 16:23-30.

Ritchie, J. C. and J. R. McHenry. 1990. Application of radioactive fallout Cesium-137 for measuring soil erosion and sediment accumulation rates and patterns: a review. *J. Environ. Qual.* 19:215-233.

Robbins, J. A. 1982. Stratigraphic and dynamic effects of sediment reworking by Great Lakes zoobenthos. *Hydrobiologia* 92:611-622.

Robbins, J. A., J. R. Krezoski, and S. C. Mozley. 1977. Radioactivity in sediments of the Great Lakes: post-depositional redistribution by deposit-feeding organisms. *Earth Planet Science Letter* 36:325-333.

Robbins, J. A., D. N. Edgington, and A. L. W. Kemp. 1978. Comparative $^{210}$Pb, $^{137}$Cs, and pollen geochronologies of sediments from Lakes Ontario and Erie. *Quat. Res.* 10:256-278.

Saarnisto, M. 1986. Annually laminated sediments. pp. 343-370. *In*: B. E. Berglund, ed., *Handbook of Holocene Palaeoecology and Palaeohydrology*. John Wiley & Sons, NY. 869 p.

Sandgren, C. D. 1991. Applications of chrysophyte stomatocysts in paleolimnology. *J. Paleolimnol.* 5:1-113.

Sandgren, P., J. Risberg, and R. Thompson. 1990. Magnetic susceptibility in sediment records of Lake Ådran, eastern Sweden: correlation among cores and interpretation. *J. Paleolimnol.* 3:129-141.

Sanger, J. E. 1988. Fossil pigments in paleoecology and paleolimnology. pp. 343-360 *In*: J. Gray, ed. *Paleolimnology Aspects of Freshwater Paleoecology and Biogeography*. Elsevier Science Publishing Co., NY. 678 p.

Sarna-Wojcicki, A. M., D. E. Champion, and J. O. Davis. 1983. Holocene volcanism in the conterminous United States and the role of silicic volcanic ash layers in correlation of latest Pleistocene and Holocene deposits. pp. 52-77. *In*: H. E. Wright, Jr., ed., *Late Quaternary Environments of the United States: The Holocene, Vol. 2.* University of Minnesota Press, Minneapolis, MN. 277 p.

Schelske, C. L., D. J. Conley, and W. F. Warwick. 1985. Historical relationships between phosphorus loading and biogenic silica accumulation in Bay of Quinte sediments. *Can. J. Fish. Aquat. Sci.* 42:1401-1409.

Schelske, C. L., D. J. Conley, E. F. Stoermer, T. L. Newberry, and C. D. Campbell. 1986. Biogenic silica and phosphorus accumulation in sediments as indices of eutrophication in the Laurentian Great Lakes. *Hydrobiologia* 143:79-86.

Schelske, C. L., H. Züllig, and M. Boucherle. 1987. Limnological investigation of biogenic silica sedimentation and silica biogeochemistry in Lake St. Moritz and Lake Zürich. *Schweiz Z. Hydrol.* 49:42-50.

Schmidt, R. 1991. Recent re-oligotrophication in Mondsee (Austria) as indicated by sediment diatom and chemical stratigraphy. *Verh. Internat. Verein.* 24:963-967.

Servant-Vildary, S. and M. Roux. 1990. Multivariate analysis of diatoms and water chemistry in Bolivian saline lakes. *Hydrobiologia* 197:267-290.

Siegenthaler, U. and U. Eicher. 1986. Stable oxygen and carbon isotope analyses. pp. 407-422. *In*: B. E. Berglund, ed., *Handbook of Holocene Palaeoecology and Palaeohydrology*. John Wiley & Sons, NY. 869 p.

Simola, H. 1979. Micro-stratigraphy of sediment laminations deposited in a chemically stratifying eutrophic lake during the years 1913-1976. *Holarct. Ecol.* 2:160-168.

Siver, P. A. and J. S. Hamer. 1990. Use of extant populations of scaled chrysophytes for the inference of lakewater pH. *Can. J. Fish. Aquat. Sci.* 47:1339-1347.

Siver, P. A. and A. Skogstad. 1988. Morphological variation and ecology of *Mallomonas crassisquama* (Chrysophyceae). *Nordic Journal of Botany.* 8:99-107.

Smol, J. P. 1987. Methods in Quaternary ecology #1. Freshwater algae. *Geoscience Canada* 14:208-217.

Smol, J. P. and M. M. Boucherle. 1985. Postglacial changes in algal and cladoceran assemblages in Little Round Lake, Ontario. *Arch. Hydrobiol.* 103:25-49.

Smol, J. P. and M. D. Dickman. 1981. The recent histories of three Canadian Shield lakes: a paleolimnological experiment. *Archiv Hydrobiol.* 93:83-108.

Smol, J. P. and J. R. Glew. 1992. Paleolimnology. pp. 551-564. *In*: W. A. Nierenberg, ed., *Encyclopedia of Earth System Science, Vol. 3*. Academic Press, Inc., CA. cirea 2,500 p.

Smol, J. P., I. R. Walker, and P. R. Leavitt. 1991. Paleolimnology and hindcasting climatic trends. *Verh. Internat. Verein. Limnol.* 24:1240-1246.

Snoeijs, P. J. M. 1990. Effects of temperature on spring bloom dynamics of epilithic diatom communities in the Gulf of Bothnia. *J. Vegetation Sci.* 1:599-608.

Stevenson, A. C. and R. J. Flower. 1991. A palaeoecological evaluation of environmental degradation in Lake Mikri Prespa, N.W. Greece. *Biol. Conserv.* 57:89-109.

Stevenson, A. C., H. J. B. Birks, R. J. Flower, and R. W. Battarbee. 1989. Diatom-based pH reconstruction of lake acidification using Canonical Correspondence Analysis. *Ambio* 18:229-233.

Stockner, J. G. and W. W. Benson. 1987. The succession of diatom assemblages in the recent sediments of Lake Washington. *Limnol. Oceanogr.* 12:513-532.

Stoermer, E. F., J. A. Wolin, C. L. Schelske, and D. J. Conley. 1990. Siliceous microfossil succession in Lake Michigan. *Limnol. Oceanogr.* 35:959-967.

Sullivan, T. J. 1990. Historical changes in surface water acid-base chemistry in response to acidic deposition. Report II. *In: NAPP State of Science and Technology*, National Acid Precipitation Assessment Program, Washington, D.C. 213 p.

Sullivan, T. J., D. F. Charles, J. P. Smol, B. F. Cumming, A. R. Selle, D. R. Thomas, J. A. Bernert, and S. S. Dixit. 1990. Quantification of changes in lakewater chemistry in response to acidic deposition. *Nature* 345:54-58.

Sullivan, T. J., R. S. Turner, D. F. Charles, B. F. Cumming, J. P. Smol, C. L. Schofield, C. T. Driscoll, B. J. Cosby, H. J. B. Birks, A. J. Uutala, J. C. Kingston, S. S. Dixit, J. A. Bernert, P. F. Ryan, and D. R. Marmorek. 1992. Use of historical assessment for evaluation of process-based model projections of future environmental change: lake acidification in the Adirondack Mountains, U.S.A. *Environ. Pollut.* 77:253-262.

Swain, E. B. 1985. Measurement and interpretation of sedimentary pigments. *Freshwat. Biol.* 15:53-75.

Sweets, P. R., R. W. Bienert, T. L. Crisman, and M. W. Binford. 1990a. Paleoecological investigations of recent lake acidification in northern Florida. *J. Paleolimnol.* 4:103-137.

ter Braak, C. J. F. 1986. Canonical correspondence analysis: a new eigenvector technique for multivariate direct gradient analysis. *Ecology* 67:1167-1179.

ter Braak, C. J. F. and L. G. Barendregt. 1986. Weighted averaging of species indicator values: its efficiency in environmental calibration. *Math. Biosci.* 78:57-72.

ter Braak, C. J. F. and C. W. N. Looman. 1986. Weighted averaging, logistic regression and the Gaussian response model. *Vegetatio* 65:3-11.

ter Braak, C. J. F. and I. C. Prentice. 1988. A theory of gradient analysis. *Adv. Ecol. Res.* 18:271-317.

ter Braak, C. J. F. and H. Van Dam. 1989. Inferring pH from diatoms: a comparison of old and new calibration methods. *Hydrobiologia* 178:209-223.

Thompson, R. and R. M. Clark. 1989. Sequence slotting for stratigraphic correlation between cores: theory and practice. *J. Paleolimnol.* 2:173-184.

Tippett, R. 1964. An investigation into the nature of the layering of deep-water sediments in two eastern Ontario lakes. *Can. J. Bot.* 42:1693-1709.

Trefry, J. H., S. Metz, R. P. Trocine, and T. A. Nelsen. 1985. A decline in lead transport by the Mississippi River. *Science* 230:439-441.

Tuchman, M. L., E. F. Stoermer, and H. J. Carney. 1984. Effects of increased salinity on the diatom assemblage in Fonda Lake, Michigan. *Hydrobiologia* 109:179-188.

Uutala, A. J. 1986. *Paleolimnological Assessment of the Effects of Lake Acidification on Chironomidae (Diptera) Assemblages in the Adirondack Region of New York.* Ph. D. dissertation. New York College of Environmental Science and Forestry, State University of New York, Syracuse. 155 p.

Uutala, A. J. 1990. Chaoborus Diptera: Chaoboridae mandibles — paleolimnological indicators of the historical status of fish populations in acid sensitive lakes. *J. Paleolimnol.* 4:139-151.

Walker, I. R. 1987. Chironomidae (Diptera) in paleoecology. *Quat. Sci. Rev.* 6:29-40.

Walker, I. R., J. P. Smol, D. R. Engstrom, and H. J. B. Birks. 1991. An assessment of chironomidae as quantitative indicators of past climatic change. *Can. J. Fish. Aquat. Sci.* 48:975-987.

Warner, B. G., ed. 1990. *Methods in Quaternary Ecology.* Geoscience Canada, Reprint Series 5, Geological Association of Canada, St. John's. 170 p.

Warwick, W. F. 1980. Palaeolimnology of the Bay of Quinte, Lake Ontario: 2800 years of cultural influence. *Can. Bull. Fish. Aquat. Sci.,* Ottawa. 206:1-117.

Watanabe, T., K. Asai, and A. Houki. 1988. Biological information closely related to the numerical index DAIpo (diatom assemblage index to organic water pollution). *Diatom* 4:49-58.

White, J. R., C. P. Gubala, B. Fry, J. Owen, and M. J. Mitchell. 1989. Sediment biogeochemistry of iron and sulfur in an acidic lake. *Geochim. Cosmochim. Acta* 53:2547-2559.

Whitehead, D. R., D. F. Charles, S. T. Jackson, J. P. Smol, and D. R. Engstrom. 1989. The developmental history of Adirondack (N.Y.) lakes. *J. Paleolimnol.* 2:185-206.

Whitmore, T. J. 1989. Florida diatom assemblages as indicators of trophic state and pH. *Limnol. Oceanogr.* 34:882-895.

Williams, J. D. H., T. P. Murphy, and T. Mayer. 1976. Rates of accumulation of phosphorus forms in Lake Erie sediments. *J. Fish. Res. Board Can.* 33:430-439.

Williams, N. E. 1988. The use of caddisflies (Trichoptera) in palaeoecology. pp. 493-500. *In*: J. Gray, ed., *Paleolimnology: Aspects of Freshwater Paleoecology and Biogeography.* Elsevier, Amsterdam. 678 p.

Williams, N. E. 1989. Factors affecting the interpretation of caddisfly assemblages from Quaternary sediments. *J. Paleolimnol.* 1:241-248.

Winter, T. C. and J. Wright. 1977. Paleohydrologic phenomena recorded by lake sediments. *EOS Transactions, American Geophysical Union* 58:188-196.

Wise, S. M. 1980. Cesium-137 and lead-210: a review of the techniques and some applications in geomorphology. pp. 109-127. *In*: R. A. Cullingford, D. A. Davidson, and J. Lewin, eds., *Timescales in Geomorphology*. John Wiley & Sons, NY. 360 p.

Wright, H. E., Jr. 1990. An improved Hongve sampler for surface sediments. *J. Paleolimnol.* 4:91-92.

Wright, H. E., Jr. 1991. Coring tips. *J. Paleolimnol.* 6:37-49.

Young, T. C. 1991. A method to assess lake responsiveness to future acid inputs using recent synoptic water column chemistry. *Water Resour. Res.* 27:317-326.

# SECTION V

# Program Considerations

# CHAPTER 14

# Biological Monitoring in the Environmental Monitoring and Assessment Program

**Steven G. Paulsen**, Environmental Research Center, University of Nevada, c/o U.S. Environmental Protection Agency, Corvallis, OR
**Richard A. Linthurst**, U.S. Environmental Protection Agency (MD-75)

Shortly after taking the position as Administrator of the U.S. Environmental Protection Agency, William Reilly was quoted as saying: "First the good news: Based on my years in the environmental movement, I think the Agency does an exemplary job of protecting the nation's public health and the quality of the environment. Now the bad news: I can't prove it." Too often, we find ourselves repeating this statement in a variety of situations. We implement a regulation or management strategy but seldom follow up to determine if we have achieved the intended result. A particular ecological system is studied, but we lack the information to determine how broadly applicable the results might be. We attack a particular environmental issue but often without sound information to determine if that problem should be our highest priority. Fundamentally, we are unable to evaluate the relative risk to ecological resources, partly because we lack an integrated approach to monitoring ecological condition and exposure to pollutants and other stressors. We generally cannot determine whether the frequency and extent of problems are increasing and whether such patterns are associated with changes in management actions or regulations. The U.S. Environmental Protection Agency, in collaboration with other agencies, has proposed a program, the

Environmental Monitoring and Assessment Program (EMAP), to alleviate many of these uncertainties in our information base. In the following pages, the rationale for EMAP's existence and the fundamentals behind its design and implementation are discussed with specific attention to the lake and stream component.

The U.S. Environmental Protection Agency (EPA) was created to ensure the protection and restoration of our Nation's ecological resources. These resources represent a heritage that we can ill afford to squander. Traditionally, we have focused our concern and efforts on the most obvious problems which often occurred at a local scale, e.g., the individual point source, a single lake or stream, or a single landfill. We have taken steps to solve many of these obvious problems and occasionally documented the success or failure of the corrective actions. The evaluation of technologies, such as wastewater treatment, and the development of chemical criteria and standards for environmental protection was frequently based on single-species laboratory test, field sampling observations from individual sites, or computer models constructed to predict pollutant transport, transformation and fate, and to assess dose-response effects of a specific pollutant on biota at a specific site. This was a rational scientific basis for determining the effects of individual pollutants as a foundation for regulatory action. Although this local scale perspective has resulted in reduced point source pollution, cumulative impacts of stresses at larger scales may lead to situations where we are "winning the battles but losing the war" as our ecological resources continually decline.

The impact of pollutants and other stresses now extends well beyond the local scale; global climate change, acidic deposition, stratospheric ozone depletion, non-point source pollution and habitat alteration threaten our ecological resources on regional and global scales. The public is increasingly concerned that these ecological resources, which provide them recreation, food, quality of life, and often an economic livelihood, may not remain sustainable. While existing policies and programs may be protecting the quality of the environment, it is not possible to demonstrate this with any certainty using available information. In 1990 alone, it is estimated that we spent $115 billion (not including the costs for wildlife conservation and other environmental programs that do not address pollution control) to clean up the environment (Science 1991). It is of increasing concern that we cannot document whether this expenditure, which is 40% of defense budget, is resulting in a healthier environment or being targeted toward the worst problems. Congressional hearings in

1983 and 1984 on the National Environmental Monitoring Improvement Act concluded that despite hundreds of millions of dollars spent annually on environmental monitoring, federal agencies could assess neither the current status of ecological resources nor the overall progress toward legally mandated goals (U.S. House of Representatives 1984).

These larger scale environmental problems require a change in our approaches to environmental protection and management. Assessing the condition of our ecological resources at the regional and national scales requires data on large geographic scales over long time periods. Meeting this need by simply aggregating data from many individual, local, short-term monitoring networks has proven difficult, if not impossible (Hren et al. 1990). Differences in monitoring objectives, methods, measurement procedures, indicator selection, length of monitoring, spatial extent, and representativeness of the sampling locations make the use of existing information to assess large-scale problems difficult. The increasing complexity, scale, and socioeconomic importance of existing and emerging environmental issues make establishing national baseline information on ecological resources, against which future changes can be documented with confidence, one of the highest priorities in ecology (Likens 1983, Roughgarden 1989).

These growing concerns prompted the EPA to adopt a proactive agenda for the future. In 1987, "Unfinished Business: A Comparative Assessment of Environmental Problems" (U.S. EPA 1987) looked at the environmental problems facing the EPA and evaluated the risk these problems posed to the environment. EPA's Science Advisory Board (SAB) recommended in 1988 that EPA reshape its strategy for addressing environmental problems in the next decade and beyond and plan, implement, and sustain a long-term research program that maintains a perspective of the environment in which the fundamental research and monitoring unit is the ecosystem (SAB 1988). Specific recommendations included:

EPA should explicitly develop and use monitoring systems to anticipate future environmental problems and recommend actions to address them.

EPA should implement a monitoring program to report on status and trends in ecological quality.

EPA should place greater emphasis on the development and use of ecological indicators.

EPA should expand its efforts to prevent/reduce environmental risk.

Fundamentally, EPA should adopt a focus on ecological risk assessment and management.

William Reilly, shortly after becoming EPA Administrator, asked the SAB to review Unfinished Business and the issue of comparative risk assessment. EPA senior managers were directed to start thinking about environmental risk and ways of reducing this risk. A quiet revolution to fundamentally change the way EPA does business in the future was set in motion (Science 1990). The future Agency focus will be on: environmental problems that pose the greatest risks rather than those that have attracted the most political attention; injecting science more prominently into the policy process; and "Managing for Results." This philosophy and direction, in part, explains why EPA initiated the Environmental Monitoring and Assessment Program (EMAP).

EMAP is part of the Office of Research and Development's (ORD) response to both the SAB recommendations (SAB 1988, 1990) and the Agency theme of "Managing for Results." EMAP represents the foundation for EPA's Ecological Risk Assessment Program.

## ENVIRONMENTAL MONITORING AND ASSESSMENT PROGRAM

EMAP represents a long-term commitment to assess and report on the condition of the nation's ecological resources. In cooperation with other agencies that share resource management and monitoring responsibilities, EMAP will provide part of the information needed to document the current condition of our ecological resources, understand why that condition exists, and assess the effects of different management alternatives. When fully implemented, EMAP will answer the following questions: What is the current status, extent and geographic distribution of our ecological resources, (e.g., estuaries, forest, streams)? What proportion of the resources are currently in good or acceptable ecological condition? What proportion of these resources are degrading or improving, in what regions, and at what rates? Are these changes associated with patterns and changes in environmental stresses? Are adversely affected resources improving in response to control and mitigation programs?

## Goals and Objectives

EMAP's goal is to monitor and report on the condition of the nation's ecological resources in order to determine the cumulative effectiveness of our regulatory and management programs and to identify emerging problems before they become widespread and irreversible. EMAP has three objectives.

1) Estimate the current status, extent, changes, and trends in indicators of the condition of the nation's ecological resources on a regional basis, with known confidence.
2) Monitor indicators of pollutant exposure and habitat condition and seek associations between human-induced stresses and ecological condition.
3) Provide periodic statistical summaries and interpretive reports on ecological status and trends to resource managers and the public.

EMAP is a long-term commitment to environmental monitoring on regional and national scales. EMAP is an interagency, interdisciplinary program in which EPA is but one participant. EMAP cannot efficiently meet its goal and objectives without the collaboration of other federal and state agencies that share responsibility for maintaining environmental quality and sustaining our ecological resources, and without the interaction of the academic research community. EMAP is designed to monitor all ecological resources and multiple environmental media (i.e., aquatic, terrestrial, and atmospheric) in order to provide an integrated picture of ecological condition.

Quantitative estimation of the status and trends in the indicators of ecological conditions requires using standardized procedures for monitoring in all regions, and using comparable methods across ecological resources. Quantitative estimates of resource condition are critical in determining the relative risk to our ecological resources and they are critical to our decision makers in order to implement effective regulatory and management decisions. EMAP has active and on-going efforts to incorporate existing data and information from other monitoring networks, both to confirm regional patterns in ecological indicators and to improve the diagnosis and assessment of large-scale environmental problems. EMAP will maintain interactive monitoring, assessment, and research activities in the program. Each of these activities is essential in maintaining a responsive, flexible, and evolving program capable of responding as new problems are identified. Monitoring, assessment, and research activities will be conducted through cooperative efforts with the scientific community. These principles are reflected in the EMAP design and approach to monitoring.

## Monitoring Approach

While there are many ways in which one could classify ecological resources, EMAP has selected one which tends to mirror the way in which the Federal government manages these resources. In EMAP, these ecological resources include: arid lands, agricultural systems, forests, lakes and streams (including the Great Lakes), inland and coastal wetlands, and estuaries and coastal waters. Several aspects of EMAP form the cornerstones of the program's unique approach. First, EMAP's focus is on ecological health rather than human health. Directly measuring indicators of the condition of ecological resources is fundamental to our ability to describe whether those biological attributes about which we are concerned are improving or degrading. Measurement of chemical and physical aspects of the environment has a role, but it is the biota that are our primary concern, and the emphasis on biological measurements in the program reflects this awareness. The second cornerstone is the reliance on a rigorous probability design for selection of sampling locations. The probability design ensures our ability to describe the entire resource based on that fraction that we actually sample. It ensures that our sample is in fact "representative" of the entire "population" about which we are concerned. It ensures that we have an estimate of the uncertainty about our descriptions of conditions and trends. A third cornerstone is the scale of interest. Currently, we have a relatively large amount of information about a variety of environmental parameters for both terrestrial and aquatic systems at the local spatial scale over fairly short temporal periods. The scales of focus within EMAP are regional and national scales of spatial resolution, with temporal scales on the order of years to decades. The indicator strategy and design are completely compatible with more intensive temporal and spatial scales but the inability to aggregate our current information to the regional and national scale has driven us to focus on these broader scales as a high priority. The final cornerstone of EMAP is its focus on interagency development and implementation. There are many federal and state agencies that have regulatory, management and assessment responsibilities for the ecological resources within the United States. EMAP is intended to fill a gap in our existing monitoring and assessment activities and in so doing it must complement rather than duplicate ongoing efforts. This will be assured by EMAP being designed and implemented by multiple federal and state agencies.

## Indicators

EMAP is faced with multiple issues: a) defining the health of ecological resources; b) selecting indicators that adequately reflect and measure that definition of health; and c) relating that condition to indicators of probable cause. We must identify the ecologically related concerns and values associated with lakes and streams as well as those aspects important to the public, aquatic biologists, and decision makers. For lakes and streams, these values and qualities historically have been expressed as designated uses and have been incorporated into the legislation and regulations embodied in the Federal Water Pollution Control Act. The designated uses include aquatic life and habitat for aquatic life, fishing, swimming, and navigation, and water for drinking, industrial, and agricultural uses. Some of these depend on ecological condition, others do so only indirectly. These designated uses are collectively referred to in the general goal of the Water Quality Act of 1987, which is to restore and maintain the physical, chemical, and biological integrity of the nation's waters. Biological integrity has been defined as a balanced, integrated, adaptive community of organisms having a species composition, diversity, and functional organization comparable to natural habitats in the region (Karr and Dudley 1981).

For surface waters in EMAP, we have selected three societal values that have ecological significance about which to make statements of condition: biotic integrity, trophic condition, and fishability. These values were chosen because of the often conflicting management programs associated with each, reflecting conflicting societal values. For example, one measure of biotic integrity in surface waters is the dominance of the fish assemblage by native species, but many programs established to improve fishability stock exotic sports species and remove native "trash" fish. Likewise, reducing enrichment of lakes in order to improve trophic state may reduce fishability by reducing productivity. Thus, for surface waters, the assessment of ecological condition needs to be made for each of these environmental attributes in order to weigh the relative impacts of our management decisions. The selection of these three environmental values is critical in EMAP because these values drive the indicators selected for monitoring.

The selection of indicators for EMAP-Surface Waters is constrained by the program's broad geographic scale. To make national- and regional-scale assessments, the program will evaluate the ecological condition of populations of lakes and streams, not of individual systems. Thus, indicators initially need to be broadly defined and as widely applicable as possible. The large number of sites and broad geographic scale also mean that the indicators must be amenable to index period sampling (i.e., adequate data can be gathered in one visit per site during a limited portion of the year).

Just as human health is not assessed by a single measurement, the status of each environmental value of concern for surface waters must be assessed by a suite of indicators. The conceptual framework for indicators used by EMAP consists of four broad categories (Figure 14.1) (Hunsaker and Carpenter 1990, Paulsen et al. 1991).

1) Response Indicators: A characteristic of the environment measured to provide evidence of the biological condition of an ecological resource at the organism, population, community, or ecosystem process level of organization (e.g., fish assemblages, chlorophyll *a* as an indicator of algal biomass, macroinvertebrate assemblages).

2) Exposure Indicators: Physical, chemical, and biological measurements that can be related to the occurrence or magnitude of contact with a physical, chemical, or biological stressor (e.g., nutrient or contaminant concentrations, tissue residues of contaminants).

3) Habitat Indicators: Physical, chemical or biological attributes that characterize the condition needed to support an organism, populations, assemblage, or ecosystem in the absence of pollutants (e.g., stream substrates, lake morphometry, suspended sediments).

## Top-Down Approach to Indicators

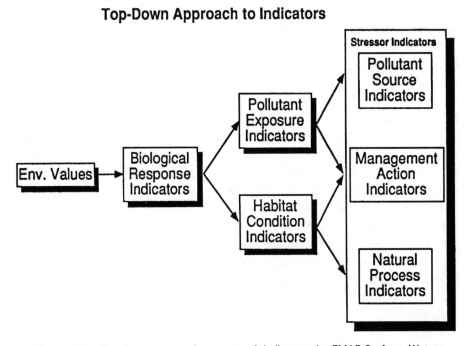

**Figure 14.1.** Top-down approach to use of indicators in EMAP-Surface Waters. Response indicators are used to describe condition for the environmental values. Exposure and habitat indicators are used to determine probable cause for condition seen in response indicators. The stressor indicators are used as confirmation of probable cause.

4) Stressor Indicators: A characteristic measured to quantify a natural process, environmental hazard, or a management action that affects changes in exposure and habitat indicators (e.g., climate, fertilizer sales and/or application rates, atmospheric deposition, fish stocking and harvesting rates).

The ability to make effective statements about condition relies primarily on our choice of biological measurements. The selected response indicators must share two critical qualities: (1) they must unambiguously and quantitatively express condition; and (2) they must relate directly to the value of concern or in an explicitly predictable way to the value of concern. Response indicators are derived from measurements that describe the biological condition at several levels of biological organization: organisms, populations, communities, or ecosystems. They serve as the focal point for describing the health of lakes and streams.

Additional general criteria for evaluating the usefulness of response indicators have been suggested by a variety of authors (Herricks and Schaeffer 1985, Hughes and Paulsen 1990, Kelly and Harwell 1990). The indicator should be biological, incorporating elements of ecosystem structure, function, and process as well as aspects of population dynamics. The selected indicators should correlate with changes in other unmonitored biological components. They must have clear connections with the three environmental values of concern (i.e., biological integrity, trophic state, and fishability) and must be responsive to a broad array of potential stressors. An ideal indicator is applicable in a broad range of surface water types across the nation and serves as an early warning of deleterious change. The response indicators must be sensitive to varying levels of stress, but not produce excessive false alarms. This includes sensitivity to important episodes that occur outside the sampling period. Useful indicators are cost effective, supplying considerable information in a limited amount of sampling time. They should be easily implemented and yield reproducible results with low (or at least measurable) sampling variability.

Habitat/exposure indicators are intended primarily for use as diagnostic tools to identify probable cause of conditions or changes in condition. Stressor indicators are intended to provide additional evidence for the cause of observed conditions and changes.

In EMAP-Surface Waters, a variety of indicators will be evaluated during the initial pilot and demonstration projects. The conceptual approach for the selection process is depicted in Figure 14.2. In lakes, several biological components are being evaluated for response indicators. These include: fish assemblages (structure and composition as well as incidence of external pathology), zooplankton assemblages, benthic macroinvertebrates, sediment cores for diatom assemblages, riparian birds, phytoplankton biomass (estimated by chlorophyll *a*), and macrophyte cover. Similar components will be evaluated in streams: fish, macroinvertebrates, periphyton, and riparian birds. Table 14.1 contains the current candidate list of indicators for EMAP-Surface Waters.

The current focus for our response indicators is on assemblage level information. This information on presence and abundance of species is desirable from a variety of perspectives. Roughgarden (1989) and Brown and Roughgarden (1990) argue for a national ecological survey.

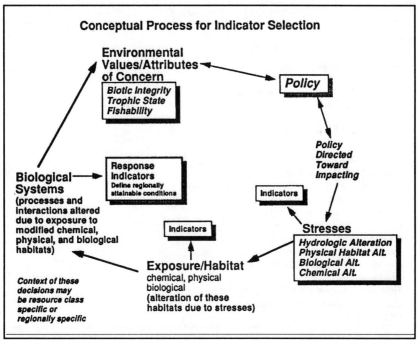

**Figure 14.2.** Conceptual process for indicator selection. The environmental values of concern must be selected based on public and scientific consideration. Description of these values is usually what shapes environmental policy. Response indicators are selected to reflect aspects of biological systems relevant to public and ecological concerns. Policy is usually directed toward one or more of a variety of stresses to ecological systems. These stresses alter the chemical, physical, and biological habitats within which biological systems must function.

EMAP collections could serve as the foundation for this effort. Preservation of the complexity of ecosystems is a desirable goal and information on the diversity and complexity of communities supported in different aquatic systems would contribute significantly to this information base. Quite often, individual species are the target of interest, either through their endangered and threatened status, or for other high-interest purposes. Whole assemblage collections do not exclude analyses of regional distributions and trends of species of particular interest. A significant amount of process- and functional-level interpretation can be obtained by analyzing the organization, or relative abundance, of different guilds.

Table 14.1. Candidate Indicators for EMAP-Surface Waters (Lakes and Streams).

**Response Indicators**
    Trophic State Index
    Assemblage Based Indices
        Fish
        Diatoms (surficial sediment cores)
        Zooplankton
        Macrobenthos
        Riparian Birds

**Exposure - Habitat Indicators**
    Physical Habitat Quality
    Chemical Habitat Quality
    Sediment Toxicity
    Chemical Contaminants in Fish

**Stressor Indicators**
    Landuse and Landcover
    Human and Livestock Population Density
    Chemical Use
    Flow and Channel Modification Structures
    Stocking and Harvesting Records
    Climatological Records

Perhaps one of the greatest challenges in EMAP, and biological monitoring in general, centers around how we intend to use the data to make statements about the condition of lakes and streams. Inevitably, we are forced to make some decisions about what we consider to be impaired (unhealthy, unacceptable, subnominal) biological conditions in water-bodies. The process for selecting these criteria or decision points is not well established or agreed upon. We know that there are really no clear thresholds; a continuum exists. Society, rather than science, will ultimately define the boundaries of acceptable and unacceptable conditions. But science can offer insights and approaches to dealing with this difficult issue. In practice, these criteria or thresholds will usually depend upon some knowledge of the acceptable range of values for the measured component (Rapport 1989, Schaeffer et al. 1988, Miller et al. 1988). This decision is often reached by evaluating indicator scores relative to some reference condition. Approaches currently being evaluated to establish this reference condition include: selecting and assessing reference sites, using historical data (might include paleo-limnological reconstructions), opinions of regional experts, pristine sites, ecological models, and present empirical distribution of indicator values.

While this challenge appears daunting, it is no different from efforts which many states have already begun in establishing biological criteria to accompany the existing chemical water quality criteria.

## Design

The primary EMAP objective is to provide unbiased estimates of status and trends in the condition of ecological resources on a regional basis with known confidence. A probability-based survey sampling design is required to satisfy this objective. The design also should enable monitoring to detect spatial and temporal patterns for each ecological resource, analyses of associations between response indicators and indicators of environmental stress, habitat and exposure, and be adequate for multiple ecological resources. The design was developed with the following considerations:

- Consistent representation of environmental reality by use of a probability sample
- Potential representation of all resources and environmental entities
- Capacity for quick response to a new question or issue
- Spatial distribution of the sample according to the distribution of the resource

Specific requirements implementing the design are:

- Explicit definition of target populations of lakes and streams and their sampling units (e.g., individual lakes, stream segments between confluences)
- Explicit definition of a frame for listing or otherwise representing all the potential sampling units within each target population (e.g., USGS 1:100,000 map series, U.S. EPA Office of Water River Reach File, version 3)
- Use of probability samples on well-defined sampling frames
- Ability to focus on subpopulations of potentially greater interest (e.g., specific types of lakes rather than all lakes)
- Ability to quantify statistical uncertainty and sources of variability for populations and subpopulations of interest

The proposed EMAP network design, which satisfies these considerations and requirements, utilizes a systematic sampling grid randomly placed over the United States. The random placement of the grid established the systematic grid sample as a probability sample so that statistically reliable regional and national estimates of current status, extent, changes, and trends in indicators of condition of ecological resources can be obtained with confidence limits. The use of a grid reflects a need to obtain good spatial representation for all resources where they exist, rather than a belief that ecological resources are distributed systematically.

The design is intended to be implemented hierarchically with four levels or tiers (Figure 14.3). The Tier I sampling units will be used to estimate the extent of the various resources classes or types (e.g., number and surface area of lakes and reservoirs, miles of streams and rivers). The information generated at Tier 1 is used to define the criteria or rules for

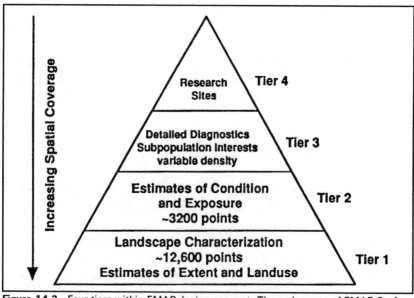

**Figure 14.3.** Four tiers within EMAP design concept. The early years of EMAP-Surface Waters will emphasize Tiers 1 and 2 because this is the primary gap in our current information.

selecting the resource sampling units on which response, exposure, and habitat indicators will be made. These resource sampling units comprise Tier 2 and represent a subsample or "double" sample of the Tier I points.

A similar approach was used to select streams for sampling in the National Surface Water Survey to assess the effects of acidic deposition on surface waters (Messer et al. 1986, Kaufmann et al. 1988). The various resource classes will be sampled annually during an index period, late summer for lakes, and early summer for streams. An index period is a period selected to provide the maximum amount of information on the ecological condition of a resource class during the year. The index period measurements provide a "snapshot" of conditions during this period. Tier 1 and 2 monitoring activities will allow the successful completion of the EMAP primary objectives. Tiers 3 and 4 will supplement and complement the information obtained as part of the Tier 1 and 2 sampling efforts.

Tier 3 sites also will be selected and used to obtain more explicit information about status, trends, and diagnostic attributes for specific subpopulations of interest, such as acid sensitive lakes in the northeast. More extensive spatial and/or intensive temporal sampling might be required for Tier 3 sites.

EMAP will rely, to a large degree, on existing programs such as the NSF Long Term Ecological Research programs, Department of Energy National Ecological Research Park network sites, and the National Park Service research sites for Tier 4 information. These research networks provide (1) important information on system characteristics, which will be useful in developing and testing indicators of ecological condition, (2) detailed process information characterizing system function that is not amenable to general monitoring, and (3) a link between spatially extensive and temporally intensive data to increase the interpretation of EMAP results. Tier 4 information should generate results that can be used to improve monitoring activities in Tiers 1–3. In turn, large-scale patterns emerging from Tier 1–3 networks can generate hypotheses to be tested at Tier 4 sites. The program will make maximum use of existing information for all four Tiers to avoid duplication and will capitalize on the experience, both successes and failures, of past efforts.

The EMAP network design for monitoring ensures broad geographic coverage, enables quantitative and unbiased estimates of ecological status and trends, facilitates analysis of associations among measurements of habitat condition, pollutant sources and exposure, and biological condition (indicators), and allows sufficient flexibility to accommodate sampling of multiple types of resources and identification of emerging environmental issues.

The emphasis over the next five years will be on implementing the monitoring network at the first two levels, Tiers 1 and 2. The current expectation is that approximately 3,200 lakes and 3,200 streams will be sampled over a four-year period. After the fourth year, samples will be taken from the systems visited during the first year. A regional demonstration project was conducted for lakes in the northeast in 1991 and will continue into the future. Figure 14.4 displays an example of what the four-year sample of lakes in the northeast would look like under the current survey design. Additional demonstration projects are planned for each new area of the country over the next five-year period. Demonstration projects will be used to: (1) test and evaluate the degree to which proposed indicators of ecological condition help distinguish polluted from unpolluted environments; (2) compile data sets that provide the information required to evaluate alternative sampling designs for assessing resource condition on regional scales; (3) identify and resolve logistical problems associated with conducting regional monitoring; and (4) test and evaluate statistical procedures for analyzing and presenting regional data. The full range of monitoring activities, from site and indicator selection, through the assessment of ecological status for regional resource classes, can be tested and evaluated through pilot and demonstration projects.

## Scale

In general, ecological studies have focused on structure and function at limited spatial scales (i.e., small areas) and short temporal scales rather than processes or patterns at broader landscape and regional scales. Karieva and Anderson (1988) found that 50% of all experimental ecological studies published between January 1970 and January 1987 were conducted on plots less than 1 m in diameter, and 25% of these studies used plots less than 25 cm in diameter. Tilman (1989) noted that 40% of ecological experiments lasted less than 1 year and only 7% lasted longer than 5 years.

However, there is a hierarchy of spatial and temporal scales of interest in ecology from organismal to ecosystem, landscape, regional, continental, and global (O'Neill et al. 1986). In addition, spatial and temporal scales are linked; short time-scale processes are coupled with

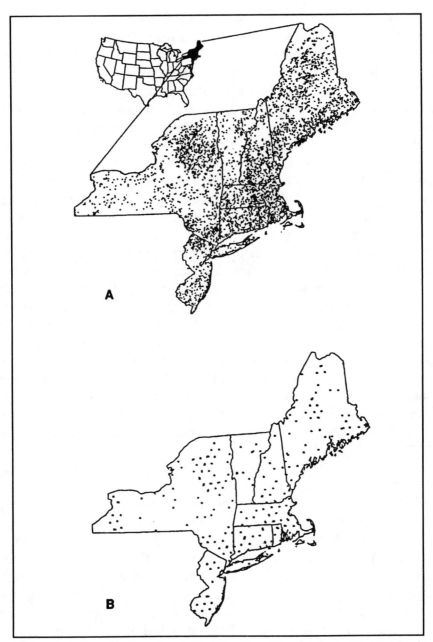

**Figure 14.4.** Example of results of the application of the design principles to lakes in the northeast. A) Distribution of lakes (only lakes 10–250 ha. plotted for clarity, all lake sizes were used in sample allocation). B) Example of a four year sample of lakes in the northeast.

small spatial scales, while processes and patterns at larger spatial scales are associated with longer time scales (Ford and Thornton 1979, O'Neill et al. 1986). Short time-scale processes such as photosynthesis or chlorophyll are measured on a space scale of meters while longer time-scale processes, such as changes in plankton community structure, are measured at space scales of kilometers. Most ecologists are familiar with the structural hierarchy of organism-population-community-ecosystem. Each level in the hierarchy is composed of components from the lower levels and becomes itself a component of higher levels. For example, populations are composed of organisms and the assemblage of populations forms a community. The behavior at one level in the hierarchy appears to be constrained or controlled by the dynamics of the next higher level (O'Neill 1988). This higher level also provides a perspective or gives significance to the phenomenon of interest. O'Neill (1988) uses the example of reproductive structures and behaviors at the organism level. These structures and behaviors are difficult to explain by studying only the single organism. It is only at the population level, or next higher level, that the significance and context of reproduction is understood. The population characteristics generally are expressed statistically to describe species density, natality, mortality, or age distribution. This same context applies for individual ecosystems. Individual systems can be intensively studied and causal mechanisms understood, but it is only when these systems are placed in a regional context that the extent and proportion of the resource impacted, or at risk from environmental problems, can be determined.

The primary spatial-temporal scales of interest in EMAP are regional and annual, respectively. These scales are compatible with monitoring and assessment for many of the environmental problems currently impacting ecological resources, such as global climate change, cumulative impacts of point and nonpoint source pollution, acidic deposition, and habitat fragmentation. These problems are subtle, with the ecological effects occurring over years to decades, not weeks or months. The ecosystem responses to these problems must be placed in a regional and global context in order to be understood and diagnosed. This regional perspective was critical in assessing the effects of acidic deposition on aquatic resources as part of the National Acid Precipitation Assessment Program (NAPAP).

Some patterns and relationships among indicators might not be observed or detected without a regional perspective for these large-scale problems. For example, in the Aquatic Effects Research Program (AERP) of NAPAP, analysis of regional patterns indicated that a strong correlation existed between regional sulfate deposition and surface water sulfate concentrations for most regions. This regional relationship, or signal, between sulfate deposition and median surface water sulfate concentration was masked at the individual watershed level and did not become apparent until data was aggregated to the sub-regional scale (Sullivan et al. 1988). Clearly, identifying appropriate ecological indicators for monitoring at regional scales represents one of the research challenges for the 1990's.

## *Collaborative Efforts*

All EMAP efforts are being conducted in collaboration with at least one other federal agency (e.g., National Oceanic and Atmospheric Administration, USDA Forest Service, U.S. Geological Survey, U.S. Fish and Wildlife Service), EPA Regions and Program Offices, states, and universities. These cooperative efforts will be expanded as EMAP moves toward full-scale, national implementation. To ensure that EMAP can meet its objectives, an active research program has been initiated in environmental statistics, indicator development, information management, landscape analysis, and ecological risk characterization. Because the program must meet the highest scientific standards, scientists in universities, national laboratories, and other agencies are active participants in the planning and implementation of EMAP. The entire program will be reviewed by the National Research Council. Although the agenda for EMAP is ambitious, the Program represents the type of monitoring program that is needed for the 1990's and beyond to address large-scale environmental problems.

## RISK ASSESSMENT

Ecological resources are potentially affected by multiple pollutants arriving through multiple media. Ecological responses to these pollutants also are influenced by biotic interactions, hydrometeorology and other natural factors, and management practices. EMAP, or any monitoring program, must acknowledge this and provide information useful in evaluating the risks to our resources. The complexity of environmental processes and subsequent effects at regional and national scales, therefore, makes it difficult to determine ecological condition based on cause-and-effect relations between pollutant concentration and ecological response or compliance with criteria or standards for individual contaminants (Messer 1990). The traditional diagnostic approaches for environmental problems focus on the cause (stressor) and develops relationships with the ecological effect or response. EMAP diagnostic approaches, however, will determine the status and trends in ecological endpoints or response indicators of ecological conditions, and then work backwards through association analyses and the process of elimination to identify plausible stressors eliciting these responses.

Suter (1990) and Fava et al. (1987) describe an epidemiological approach to ecological risk assessment. It is this epidemiological effects-driven assessment that is under development in EMAP. In effects-driven assessments, an effect is observed, and one then works toward identification of the hazard or stressors and exposure information to determine possible causes of the ecological response or effect. The first step is the identification and selection of assessment endpoints. Assessment endpoints are formal expressions of the actual environmental value or attribute that is to be protected or managed (Suter 1990). In many instances, assessment endpoints can not be quantitatively established or defined. Quantitative measurements, however, can be taken that characterize or relate to the assessment endpoint. These measurements are referred to as the measurement endpoints (Suter 1990). Therefore, measurement endpoints are ideal ecological response indicators. The impact of this perspective was the foundation of the EMAP indicator selection process described above.

## ADAPTING TO FUTURE NEEDS

One of the greatest challenges in any monitoring program is the ability to provide consistent information to describe trends, yet maintain flexibility to adapt to emerging issues. It was recommended that EPA improve its capability to anticipate environmental problems so that early actions can be taken to ameliorate these problems (SAB 1988). Anticipatory monitoring implies future prediction or forecasting of problems. Anticipatory monitoring forecasting future change is exceptionally difficult, if not impossible. Based on the large spatial extent and long time frames associated with those environmental problems posing the greatest risk to ecological systems, forecasting or anticipating environmental problems is less important than identifying and responding to these problems before they become widespread and irreversible. Currently there are no networks monitoring multiple ecological resources at regional scales. EMAP provides a nationally designed network capable of being adapted to a variety of spatial and temporal scales, and thus, is well suited to use as a platform from which to launch relatively rapid evaluations of the extent of recently recognized problems.

The integrated use of existing monitoring networks to obtain a regional perspective of environmental condition is extremely difficult, if not impossible. Recent studies on the potential use of local and state water quality data for regional assessments illustrate some of the problems associated with this approach: sampling sites were clustered in areas with known or suspected problems, resulting in a biased estimate of regional conditions or trends; samples were collected to meet permit or effluent requirements rather than evaluate ambient stream conditions; the monitoring stations did not have sufficient periods of record to evaluate trends, or the records were fragmented; samples collected from these stations could not be verified as representative; and monitoring records did not have adequate documentation to determine data quality (Hren et al. 1990). Similar problems have been noted for other monitoring networks over the past two decades (GAO 1981, US Congress 1983, 1984, U.S. EPA 1987). EMAP has focused on designing a flexible, adaptive network that is responsive to emerging environmental issues in order to overcome some of these monitoring problems. Some of the Program features that facilitate this responsiveness include a flexible design, methodology, continual development of indicators, and better data analysis procedures.

The four tiered approach to the EMAP network design, discussed in previous sections, in conjunction with other design characteristics results in a sampling frame adapted for global to subregional scales of resolution and from biomes to specific subpopulations of interest. The triangular arrangement of grid points provides flexibility to increase or decrease the density of the grid by three, four, or seven-fold, or by multiples of these values. This flexibility can provide resolution at the state or landscape level, if desired, as well as sampling at densities appropriate for scarce resource or specific subpopulations, such as the redwood forests or coastal wetlands.

A strategic plan for indicator development has been prepared so that better diagnostic and response indicators will continually be identified, tested, and evaluated through demonstration projects and implemented in EMAP (Charles et al. 1990, Knapp et al. 1991). This will permit EMAP to continually evolve and address emerging environmental problems as well as develop better indicators for diagnosing these problems.

Better analysis and presentation procedures also will be developed to improve communication with decision makers and the public. In general, most scientific and technical information is presented in a format that is not useful for decision makers (NRC 1989). These presentations typically use unfamiliar language, complex graphics that are difficult to interpret, and do not relate scientific results to policy issues. Innovative approaches for displaying and presenting ecological effects must be an integral part of indicator development research.

## LIMITATIONS

In describing any program it is critical to describe its limitations as well as its capabilities. EMAP will make a significant contribution to assessing the risk to ecological resources from large-scale problems but some of the caveats associated with the approach must be understood. First, this effort requires a long-term commitment of funds and personnel. Long periods of record are required to detect statistically reliable trends. Second, multiple objectives can lead to multiple conflicting expectations. Efforts are on-going to identify questions that EMAP will, and will not, be able to address when fully implemented. Third, ecological condition or health cannot be unambiguously defined. Ecological health will be operationally defined based both on scientific understanding and on desired socioeconomic use of the ecological resource. These uses, in many instances, can also be conflicting. EMAP will provide the information to determine the ecological condition of the resource and whether desired use can be achieved, but it will not unconditionally define ecosystem health. Finally, monitoring information is useful in correlating or associating ecological response indicators with indicators of environmental stress, but it does not establish cause and effect. EMAP information can be used to formulate hypotheses for testing and evaluation and identify resources that might be at risk, but it is not causally oriented. These risks must be quantified throughout the ecological risk assessment process.

# CONCLUSIONS

The increasing complexity of environmental problems dictates that we develop new approaches for monitoring these problems and our effectiveness in addressing them. The use of biological indicators should be the foundation of any new efforts. The Environmental Monitoring and Assessment Program (EMAP) is being designed to monitor biological indicators of ecological condition at a regional scale of resolution. The EMAP focus on regional spatial scales and annual temporal scales is compatible with the spatial-temporal scales of emerging environmental problems, such as global climate change, acidic deposition, nonpoint source pollution, and habitat fragmentation. The incorporation of biological measures in EMAP, with a hierarchical network design, will permit unbiased estimates of the current status, extent, changes, and trends in the condition of the nation's ecological resources on a regional basis with known confidence. EMAP will contribute to the ecological risk assessment process through an epidemiological effects-driven approach utilizing indicators of condition in conjunction with indicators of pollutant exposure and habitat alteration.

# REFERENCES

Charles, D. P., C. M. Knapp, D. R. Marmorek, J. P. Baker, K. W. Thornton, J. M. Klopatek. 1990. *Indicator Development Strategy for Ecological Indicators*. U.S. EPA.

Eagleson, P. S. 1986. The emergence of global-scale hydrology. *Water Resources Research* 22(9): 6S-14S.

Fava, J. A., W. J. Adams, R. J. Larson, G. W. Dickson, K. L. Dickson, and W. E. Bishop. 1987. *Research Priorities in Environmental Risk Assessment*. Society of Environmental Toxicology and Chemistry, Washington, D.C.

Ford, D. E. and K. W. Thornton. 1979. Time and length scales for the one-dimensional assumption and its relation to ecological models. *Water Resource Research* 15:113-120.

General Accounting Office. 1981. *Better Monitoring Techniques Are Needed to Assess the Quality of Rivers and Streams. Vol. I*. United States General Accounting Office, CED-81-30.

Gray, W. M. 1990. Strong association between West African rainfall and U.S. landfall of intense hurricanes. *Science* 249: 1251-1256.

Herricks, E. E. and D. J. Schaeffer. 1985. Can we optimize biomonitoring? *Environmental Management* 9:487-492.

Hren, J., C. J. Oblinger Childress, J. M. Norris, T. H. Chaney, and D. N. Myers. 1990. Regional water quality: evaluation of data for assessing conditions and trends. *Environ. Sci. Technol.* 24(8):1122-1127.

Hughes, R. M. and S. G. Paulsen. 1990. Indicator strategy for inland surface waters. pp. 4.1-4.20. *In*: C. T. Hunsaker, and D. E. Carpenters, eds., *Ecological Indicator Report for the Environmental Monitoring and Assessment Program.* U.S. Environmental Protection Agency, Research Triangle Park, NC.

Hunsaker, C. T. and D. E. Carpenter, eds. 1990. *Environmental Monitoring and Assessment Program Ecological Indicators.* EPA/600/3)90/060. Office of Research and Development. U.S. Environmental Protection Agency, Washington, D.C.

Kareiva, P. and M. Anderson. 1988. Spatial aspects of species interactions: the wedding of models and experiments. pp. 38-54. *In*: A. Hastings, ed., *Community Ecology.* Springer Verlag, NY.

Karr, J. R., and D. R. Dudley. 1981. Ecological perspective on water quality goals. *Environmental Management* 5:55-68.

Kaufmann, P. R., A. T. Herlihy, J. W. Elwood, M. E. Mitch, W. S. Overton, M. J. Sale, J. J. Messer, K. H. Reckhow, K. A. Cougan, D. V. Peck, A. J. Kinney, S. J. Christie, D. D. Brown, C. A. Hagley, and H. I. Jager. 1988. *Chemical Characteristics of Streams in the Mid-Atlantic and Southeastern United States. Volume 1: Population Descriptions and Physico-Chemical Relationships.* EPA 600/3-88-021a. U.S. Environmental Protection Agency, Washington, D.C.

Kelly, J. R. and M. A. Harwell. 1990. Indicators of ecosystem recovery. *Environmental Management* 14:527-545.

Knapp, C. M., D. R. Marmorek, J. P. Baker, K. W. Thornton, and J. M Klopatek. 1991. *The Indicator Development Strategy for the Environmental Monitoring and Assessment Program* (Draft). U.S. Environmental Protection Agency, Environmental Research Laboratory, Corvallis, OR.

Landers, D. H., J. M. Eilers, D. F. Brakke, W. S. Overton, P. E. Kellar, M. E. Silverstein, R. D. Schonnbrod, R. E. Crowe, R. A. Linthurst, J. M. Omernik, S. A. Teague, and E. P. Meier. 1987. *Characteristics of Lakes in the Western United States. Vol. I. Population Descriptions and Physico-Chemical Relationships.* EPA/600/3-86-/045a. U.S. EPA, Washington, D.C. 176 p.

Likens, G. E. 1983. A priority for ecological research. *Bull. Ecol. Soc. Am.* 64:234-243.

Linthurst, R. A., D. H. Landers, J. M. Eilers, D. F. Brakke, W. S. Overton, E. P. Meier, and R. E. Crowe. 1986. *Characteristics of Lakes in the Eastern United States. Volume 1: Population Descriptions and Physico-Chemical Relationships.* EPA/600/4-86/007a. U.S. EPA, Washington, D.C.

Messer, J. J. 1990. Indicators concepts. Section 2. *In*: C. T. Hunsaker and D. E. Carpenter, eds., *Environmental Monitoring and Assessment Program Ecological Indicators*. EPA/600/3-90/060. U.S. Environmental Protection Agency, Office of Research and Development, Washington, D.C.

Messer, J. J., C. W. Ariss, J. R. Baker, S. K. Drouse, K. N. Eshleman, A. J. Kinney, W. S. Overton, J. J. Sale, and R. D. Schonbrod. 1986. Stream chemistry in the Southern Blue Ridge: feasibility of a regional synoptic sampling approach. *Water Resource Bull.* 24:821-829.

Miller, D. L., P. M. Leonard, R. M. Hughes, J. R. Karr, P. B. Moyle, L. H. Scharader, B. A. Thompson, R. A. Daniels, K. D. Fausch, G. A. Fitzhugh, J. R. Gammon, D. B. Halliwell, P. L. Angermeier, and D. J. Orth. 1988. Regional applications of an index of biotic integrity for use in water resource management. *Fisheries* 13:12-20.

National Acid Precipitation Assessment Program (NAPAP). 1990. *Integrated Assessment* (Volume 1: Questions 1 & 2; Volume 2: Question 3; Volume 3: Questions 4 & 5, External Review Drafts). Washington, D.C.

National Research Council. 1989. *Improving Risk Communication*. National Academy Press, Washington, D.C.

Neilson, R. P. 1986. High resolution climatic analysis and Southwest biogeography. *Science* 232:27-34.

Neilson, R. P. 1987. Biotic regionalization and climatic controls in western North America. *Vegetation* 70:135-147.

O'Neill, R. V. 1988. Hierarchy theory and global change. pp. 29-45. *In*: T. Rosswall, R. G. Woodmansee, and P. G. Risser, eds., *Scales and Global Change: Spatial and Temporal Variability in Biospheric and Geospheric Processes*. John Wiley & Sons, NY. 355 p.

O'Neill, R. V., D. L. DeAngelis, J. B. Waide, and T. F. H. Allen. 1986. *A Hierarchical Concept of Ecosystems*. Princeton University Press, Princeton, NJ.

Paulsen, S. G., D. P. Larsen, P. R. Kaufmann, T. R. Whittier & others. 1991. *EMAP-Surface Waters Monitoring and Research Strategy — Fiscal Year 1991*. EPA/600/3-91/022. U.S. Environmental Protection Agency, Corvallis, OR. 183 p.

Rapport, D. J. 1989. What constitutes ecosystem health? *Perspect. Biol. Med.* 33:120-132.

Reynoldson, T. B., D. B. Schloesser, and B. A. Manny. 1989. Development of a Benthic Invertebrate Objective for Mesotrophic Great Lakes Waters. *J. Great Lakes Res.* 15:669-686.

Roughgarden, J. 1989. The United States needs an ecological survey. *Bioscience.* 39:5.

Ryder, R. A. and C. J. Edwards. 1985. *A Conceptual Approach for the Application of Biological Indicators of Ecosystem Quality in the Great Lakes Basin*. Report to the Great Lakes Science Advisory Board, IJC, Windsor, Canada.

Schaeffer, D. J., E. E. Herricks, and H. W. Kerster. 1988. Ecosystem health: I. Measuring ecosystem health. *Environmental Management* 12:445-455.

Science. 1990. *Science* 429:616.

Science. 1991. Costs of a Clean Environment. *Science* 251:1182.

Science Advisory Board. 1988. *Future Risk: Research Strategies of the 1990s*. SAB-EC-88-040. U.S. Environmental Protection Agency, Washington, D.C.

Science Advisory Board. 1990. *Reducing Risk: Setting Priorities and Strategies for Environmental Protection*. SAB-EC-90-021. U.S. Environmental Protection Agency, Washington, D.C.

Strachan, W. M. J. and M. G. Henry. 1990. *Final Report of the Ecosystem Objectives Committee*. Report to the Great Lakes Science Advisory Board. International Joint Commission, Windsor, Canada.

Sullivan, T. J., J. M. Eilers, M. R. Church, D. J. Blick, Jr., K. N. Eshleman, D. H. Landers, and M. S. DeHaan. 1988. Atmospheric wet sulfate deposition and lake water chemistry. *Nature* 331:607-609.

Suter, G. W., II. 1990. Endpoints for regional ecological risk assessments. *Environmental Management* 14:9-23.

Swetman, T. W. and J. L. Betancourt. 1990. Fire-southern oscillation relations in the southwestern United States. *Science* 249:1017-1020.

Tilman, D. 1989. Ecological experimentation: Strengths and conceptual problems. pp. 136-157. *In*: G. E. Likens, ed., *Long-Term Studies in Ecology: Approaches and Alternatives*. Springer Verlag, NY.

U.S. Environmental Protection Agency. 1987. *Unfinished Business: A Comparative Assessment of Environmental Problems. Overview Report*. Office of Policy Analysis and Office of Policy, Planning, and Evaluation, Washington, D.C.

U.S. Environmental Protection Agency. 1987. *Surface Water Monitoring: A Framework for Change*. Office of Water and Office of Policy Planning, and Evaluation, Washington, D.C.

U.S. House of Representatives Committee on Science and Technology. 1983. *National Environmental Monitoring*. Hearings before the Subcommittee on Natural Resources, Agriculture Research, and the Environment of the Committee on Science and Technology. May 19, 26; June 2, 1983.

U.S. House of Representatives Committee on Science and Technology. 1984. *Environmental Monitoring Improvement Act*. Hearings before the Subcommittee on Natural Resources, Agriculture Research, and the Environment of the Committee on Science and Technology. March 28, 1984.

# CHAPTER 15

# Design of Biological Components of the National Water-Quality Assessment (NAWQA) Program

**Martin E. Gurtz**, U.S. Geological Survey, Raleigh, North Carolina

## INTRODUCTION

The National Water-Quality Assessment (NAWQA) Program is a long-term program of the U.S. Geological Survey (USGS) designed to describe the status of, and trends in, the quality of the Nation's surface- and ground-water resources and to provide an understanding of the natural and human factors that affect the quality of these resources (Hirsch et al. 1988, Leahy et al. 1990). The program will evaluate water quality at a wide range of spatial scales, from local to national, and will focus on persistent water-quality conditions that affect large areas of the Nation or occur commonly within small areas. The NAWQA Program is an integrated assessment of water quality that incorporates physical, chemical, and biological components. This approach is designed to provide new insights into the status and trends of national water quality, and "more importantly . . . improve our understanding of physical, chemical, and biological processes and causal relationships" (National

Research Council 1990). Among the reasons for including biological components in NAWQA are the following:

1) Biota respond to a wide variety of natural and anthropogenic environmental influences, including: stresses from point and nonpoint sources, toxic effluents, enriched organic effluents, extreme flows, and habitat degradation.

2) Biota integrate over space. Stream organisms can serve as monitors for disturbances in the upstream landscape (Hynes 1975, Hynes 1994). Thus, an examination of biological communities in a stream provides a means to evaluate effects of human activities throughout the catchment.

3) Biota integrate over time. Biota that have aquatic stages for most or all of their life cycles integrate the effects of temporal variability in the aquatic environment. Fixed-interval monitoring programs (for example, monthly water samples) may not detect the water-quality effects of episodic events such as pulsed effluent discharges or high flows. Such events can have long-term effects on aquatic biota that can be determined by tissue analysis or changes in the structure or function of aquatic communities.

4) Biota, in many cases, provide a more sensitive indication of environmental change than other media. For example, biological tissues may concentrate some chemicals at levels that are easier to detect than in water or sediment. Such bioconcentration may result in tissue concentrations that are high enough to be of concern to human or ecological health (Phillips 1980).

5) Public concerns about water quality often focus on biological issues (Karr 1991). Protection or restoration of natural biological characteristics of streams and rivers—variously expressed as biological integrity, biodiversity, or ecological condition—has received increasing attention by both legislators and the general public (Plafkin et al. 1989, Hunsaker and Carpenter 1990, U.S. Environmental Protection Agency 1990). Public-health considerations of recreational activities or fish consumption also require biologically based information.

Objectives of various biological monitoring and research programs include: (1) evaluation of response to (or recovery from) a specific stress or disturbance, (2) protection of biological diversity or endangered species or habitats, and (3) evaluation of compliance with specific regulations, or development of those regulations. The biological components of NAWQA have broader objectives reflecting an emphasis on the development of an improved understanding of relations among physical, chemical, and biological characteristics of streams as an integral part of interpreting water-quality status and trends. The approach of NAWQA is to describe spatial and temporal patterns in selected environmental settings and to develop and test hypotheses about the causes of the observed patterns.

The NAWQA Program will complement programs of other Federal and State agencies through an integrated assessment of water quality based on studies of entire river basins and major aquifer systems. The river-basin approach to research and management of surface-water resources allows knowledge of the hydrology of the system to contribute to an understanding of sources and transport of constituents. Biological studies of entire river basins can incorporate important ecological principles, for example, the river continuum concept (Vannote et al. 1980) and prior knowledge of regional differences (e.g., ecoregions) (Omernik 1987), while encompassing multiple political units (Tyus 1990). NAWQA will contribute to the spatially and temporally extensive and consistent data bases that are needed for regional risk assessments (Suter 1990), as well as to the development and testing of emerging theories of stream ecology (Gore et al. 1990). Regional differences in species composition, habitat characteristics, thermal and hydrologic regimes, and water-quality problems must be taken into consideration in the design of a nationally consistent biological-assessment program.

This chapter presents the conceptual design of the NAWQA Program and the design considerations associated with the ecological and tissue components of the surface-water design. Ground-water components of NAWQA are not discussed.

## CONCEPTUAL DESIGN OF THE NATIONAL WATER-QUALITY ASSESSMENT PROGRAM

The building blocks of the NAWQA Program are study-unit investigations, which will be conducted in about 60 hydrologic systems that consist of major river basins and large parts of aquifers or aquifer systems distributed throughout the Nation (Figure 15.1; Table 15.1). The study units are large, ranging in size from about 1,550 square kilometers

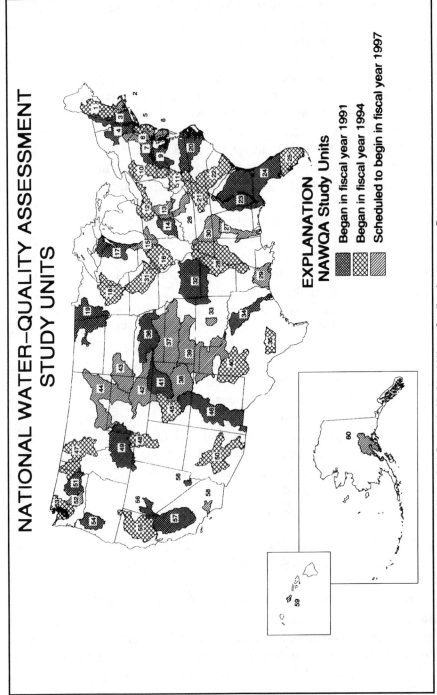

**Figure 15.1.** Study units in the U.S. Geological Survey's National Water-Quality Assessment Program.

**Table 15.1.** Names of study units in the National Water-Quality Assessment Program. Numbers refer to Figure 15.1. Names in bold and italic are study units that started in 1991.

| Map No. | Study Unit Name |
|---------|-----------------|
| 1 | New Hampshire-Southern Maine Basins |
| 2 | Southeastern New England |
| 3 | *Connecticut, Housatonic and Thames River Basins* |
| 4 | *Hudson River Basin* |
| 5 | Long Island and New Jersey Coastal Plain |
| 6 | Delaware River Basin |
| 7 | *Lower Susquehanna River Basin* |
| 8 | Delmarva Peninsula |
| 9 | *Potomac River Basin* |
| 10 | Allegheny and Monongahela Basins |
| 11 | Kanawha-New River Basin |
| 12 | Lake Erie-Lake Saint Clair Drainage |
| 13 | Great and Little Miami River Basins |
| 14 | *White River Basin* |
| 15 | Upper Illinois River Basin |
| 16 | Southern Illinois |
| 17 | *Western Lake Michigan Drainages* |
| 18 | Upper Mississippi River Basin |
| 19 | *Red River of the North* |
| 20 | *Albemarle-Pamlico Drainage* |
| 21 | Upper Tennessee River Basin |
| 22 | Santee Basin and Coastal Drainage |
| 23 | *Apalachicola-Chattahoochee-Flint River Basin* |
| 24 | *Georgia-Florida Coastal Plain* |
| 25 | Southern Florida |
| 26 | Kentucky River Basin |
| 27 | Mobile River and Tributaries |
| 28 | Mississippi Embayment |
| 29 | Chicot-Evangeline |
| 30 | Lower Tennessee River Basin |
| 31 | Eastern Iowa Basins |
| 32 | *Ozark Plateaus* |
| 33 | Central Oklahoma Aquifer |
| 34 | *Trinity River Basin* |
| 35 | South Central Texas |
| 36 | *Central Nebraska Basins* |
| 37 | Kansas River Basin |
| 38 | Upper Arkansas River Basin |
| 39 | Central High Plains |
| 40 | Southern High Plains |
| 41 | *South Platte River Basin* |
| 42 | North Platte River Basin |
| 43 | Cheyenne and Belle Fourche Basins |
| 44 | Yellowstone Basin |
| 45 | Upper Colorado Basin |
| 46 | *Rio Grande Valley* |
| 47 | Northern Rockies Intermontane Basins |
| 48 | Great Salt Lake Basin |

**Table 15.1.** Continued

| Map No. | Study Unit Name |
| --- | --- |
| 49 | *Upper Snake River Basin* |
| 50 | Southern Arizona |
| 51 | *Central Columbia Plateau* |
| 52 | Yakima River Basin |
| 53 | Puget Sound Drainages |
| 54 | *Willamette Basin* |
| 55 | Sacramento Basin |
| 56 | *Nevada Basin and Range* |
| 57 | *San Joaquin-Tulare Basins* |
| 58 | Santa Ana Basin |
| 59 | Oahu |
| 60 | Cook Inlet Basin |

(600 square miles) to more than 155,000 square kilometers (60,000 square miles). The NAWQA study units collectively cover a large part of the United States, encompass a large percentage of the Nation's population and water use, and include diverse hydrologic systems that differ widely in the natural and human factors that affect water quality. The distribution of study units ensures that the most important regional and national water-quality issues can be addressed by comparative studies between and among study units. Surface water, ground water, and their interactions will be examined to varying degrees in each study unit, depending on their relative importance to the resource.

Study-unit investigations consist of intensive assessment activity for 4 to 5 years, followed by 5 years of low-intensity assessment, and then repeating the cycle (Figure 15.2). A typical study-unit investigation begins with a 2-year effort focused on planning, analysis of available data, and reconnaissance-level field investigations. This is followed by 3 years of intensive data collection to document current water-quality conditions, establish a baseline for evaluating temporal trends, and progressively build an understanding of the factors affecting water-quality conditions in the basin. The intensive data-collection phase is followed by report preparation and selected low-intensity assessment activities. Intensive assessment in each of the study units is thus done on a rotational rather than a continuous basis, with one-third (20) of the study units studied in detail at a given time. Assessment activities for the first 20 study units (Figure 15.1) began in 1991. The first cycle of study-unit investigations is scheduled to be completed in 12 years (1991-2002).

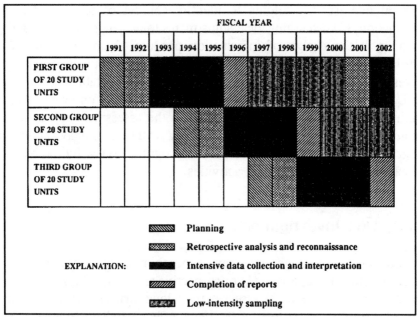

**Figure 15.2.** Schedule of first cycle of study-unit investigations, by dominant activity, for the National Water-Quality Assessment Program. Fiscal year begins October 1.

The NAWQA Program emphasizes a multidisciplinary approach using physical, chemical, and biological measurements to provide multiple lines of evidence with which to evaluate water quality. The program focuses on a broad spectrum of constituents and sampling approaches for surface water, including information on: (1) hydrology, (2) inorganic constituents (major ions, trace elements, nutrients), physical measurements (suspended sediment, specific conductance, temperature), radionuclides, and organic contaminants in water, (3) trace elements and organic contaminants in bed material and aquatic biota, (4) ecological information (fish, benthic invertebrate, and algal communities), and (5) stream habitat characterization.

Communication and coordination with other Federal, State, and local agencies, universities, and the public are important to the NAWQA Program. Each study-unit investigation has a liaison committee to help ensure that the scientific information produced by the program is relevant to local and regional interests. The committees are composed of non-USGS members who represent additional technical and management interests and expertise. Specific activities of each liaison committee include: (1) exchanging information about water-quality issues of local and regional interest; (2) helping to identify sources of relevant data and information; (3) discussing local adjustments to project guidelines; and (4) assisting in the design of information products from the projects. In addition to liaison committees, a Federal Advisory Council consisting of representatives of other Federal agencies has been established to advise the USGS on the needs of water-information users and on procedures for making the data and the information from the assessment appropriate to planning needs and available in a timely manner. Collaboration with programs of other Federal agencies and program review by the National Research Council are ongoing activities.

## Study-Unit Investigations

Study-unit investigations consist of four main components with different, but related, goals: (1) retrospective analysis; (2) occurrence and distribution assessment; (3) assessment of long-term trends and changes; and (4) source, transport, fate, and effect studies (Figure 15.3). Reconnaissance represents an intermediate step in the planning of the occurrence and distribution assessment. In cooperation with other agencies, potential implications of study results from each component are assessed for management priorities and information requirements for local, regional, and national water-quality issues.

The retrospective analysis is an interpretation of existing data in order to address NAWQA objectives to the extent practicable and to aid in the design of NAWQA studies. These analyses provide a historical perspective on water quality in the study unit, strengths and deficiencies of available information, and implications for water-quality issues, study priorities, and study design. The purposes of these analyses are to: (1) develop an improved conceptual model of factors affecting spatial and temporal patterns of water quality within the study unit; (2) guide additional data collection; (3) contribute to early national-synthesis products; and (4) document findings for future NAWQA studies.

A water-quality reconnaissance is conducted in each study unit to: (1) improve the initial conceptual model of factors governing water quality in the study unit; (2) evaluate and select sampling sites for the occurrence and distribution assessment; and (3) familiarize project personnel with field techniques and study-unit characteristics. This effort comprises

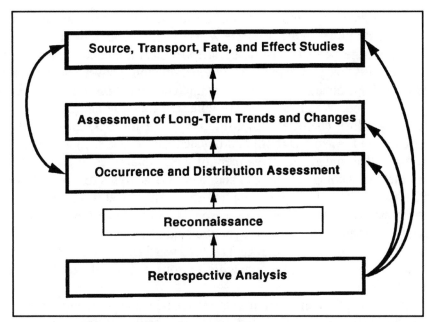

**Figure 15.3.** The main components of study-unit investigations in the National Water-Quality Assessment Program. From bottom to top, activities progress from large to small spatial scale and from low to high contribution to understanding causal relations. Arrows show primary directions of information flow in the design.

the first field sampling effort and is an integral part of the study-unit design process. Each candidate sampling site is visited first as part of an initial reconnaissance that includes a rapid visual assessment of site accessibility, general habitat conditions, the proximity of natural or anthropogenic sources of contamination, conditions for sampling, and the methods and equipment appropriate for that location. This information serves as the basis for selecting sampling locations for detailed on-site assessments in which field water-quality measurements are made, including qualitative assessments of habitat conditions and biological communities.

The primary objective of the occurrence and distribution assessment is to characterize geographic and seasonal distributions of water-quality conditions in relation to major contaminant sources and background conditions. Sampling activities during the 3-year intensive data-collection phase include collection of physical, chemical, and biological information at a set of fixed sites that represent important environmental settings (for example, relatively homogeneous land use and physiographic conditions) in each study unit, or that integrate water-quality conditions at key points in the drainage network. These basic fixed sites are continuously gaged for stream discharge to support analyses of relations between water quality and hydrologic conditions; specific conductance and water temperature are also monitored. Water samples are collected monthly and at high flow, and analyzed for chemical composition (e.g., major ions, trace elements, nutrients, and organic carbon) and physical properties such as suspended sediment. The integrated assessment of water-quality conditions at these sites includes ecological surveys and chemical analyses of bed sediments and tissues. More-intensive sampling is conducted at a subset of these sites to assess the occurrence and seasonal patterns of pesticides and other dissolved organic constituents. Synoptic studies will be conducted in each study unit to investigate the geographic distribution of selected water-quality characteristics during specific seasons and in relation to sources and transport.

Assessment of long-term trends and changes in selected water-quality characteristics will be designed from the results of the retrospective analyses, reconnaissance, occurrence and distribution assessment, and the concurrent development of information on the environmental framework. Source, transport, fate, and effect studies are detailed case studies designed to investigate causes and governing processes of water-quality conditions and trends for specific situations within the study units that have the greatest relevance to characterizing and managing water quality.

## Environmental Framework

A common set of important natural and anthropogenic factors provides a unifying environmental framework for making comparative assessments of water quality among hydrologic systems across the Nation. Such factors include landscape features and hydrogeologic setting. Landscape features include ecological regions such as "ecoregions" (Omernik 1987, Gallant et al. 1989, Hughes et al. 1994), potential natural vegetation (Kuchler 1970), geology, physiography, soil and crop types, wetland types and distributions, and land use and land cover. The hydrogeologic setting of the study unit also includes major surface-water hydrologic features and manmade alterations to the natural flow system, as well as spatial properties of the groundwater system.

Characterization of the environmental framework contributes to the initial conceptual model of dominant factors that influence water quality in the study unit. This broad geographic description of natural and anthropogenic features provides a general stratified structure for initial study planning in that each unique combination of categories of natural features and land use represents a relatively homogeneous condition, or environmental setting, for this scale of water-quality assessment. Initial study priorities can be assessed and the scope of assessment narrowed on the basis of the relative importance of each setting to water quality and its areal extent. A long-term goal is the systematic documentation of environmental setting characteristics in a nationally consistent fashion so that each setting can be evaluated in the context of contributing to a national perspective on water-quality conditions.

## National Synthesis

NAWQA national-synthesis projects will combine key findings of study-unit investigations with existing information from other programs, and with studies of the USGS and other agencies and researchers, in order to produce regional and national-scale assessments for priority water-quality issues. National synthesis will initially focus on priority issues related to pesticides and nutrients.

Whereas the study-unit framework characterizes specific patterns in factors that govern water quality within study-unit boundaries, the national-level framework characterizes broad patterns in factors that govern water quality and it defines the role of each study unit in representing a particular set of natural and anthropogenic features of regional or national significance. The distribution of specific types of potential sources of contaminants, such as specific agricultural activities or urban areas, are evaluated in relation to natural features, such as climate, soils, and aquifers, that may affect water-quality conditions. National synthesis will consist of an accumulation of analyses of the most significant influences, which often are regional in nature, evaluated in a national context. Overlapping distributions of major source activities within particular hydrologic settings will provide opportunities to compare the significance of the different activities, for example, urban land use compared to corn and soybean cultivation, with respect to distribution and levels of pesticides and nutrients.

## DESIGN OF BIOLOGICAL COMPONENTS

One of the key goals of the NAWQA Program is to develop a better understanding of the linkages, relative to status and trends, among physical, chemical, and biological characteristics of streams in selected environmental settings across the Nation. Biological methods for describing status and trends in the quality of surface-water resources must be well integrated with physical and chemical approaches in a common sampling design. However, few biological studies or monitoring programs have been implemented at a spatial scale as large as that of NAWQA, and few biological methods have been tested in such a variety of geographical settings. The need for biological tools that have low natural variability, but high sensitivity to a wide range of stresses, is a tradeoff among levels of biological organization. Variability tends to decrease with aggregation (individual to ecosystem-level phenomena), whereas sensitivity tends to be highest at the species level (Kratz et al. 1994).

Biological approaches most appropriate for meeting NAWQA objectives include ecological studies and tissue studies. Ecological studies include surveys that focus on community structure and function, and habitat characteristics to assess water quality. This approach represents a compromise between the greater sensitivity of species indicators or physiological responses to individual stresses and the lower variability, but broader response, of ecosystem processes. Chemical analyses of tissues provide the means to assess the occurrence and distribution of potentially toxic trace elements and organic compounds. Other biological approaches that may be used in some study units include: (1) synoptic surveys of *Escherichia coli* bacteria to assess occurrence and distribution of stream reaches that may be heavily affected by fecal contamination from human or animal wastes; (2) toxicity studies (Elder 1989), biomarkers, and other stress indicators to address specific questions related to causes and effects of selected contaminants; and (3) studies of biogeochemical processes to examine interactions between surface water and ground water.

## Ecological Studies

The primary goal of ecological studies in NAWQA is to collect information about biological communities that contributes to the conceptual model of factors affecting water quality, and to improved understanding of the relations among physical, chemical, and biological characteristics of streams. The following are examples of ecological questions that can be addressed by the NAWQA design:

1) How do biological communities differ among selected environmental settings in each study unit?
2) What are the primary physical and chemical factors influencing biological communities in selected environmental settings?
3) How are biological community characteristics affected by physical and chemical stream characteristics at different spatial scales?
4) How do biological communities affect physical and chemical stream characteristics spatially and temporally?

Three taxonomic groups—fish, invertebrates, and algae—will be investigated because each responds differently to natural or anthropogenic disturbances due to differences in habitat, food, mobility, physiology, and life history. Use of a multiple-community approach adds additional power to the design; agreement, or lack thereof, among these sets of community data can be very instructive (e.g., Stewart and Loar 1994). A multiple-community approach is especially valuable in broad-scope water-quality programs.

For all NAWQA ecological studies, taxonomic identifications are made to the lowest practicable level, preferably species. Advantages of identifications at the genus or species level, compared with family- or higher-level taxonomy, were reviewed by Lenat and Barbour (1994). Genus or species identifications increase the precision of site classifications and detection of subtle changes in water quality, resulting in improved interpretations of spatial patterns — for example, contrasting sites in different environmental settings, examining longitudinal changes within a basin, or comparing communities among river basins regionally or nationally. Biotic indices based on species identifications are more accurate than those based on family-level identifications (Hilsenhoff 1988). The integrated NAWQA data base of taxonomic information, coupled with physical and chemical water-quality characteristics, will be useful for a variety of biological community analyses as well as development of improved biological indices or component metrics.

Habitat characterization is an essential component of many water-quality assessment programs (Osborne et al. 1991) and can have a primary role in monitoring programs targeted toward protection of anadromous species (Conquest et al. 1994). In NAWQA, habitat descriptions of channel, riparian, and flood-plain features known to have important structural influences on biological communities will also be useful in explaining physical or chemical characteristics of surface waters. The challenge in habitat assessment is to select and prioritize measurements from a large array of channel and riparian features that influence stream communities and processes. There is no set of widely accepted techniques for use on a national scale; existing manuals on habitat descriptions (e.g., Armour et al. 1983, Hamilton and Bergersen 1984, Platts et al. 1987) often have a regional or single-purpose focus. The framework for habitat characterization should incorporate different spatial scales because of corresponding differences in geomorphic processes and in the role of habitats as refugia in recovery from disturbances (Sedell et al. 1990).

Habitat descriptions in NAWQA follow a spatial hierarchy modified from that of Frissell et al. (1986), incorporating four spatial scales—basin, stream segment, stream reach, and microhabitat (Meador et al., unpublished data). Basin descriptors, such as physiographic province, geology, and climate, contribute to an understanding of long-term evolution of habitats, cumulative effects of human activities, and comparative biogeographic patterns in biological communities (Biggs et al. 1990). Much of the data at the basin scale is obtained from geographic information system data bases as part of the environmental framework description for each NAWQA study unit. A nationally consistent scale of resolution facilitates regional comparisons. Non-digital coverages, or local data bases of greater resolution than the national coverage, are used at the study-unit level. At the stream-segment scale, characteristics such as segment slope, stream order, and channel sinuosity generally are described using 7.5-minute topographic maps. Habitat description at the stream-reach scale includes classification of geomorphic channel units (e.g., pool, riffle, or run). Channel features include stream type, for example, meandering, braided, channelized, or pool/riffle (Leopold and Wolman 1957). Instream cover (e.g., woody debris or undercut banks), macrophyte cover, and channel-substrate characteristics are also included. Characteristics of the riparian zone, such as riparian vegetation and bank stability, are important descriptors because of the strong influences of riparian systems on aquatic communities. Microhabitat descriptors include physical variables, such as water depth, current velocity, and substrate particle size, that are measured concomitantly with the collection of biological-community samples.

Ecological studies in NAWQA are designed according to the main components of NAWQA as shown in Figure 15.3. The primary mechanism for addressing the questions stated above is the occurrence and distribution assessment, with retrospective and reconnaissance activities contributing to the design. Other objectives will be addressed that are consistent with the assessment of long-term trends and changes, and case studies of sources, transport, fate, and effects. Integration of physical, chemical, and biological investigations of surface-water quality is emphasized in each of these activities.

## Retrospective Analysis of Biological Information

A review of existing biological information is an important part of the retrospective analysis in each study-unit investigation. This activity contributes to the conceptual model of the natural and anthropogenic factors contributing to spatial and temporal patterns of water quality (physical, chemical, and biological) within and among study units. An improved conceptual model will guide the study design in terms of selection of sites and constituents most important for understanding the local river basins and for complementing existing knowledge. A variety of sources of biological information can be used as part of this analysis, including: the U.S. Geological Survey's National Water Information System; the U.S. Environmental Protection Agency's STORET data base and the National Bioaccumulation Study; the U.S. Fish and Wildlife Service's National Contaminant Biomonitoring Program; and data bases of the Nature Conservancy and various State natural heritage programs. Additional sources of information on biological characteristics of study units include reports of State and other agency monitoring programs, environmental impact statements, taxonomic publications on local fauna and flora, and investigations of university researchers and others. Species introductions (intentional or otherwise) and historical, ongoing, or planned programs of stocking or reintroducing native species should be documented. Products of the retrospective analysis may include maps, showing sampling locations of major studies and water-quality sampling programs where biological data have been collected. Synthesis of information from multiple sources may provide valuable insight into water-quality relations when interpreted from the perspective of a whole basin or comparisons among environmental settings within a study unit.

## Reconnaissance

Reconnaissance of candidate sampling locations is a two-phase activity. An initial reconnaissance of a large number of sites, providing a quick overview of study-unit characteristics, is followed by a more intensive on-site assessment of a subset of these sites. The primary ecological information recorded in the initial reconnaissance includes the general habitat conditions of the stream, for example, geomorphic channel units (pool, riffle, run), channel features, bank stability, and riparian vegetation. Observations of local land use or riparian disturbances, dominant channel substrates, and are useful for comparing and contrasting among candidate sites for the on-site assessment. Other elements of the initial reconnaissance may include general access considerations, wadeability of the channel, availability of a boat ramp, and potential choices for sampling gear.

The on-site assessment involves a larger crew and more time per site than the initial reconnaissance. In addition to field measurements of stream discharge and selected physical and chemical water-quality constituents, suitability of different types of sampling gear are assessed, and initial efforts to describe the occurrence and relative abundance of fish, benthic invertebrates, and aquatic vegetation are accomplished by field identification of specimens. No samples are collected for quantitative processing, but some may be retained for purposes of training and familiarization of study-unit staff. Habitat description in the on-site assessment depends on study-unit objectives, but may incorporate an objective index approach such as the Ohio Environmental Protection Agency's Qualitative Habitat Evaluation Index (QHEI) (Rankin 1989), with local modifications as necessary. This index includes a qualitative classification of habitat characteristics based on six interrelated factors—substrate, instream cover, channel morphology, riparian condition, pool and riffle quality, and gradient.

## Occurrence and Distribution Assessment

The sampling design for ecological studies in the occurrence and distribution assessment emphasizes the integration of physical, chemical, and biological approaches to assessing surface-water resources. Collections of biological communities (fish, invertebrates, and algae) and descriptions of riparian and instream habitat conditions are made at each fixed site, where physical and chemical characteristics are also assessed. Nationally consistent methods are used to maximize the ability to compare selected ecological characteristics among important environmental settings, or among selected mainstem sites, at study-unit, regional, and national scales. Examinations of these descriptive data will lead to development of hypotheses about relations among physical, chemical, and biological characteristics and influences of land use and other human activities.

Each collection represents a composite of samples distributed throughout a sampling reach, the length of which is defined at each site by a combination of factors, including stream geomorphology and meander wavelength. Criteria for minimum and maximum length are used to provide a minimum sampling reach length sufficient to ensure the collection of representative samples (e.g., of the fish community), and to limit the sampling reach length to a distance that prevents unnecessary sampling and minimizes crew fatigue. Sampling reaches are located so that there are no major discontinuities, relative to geomorphology or other major natural or human influences on water quality, between the sampling reach and the fixed site (e.g., no intervening major point sources).

Fish-community sampling incorporates multiple collection methods to provide information on the presence and relative abundance of species at each site (Meador et al. 1993). The primary method of collection is electrofishing (backpack or boat units, depending on the site). Electrofishing is supplemented in some cases by seining, especially in streams where low specific conductance or high turbidity may preclude efficient use of electrofishing gear. Data recorded in the field include length and weight information and the presence of externally visible skin or subcutaneous disorders, or parasites. Species identifications are made in the field, to the extent possible.

For benthic invertebrates, three types of samples are collected (Cuffney et al., unpublished data). A semi-quantitative (relative abundance) sample is collected from the habitat expected to support the faunistically richest community within the reach. In most wadeable streams, such habitats include riffles; however, samples in some streams may be collected from woody snags or macrophyte beds. A second semi-quantitative sample is collected from a depositional habitat (e.g., pool). In addition, a qualitative (presence or absence) sample is collected that encompasses the variety of instream habitats present throughout the reach. In each case, samples are composites collected throughout the entire reach.

Algal community sampling follows an approach analogous to that for benthic invertebrates, with semi-quantitative collections from one or more habitats supplemented by a composite multi-habitat sample (Porter et al., unpublished data). In wadeable streams, semi-quantitative samples are generally collected by scraping rocks or other stable surfaces. Qualitative samples are collected by a combination of scraping, brushing, siphoning, or other methods appropriate for the types of habitat present in the sampling reach.

Habitat measurements in the field emphasize reach characteristics such as bank stability, channel form and substrate, geomorphic channel units, and riparian woody vegetation (Meador et al., unpublished data). Information of this type provides insight into geomorphic processes and major structural features. Physical characteristics such as water depth, current velocity, and substrate particle size are also noted to document conditions for each sample. The primary limitation of the amount of habitat information that can be obtained in these studies is the field time and personnel required.

Sampling of biological communities will be conducted during hydrologic and seasonal conditions that are comparable among all sites within the study unit. Scheduling of the survey takes into consideration factors such as variability and extremes of flow at different times of the year, life cycles of aquatic taxa (e.g., emergence of important insect species, periods of fish migration or spawning), accessibility of sites, and timing of major human activities.

At a subset of the fixed sites in each study unit, intensive ecological assessments are conducted to provide essential information on spatial and temporal variability. Sample variance is estimated at these sites by sampling multiple reaches (minimum of three) located so that each represents the same water-quality conditions as the fixed site; information from these sites will help establish confidence levels for site comparisons over space or time. Multiple-year collections (annually during each year of the 3-year intensive data-collection phases) at these sites provide estimates of year-to-year variability; these data will help define present biological conditions and establish baseline conditions for comparisons against future long-term (decadal) changes.

In addition to fixed-site sampling, ecological synoptic studies are also conducted to: (1) provide improved spatial resolution for ecological characteristics important at the study-unit scale, or (2) evaluate the spatial distribution of selected ecological characteristics in relation to causative factors, such as land uses or contaminant sources and instream habitat conditions. Design of synoptic surveys is flexible to address questions that are study-unit driven (in the context of regional and national priorities), but nationally consistent methods are used to the extent possible to facilitate regional and national comparisons. Examples of study-unit objectives for ecological synoptic surveys include: (1) spatial distribution of selected biological community characteristics in parts of the study unit (e.g., to compare selected environmental settings), (2) characterization of regional reference conditions, and (3) comparisons of biological responses along a gradient of physical or chemical conditions such as riparian habitat, nutrients, or pesticides. Investigators in some study units may choose to fill in gaps of other ongoing programs, either in terms of spatial coverage or taxonomic groups studied, or to collaborate with other agencies to achieve mutually beneficial objectives.

## Assessment of Long-Term Trends and Changes

Assessment of long-term trends and changes in biological communities and habitat characteristics will include comparisons with changes in physical and chemical constituents, landscape features, and human activities. At the intensively studied subset of fixed sites, ecological studies will be conducted annually during each 3-year intensive phase. Long-term changes can be evaluated by comparing these discrete 3-year data sets at 9-year intervals. Annual sampling of biological communities may be continued at some sites during the low-intensity sampling period to assess both natural variability and long-term trends due to human activities. Sampling of multiple reaches at some sites will allow partitioning of sources of variability, for example, among reaches in the same stream segment.

Interpretation of temporal patterns in biological communities requires an understanding of temporal variability induced by natural patterns, such as seasonal changes in food resources, water temperature, or discharge.

Hydrologic characteristics, for example, have a strong influence on temporal variability of biological phenomena. Extremes in stream discharge may cause scour, disruption and export of debris dams, or establishment of riparian woody species, and may have severe and potentially long-lasting consequences on communities through drastic reductions in population size or modifications to channel and riparian characteristics. Long-term characteristics of the hydrologic regime, such as flow predictability, may also have significant effects on community structure, life-history characteristics, and response to disturbance (Resh et al. 1988). For example, biological communities may be most predictable in highly constant or highly variable flow regimes (Connell 1978, Moyle 1994). The NAWQA Program provides a framework for examining influences of hydrologic regime on aquatic biota and their responses to various water-quality stresses. Temporal variability can be interpreted in the context of spatial variability at different scales (see Resh and Rosenberg 1989, Meador and Matthews 1992).

To minimize the effects of some of the natural sources of temporal variability, samples of biological communities will be collected during a similar range of seasonal and flow conditions each year for each site selected for multiple-year sampling. Development of criteria for sampling during a specified range of degree-day accumulation, flow exceedance probabilities, or time since a scouring flood would make these decisions more objective. Timing of sampling to coincide with a period of minimal flow variability will help reduce some of the year-to-year variability.

## Source, Transport, Fate, and Effect Studies

Intensive case studies that examine the source, transport, fate, and effects of selected water-quality constituents will be conducted to address specific cause-and-effect questions and to help explain observed spatial and temporal patterns in physical, chemical, and biological water-quality data. One objective for such studies is to identify the primary physical and chemical factors influencing biological communities in selected environmental settings and to evaluate potential effects on biological communities related to land use and natural features within the study unit. Priority will be given to studies that have: (1) strong potential for technical integration of ecological studies with physical and chemical properties of surface water and ground water; (2) potential for long-term study; (3) availability of high-quality ancillary data such as detailed landscape characterization; and (4) participation by other programs and researchers within and external to the USGS. Data from some study units may be used to test large-scale and long-term hypotheses that would lead to better: (1) design of biological sampling protocols; (2) interpretation of biological water-quality data at local, regional, or national levels; and (3) understanding of the functioning of river systems.

Case studies incorporating ecological issues may include more-detailed habitat measurements, biological process studies, detailed mapping of macrophyte beds, or toxicological studies. These studies could address questions such as: How does a particular activity or land use affect biological community structure or function? What factors cause seasonal differences in the response of biological communities to a particular stress? What are the effects of stream restoration or changes in land-management practices on community structure and function? Other opportunities for ecologically focused case studies exist in conjunction with NAWQA ground-water flowpath studies, where interactions between surface water and ground water can be studied.

## Data Analysis

The basin orientation of NAWQA surface-water studies facilitates interpretation of ecological information at different spatial scales. Integrated assessments of water quality (physical, chemical, and biological) at a relatively small number of fixed sites in each study unit provide, first, descriptions of environmental settings considered important from both study-unit and national perspectives. These descriptions are then interpreted in the context of conceptual models concerning interrelations among factors affecting water quality at different spatial scales. Within a study unit, ecological synoptic studies may provide a broader spatial context for interpreting fixed-site data. Interpretation of differences among sites must also take into consideration regional differences within and among study units relative to geology, land form, land use, and vegetation (Omernik 1987, Hughes et al. 1990, Hughes et al. 1994), as well as expected changes along a longitudinal continuum from headwaters to larger-order downstream reaches within the same river basin (Vannote et al. 1980). Data from sites representative of "reference" conditions (Hughes et al. 1986) — where human influences on water quality are expected to be minimal and where the probability of major change in land use is low — will aid interpretation of changes due to human activities and contribute to the understanding of spatial and temporal variability in natural systems; reference sites are included in study-unit designs wherever possible.

Historically, research on spatial variation in natural systems has focused on finer-scale, habitat-level influences on biota, rather than on larger-scale regional influences. Thus, comparisons of ecological properties of streams across large areas present additional challenges. For example, the dynamics of recovery from disturbance varies among spatial scales (Reice et al. 1990). National synthesis of ecological data will include exploratory analyses to search for patterns in biological community characteristics or responses to environmental conditions at scales larger than the study unit. These analyses will focus initially on priority issues associated with pesticides and nutrients.

Comparisons will be made of biological community characteristics to distributions of physical, chemical, land-use, geomorphic, and regional characteristics within and among study units. These comparisons will aid in development of hypotheses about causes of observed spatial and temporal patterns. Further, comparisons can be made among different biological communities to evaluate their relative sensitivities to changes in water quality.

Biological community data can be summarized and interpreted in many different ways, from simple measurements such as taxa richness to complex indices of biological integrity or multivariate analyses (see review by Ford 1989). Often, the best choice depends on the specific stress or water-quality issue. Approaches such as the Index of Biotic Integrity (Fausch et al. 1984, Karr et al. 1986, Karr 1991) and analogous methods for invertebrates (Ohio Environmental Protection Agency 1988, Shackleford 1988), or other index approaches (e.g., Hilsenhoff 1982), will be used where sufficient information is available locally to provide for necessary calibration with stream size and reference conditions. However, appropriate baseline data are not presently available for most regions or biological community types (Suter 1990). Hence, such approaches initially will not be used for nationwide comparisons. As the NAWQA database grows, the ability to develop and use indices (or component metrics) for evaluating different kinds of biological stresses will improve, both for NAWQA and for other programs.

# Tissue Studies

Analyses of the chemical composition of biological tissues provide information about the occurrence and distribution of contaminants in aquatic systems. Some chemicals may be more highly concentrated in tissues, thus making their detection in tissues easier than in water or sediments. Tissue analyses provide a time-averaged assessment of contaminants as well as a direct measurement of the bioavailability of potential toxicants.

Tissue sampling is applicable only to persistent chemicals that accumulate in tissues. Factors considered in the selection of target chemicals for analysis in biological tissues include: availability of analytical methods, toxicity, bioaccumulative potential, and the capacity of target taxa to metabolize the chemicals. Target chemicals include trace elements and organic contaminants such as organochlorine pesticides, polynuclear aromatic hydrocarbons, and polychlorinated biphenyls. Chemicals analyzed in each study unit include national target constituents, plus other chemicals chosen based on information acquired during the retrospective analysis.

Resident taxa collected for tissue analyses are chosen according to nationally consistent guidelines from a list of National Target Taxa by use of a decision tree. This helps ensure maximum comparability among sites and over time. Species are avoided if they are known to be especially sensitive to certain chemicals, and thus usually absent in the streams where concentrations are highest (Nehring 1976). Different decision trees are used for trace elements and for organic contaminants. For trace element analysis, the top taxon priority is the introduced bivalve mollusk *Corbicula fluminea*. If *C. fluminea* cannot be found in adequate abundance, the decision progresses to other groups of target taxa, with the order of preference being insects, livers of fish, and vascular plants. Taxa are prioritized within each of these groups of target taxa. For analysis of organic contaminants, *C. fluminea* is again the first choice, followed by fish. Variability of concentrations in organisms due to size, age, or reproductive stage of the organism is minimized by specifying criteria for each. Decisions on taxa and other criteria will be made based on information gathered during the retrospective and reconnaissance activities.

The first objective in NAWQA tissue studies is to define occurrence of significant contaminants in the study unit (Crawford and Luoma 1993).

An initial sampling survey of tissues is conducted in conjunction with collections of bed sediments for chemical analyses at a small number of sites (commonly 15-20) selected to maximize the probability of assessing contaminants that commonly occur and are important in the study unit. This is followed by a spatial-distribution survey to add improved geographic coverage, with particular emphasis on the priority constituents whose occurrences were detected in the initial survey. In addition, several sites thought to be minimally affected by human activity are included for comparisons with contaminated sites and for identification of widely dispersed contaminants. Timing of the tissue-analysis surveys is determined according to local criteria, including seasonal, hydrologic, and biological conditions, and inputs from known sources of contaminants. Most tissue sampling will be conducted at low flow.

Samples for each taxon are composited for each site to average individual variability and to meet a minimum mass requirement for the intended analyses. At most sites, one composite sample is taken for analysis of organic contaminants and one for analysis of trace elements. The collection of replicate samples and multiple species, preferably one fish and one invertebrate species, is recommended for a subset of the spatial-distribution survey sites, especially for analysis of trace-element concentrations. Replicate sampling should be adequate in order to statistically differentiate critical reaches in the study unit after this stage of study. Because no single taxon is likely to occur in all streams and rivers of a given study unit, spatial distributions of constituents will be derived from comparisons of streams and river segments with similar taxa. Water and sediment chemistry data can also be considered in such comparisons.

The product of the occurrence and distribution assessments of tissues will be a conceptual model of the distribution of locally important trace elements and organic contaminants in each study unit. Products may include maps of selected constituent concentrations in association with land uses and other potential sources, or reference conditions. In addition, an assessment will be made of locations where concentrations exceed levels that are potentially toxic to humans or aquatic organisms based on established regulations and guidelines.

A few sites will be selected in each study unit to establish a data base for long-term trend detection of contaminants in tissues, with greater sampling intensity at these sites. As with ecological surveys, temporal trends in tissue data will be appraised primarily by comparing changes between 3-year discrete periods in subsequent NAWQA study cycles.

National trend analysis questions will have the highest priority in determining specific sites and chemicals targeted for trend analysis in each study unit. Results of the occurrence and distribution assessments may suggest potential case studies in order to investigate local influences of land uses, point-source effects, or effects of hydrologic processes or features on the distribution and effects of trace elements and organic contaminants.

## IMPLEMENTATION CHALLENGES AND OPPORTUNITIES

Development of sampling design and protocols for the biological components of NAWQA has been an iterative process. Site-selection strategies and field and laboratory methods were tested in pilot studies, and protocols were revised based on these experiences. Implementation of NAWQA or any other large-scale water-quality program presents certain challenges as well as opportunities.

Large (non-wadeable) rivers provide a major challenge because most standardized sampling methods for biological monitoring have been developed for shallow wadeable parts of streams and rivers. In addition, large rivers are typically more complex than smaller rivers and streams because of their wide flood plains, large lateral components to nutrient and organic matter fluxes, and adaptations of many species to transient aquatic-terrestrial environments (Junk et al. 1989, Sparks et al. 1990). In NAWQA, sampling strategies for large rivers follow the same general approaches as for smaller streams, with modifications as necessary to adjust reach length and to select appropriate targeted habitats for sampling. For example, shoreline woody snags may support the highest diversity of invertebrates compared with other habitats in some large rivers (e.g., Lenat and Barbour 1994), and methods must be used that adequately sample that habitat type.

Storage and management of biological data is a challenge previously faced by few agencies on a national scale. A revision of the U.S. Geological Survey's National Water Information System (NWIS-II) has been designed to handle biological and habitat data collected by NAWQA. The new system provides for storage of physical, chemical, and biological characteristics of water quality in the same data base, facilitating integrated analyses. NWIS-II will also have the capability to associate taxonomic information with a code structure that facilitates tracking of name changes so that nomenclature is consistent among NAWQA study units and among NAWQA cycles.

Quality assurance is an essential component of water-quality monitoring programs and ecological studies. Elements of quality assurance include collection of replicate samples in a design that allows quantification of sources of variance at some sites. Field personnel receive standardized training on methods for collecting and processing samples. Published sampling protocols will be used consistently; any deviations will be noted in the data base. A Biological Quality-Assurance Unit (in the USGS's National Water-Quality Laboratory) will closely monitor taxonomic data, including coordinating verification of identifications, establishing taxonomic voucher collections, developing and maintaining computer databases, and collaborating with other agencies in sharing taxonomic data and coordinating taxonomic databases (Cuffney et al., unpublished data).

The implementation of this large program is mainly by local study-unit teams, each of which includes a biologist. The local team designs, executes, interprets, and publishes the results of the study-unit investigation. This is a key element in the design of the national program and ensures that the data are interpreted first by scientists who are most familiar with the study area. In addition, each of the four USGS regions has a team of biologists to advise project staffs and oversee the design, sampling, and interpretation of the biological aspects of all studies in their respective regions. National Biological Survey personnel are participating in NAWQA at headquarters and regional levels. National-synthesis teams contribute to the design and coordination of study-unit activities to facilitate interpretation of study-unit data at regional and national scales.

NAWQA provides excellent opportunities for collaborative research. The large-scale and long-term nature of the NAWQA Program will provide data that are useful in addressing important ecological research questions or providing a framework upon which additional studies can build. For example, the Sustainable Biosphere Initiative (SBI) of the Ecological Society of America (Lubchenco et al. 1991) presented recommendations for a research agenda for the coming decade that focused on three areas—global processes, biological diversity, and sustainability of ecological systems. Some of the critical issues identified by the SBI include effects of spatial and temporal scales on aquatic communities, effects of multiple factors on ecological systems, and the role of environmental variability. These issues cut across basic and applied fields of aquatic ecology and require long-term interdiscipinary studies, perhaps building on the base of information provided by programs such as NAWQA. Collaborative studies could be designed, for example, to evaluate the magnitude and specific action of natural and anthropogenic disturbance, and interaction with other biotic and abiotic factors, on structure and function of ecological systems at different spatial and temporal scales.

## ACKNOWLEDGMENTS

Many people have contributed to the ideas expressed in this paper and to the approach to integrating biology in the National Water-Quality Assessment Program. Among these people, the author especially thanks the following, all with the U.S. Geological Survey: J. Kent Crawford, Thomas F. Cuffney, Robert J. Gilliom, Harry V. Leland, Samuel N. Luoma, Diane M. McKnight, Michael R. Meador, David A. Rickert, Marc A. Sylvester, and William G. Wilber.

## REFERENCES

Armour, C. L., K. P. Burnham, and W. S. Platts. 1983. *Field Methods and Statistical Analyses for Monitoring Small Salmonid Streams.* U.S. Fish and Wildlife Service Publication FWS/OBS-83/33, Fort Collins, Colorado. 200 p.

Biggs, B. J. F., M. J. Duncan, I. G. Jowett, J. M. Quinn, C. W. Hickey, R. J. Davies-Colley, and M. E. Close. 1990. Ecological characterisation, classification, and modelling of New Zealand rivers: an introduction and synthesis: *New Zealand Journal of Marine and Freshwater Research* 24:277-304.

Connell, J. H. 1978. Diversity in tropical rain forests and coral reefs. *Science* 199:1302-1310.

Conquest, L. L., S. C. Ralph, and R. J. Naiman. 1994. Implementation of large-scale stream monitoring efforts: sampling design and data analysis issues. *In*: S. L. Loeb and A. Spacie, eds., *Biological Monitoring of Aquatic Systems.* Lewis Publishers, Boca Raton, FL.

Crawford, J. K. and S. N. Luoma. 1993. Guidelines for studies of contaminants in biological tissues for the National Water-Quality Assessment Program. *U.S. Geological Survey Open-File Report* 92-494. 69 p.

Cuffney, T. F., M. E. Gurtz, and M. R. Meader. (unpublished data). Guidelines for processing and quality assurance of benthic invertebrate samples collected as part of the National Water-Quality Assessment Program. *U.S. Geological Survey Open-file Report* 93-407.

Cuffney, T. F., M. E. Gurtz, and M. R. Meador. (unpublished data). Methods for collecting benthic invertebrate samples as part of the National Water-Quality Assessment Program. *U.S. Geological Survey Open-File Report* 93-406.

Elder, J. F. 1989. Applicability of Ambient Toxicity Testing to National or Regional Water-Quality Assessment. *U.S. Geological Survey Open-File Report* 89-55. 102 p.

Fausch, K. D., J. R. Karr, and P. R. Yant. 1984. Regional application of an index of biotic integrity based on stream fish communities. *Transactions of the American Fisheries Society* 113:39-55.

Ford, J. 1989. The effects of chemical stress on aquatic species composition and community structure. pp. 99-144. *In*: S. A. Levin, M. A. Harwell, J. R. Kelly, and K. D. Kimball, eds., *Ecotoxicology: Problems and Approaches*. NY. 547 p.

Frissell, C. A., W. J. Liss, C. E. Warren, and M. D. Hurley. 1986. A hierarchical framework for stream habitat classification: viewing streams in a watershed context. *Environmental Management* 10:199-214.

Gallant, A. L., T. R. Whittier, D. P. Larsen, J. M. Omernik, and R. M. Hughes. 1989. *Regionalization as a Tool for Managing Environmental Resources*. U.S. Environmental Protection Agency Publication EPA 600/3-89-060, Corvallis, Oregon. 152 p.

Gore, J. A., J. R. Kelly, and J. D. Yount. 1990. Application of ecological theory to determining recovery potential of disturbed lotic ecosystems: research needs and priorities. *Environmental Management* 14:755-762.

Hamilton, K. and E. P. Bergersen. 1984. *Methods to Estimate Aquatic Habitat Variables*. Bureau of Reclamation, Denver, CO.

Hilsenhoff, W. L. 1982. Using a biotic index to evaluate water quality in streams. *Wisconsin Department of Natural Resources Technical Bulletin*, Madison, WI. Number 132.

Hilsenhoff, W. L. 1988. Rapid field assessment of organic pollution with a family-level biotic index. *Journal of the North American Benthological Society* 7:65-68.

Hirsch, R. M., W. M. Alley, and W. G. Wilber. 1988. Concepts for a National Water-Quality Assessment Program. *U.S. Geological Survey Circular* 1021. 42 p.

Hughes, R. M., S. A. Heiskary, W. J. Matthews, and C. O. Yoder. 1994. Use of Ecoregions in Biological Monitoring. *In:* S. L. Loeb and A. Spacie, eds., *Biological Monitoring of Aquatic Systems.* Lewis Publishers, Boca Raton, FL.

Hughes, R. M., D. P. Larsen, and J. M. Omernik. 1986. Regional reference sites: a method for assessing stream potentials. *Environmental Management* 10:629-635.

Hughes, R. M., T. R. Whittier, C. M. Rohm, and D. P. Larsen. 1990. A regional framework for establishing recovery criteria. *Environmental Management* 14:673-683.

Hunsaker, C. T. and D. E. Carpenter, eds. 1990. *Ecological Indicators for the Environmental Monitoring and Assessment Program.* U.S. Environmental Protection Agency Publication EPA 600/3-90-060, Office of Research and Development, Research Triangle Park, NC.

Hynes, H. B. N. 1994. Historical perspective and future direction of biological monitoring of aquatic ecosystems. *In:* S. L. Loeb and A. Spacie, eds., *Biological Monitoring of Aquatic Systems.* Lewis Publishers, Boca Raton, FL.

Hynes, H. B. N. 1975. The stream and its valley. *Verhandlungen Internationale Vereinigung fur Theoretische und Angewandte Limnologie* 19:1-15.

Junk, W. J., P. B. Bayley, and R. E. Sparks. 1989. The flood pulse concept in river-floodplain systems. *In:* D. P. Dodge, ed., *Proceedings of the International Large River Symposium.* Canadian Special Publications in Fisheries and Aquatic Sciences 106:110-127.

Karr, J. R. 1991. Biological integrity: a long-neglected aspect of water resource management. *Ecological Applications* 1:66-85.

Karr, J. R., K. D. Fausch, P. L. Angermeier, P. R. Yant, and I. J. Schlosser. 1986. Assessing biological integrity in running waters: a method and its rationale. Champaign, Illinois, *Illinois Natural History Survey Special Publication* 5. 28 p.

Kratz, T. K., J. J. Magnuson, S. R. Carpenter, and T. M. Frost. 1994. Landscape position, scaling, and the spatial temporal variability of ecological parameters: considerations for biological monitoring. *In:* S. L. Loeb and A. Spacie, eds., *Biological Monitoring of Aquatic Systems.* Lewis Publishers, Boca Raton, FL.

Kuchler, A. W. 1970. Potential natural vegetation: map (scale 1:7,500,000). *In: The National Atlas of the United States of America.* U.S. Geological Survey, Washington, D.C., Plates 89-91.

Leahy, P. P., J. S. Rosenshein, and D. S. Knopman. 1990. Implementation plan for the National Water-Quality Assessment Program. *U.S. Geological Survey Open-File Report* 90-174. 10 p.

Lenat, D. R. and M. T. Barbour. 1994. Using benthic macroinvertebrate community structure for rapid, cost-effective, water-quality monitoring: rapid bioassessment. In: S. L. Loeb and A. Spacie, eds., *Biological Monitoring of Aquatic Systems.* Lewis Publishers, Boca Raton, FL.

Leopold, L. B. and M. G. Wolman. 1957. River channel patterns: braided, meandering, and straight. pp. 39-85. *In: U.S. Geological Survey Professional Paper* 282-B. 210 p.

Lubchenco, J., A. M. Olson, L. B. Brubaker, S. R. Carpenter, M. M. Holland, S. P. Hubbell, S. A. Levin, J. A. MacMahon, P. A. Matson, J. M. Melillo, H. A. Mooney, C. H. Peterson, H. R. Pulliam, L. A. Real, P. J. Regal, and P. G. Risser. 1991. The Sustainable Biosphere Initiative: an ecological research agenda. *Ecology* 72:371-412.

Meador, M. R., T. F. Cuffney, and M. E. Gurtz. 1993. Methods for sampling fish communities as part of the National Water-Quality Assessment Program. *U.S. Geological Survey Open-File Report* 93-104. 40 p.

Meador, M. R., C. R. Hupp, T. F. Cuffney, and M. E. Gurtz. (unpublished data). Methods for characterizing stream habitat as part of the National Water-Quality Assessment Program. *U.S. Geological Survey Open-File Report* 93-408.

Meador, M. R. and W. J. Matthews. 1992. Spatial and temporal patterns in fish assemblage structure of an intermittent Texas stream. *American Midland Naturalist* 127:106-114.

Moyle, P. B. 1994. Biodiversity, biomonitoring, and the structure of stream fish communities. *In:* S. L. Loeb and A. Spacie, eds., *Biological Monitoring of Aquatic Systems.* Lewis Publishers, Boca Raton, FL.

National Research Council. 1990. *A Review of the U.S.G.S. National Water Quality Assessment Pilot Program.* National Academy Press, Washington, D.C. 153 p.

Nehring, R. B. 1976. Aquatic insects as biological monitors of heavy metal pollution. *Bulletin of Environmental Contamination and Toxicology* 15:147-154.

Ohio Environmental Protection Agency. 1988. *Biological criteria for the protection of aquatic life, Volume I-III.* Division of Water Quality Monitoring, Surface Water Section, Columbus. 351 p.

Omernik, J. M. 1987. Ecoregions of the conterminous United States. *Annals of the Association of American Geographers* 77:118-125.

Osborne, L. L., B. Dickson, M. Ebbers, R. Ford, J. Lyons, D. Kline, E. Rankin, D. Ross, R. Sauer, P. Seelbach, C. Speas, T. Stefanavage, J. Waite, and S. Walker. 1991. Stream habitat assessment programs in states of the AFS Central Division. *Fisheries* (Bethesda) 16:28-35.

Phillips, D. J. H. 1980. *Quantitative Aquatic Biological Indicators.* Applied Science Publishers Ltd., London. 488 p.

Plafkin, J. L., M. T. Barbour, K. D. Porter, S. K. Gross, and R. M. Hughes. 1989. *Rapid Bioassessment Protocols for Use in Streams and Rivers: Benthic Macroinvertebrates and Fish*, U.S. Environmental Protection Agency Publication EPA 444/4-89-001, Office of Water, Washington, D.C.

Platts, W. S., C. Armour, G. D. Booth, M. Bryant, J. L. Bufford, P. Cuplin, S. Jensen, G. W. Lienkaemper, G. W. Minshall, S. B. Monsen, R. L. Nelson, J. R. Sedell, and J. S. Tuhy. 1987. *Methods for Evaluating Riparian Habitats with Applications to Management.* General Technical Report INT-221: U.S. Department of Agriculture, Forest Service, Ogden, Utah. 177 p.

Porter, S. D., T. F. Cuffney, M. E. Gurtz, and M. R. Meador. (unpublished data). Methods for collecting algal samples as part of the National Water-Quality Assessment Program. *U.S. Geological Survey Open-File Report* 93-409.

Rankin, E. T. 1989. *The Qualitative Habitat Evaluation Index (QHEI): Rationale, Methods, and Approach.* Ohio Environmental Protection Agency, Columbus, Ohio. 54 p.

Reice, S. R., R. C. Wissmar, and R. J. Naiman. 1990. Disturbance regimes, resilience, and recovery of animal communities and habitats in lotic ecosystems. *Environmental Management* 14:647-659.

Resh, V. H., A. V. Brown, A. P. Covich, M. E. Gurtz, H. W. Li, G. W. Minshall, S. R. Reice, A. L. Sheldon, J. B. Wallace, and R. C. Wissmar. 1988. The role of disturbance in stream ecology. *Journal of the North American Benthological Society* 7:433-455.

Resh, V. H. and D. M. Rosenberg. 1989. Spatial-temporal variability and the study of aquatic insects. *Canadian Entomologist* 121:941-963.

Sedell, J. R., G. H. Reeves, F. R. Hauer, J. A. Stanford, and C. P. Hawkins. 1990. Role of refugia in recovery from disturbances: modern fragmented and disconnected river systems. *Environmental Management* 14:711-724.

Shackleford, B. 1988. *Rapid Bioassessments of Lotic Macroinvertebrate Communities: Biocriteria Development.* Arkansas Department of Pollution Control and Ecology, Little Rock, Arkansas. 45 p.

Sparks, R. E., P. B. Bayley, S. L. Kohler, and L. L. Osborne. 1990. Disturbance and recovery of large floodplain rivers. *Environmental Management* 14:699-709.

Stewart, A. J. and J. W. Loar. 1994. Spatial and temporal variability in biological monitoring data. *In:* S. L. Loeb and A. Spacie, eds., *Biological Monitoring of Aquatic Systems.* Lewis Publishers, Boca Raton, FL.

Suter, G. W., II. 1990. Endpoints for regional ecological risk assessments: *Environmental Management* 14:9-23.

Tyus, H. M. 1990. Effects of altered stream flows on fishery resources. *Fisheries* (Bethesda) 15:18-20.

U.S. Environmental Protection Agency. 1990. *Biological criteria: National Program Guidance for Surface Waters.* U.S. Environmental Protection Agency Publication EPA 440/5-90-004, Office of Water Regulations and Standards, Washington, D.C. 57 p.

Vannote, R. L., G. W. Minshall, K. W. Cummins, J. R. Sedell, and C. E. Cushing. 1980. The river continuum concept. *Canadian Journal of Fisheries and Aquatic Sciences* 37:130-137.

# SECTION VI

# Conclusion

# CHAPTER 16

# Biological Monitoring: Challenges for the Future

**James R. Karr,** Institute for Environmental Studies, University of Washington, Seattle, WA

The development of an effective integrated biological monitoring program is essential to achieve the primary objective of the Clean Water Act: "To restore and maintain the physical, chemical, and biological integrity of the Nation's waters" (U.S. EPA 1990a). Recent advances in the development of approaches to biological monitoring, coupled with widespread recognition that conventional chemical and physical approaches have not protected water resources, have stimulated interest in biological monitoring. The road toward implementing biological monitoring is littered with many challenges. My goal in this paper is to outline the evidence that biological monitoring is needed, review historical impediments to use of biological monitoring, discuss principles central to the protection of the quality of water resources, and, finally, review major challenges that lie ahead.

## THE NEED FOR BIOLOGICAL MONITORING

**What is pollution?** The Clean Water Act of 1987 defines pollution as "manmade or man-induced alteration of the chemical, physical, biological, or radiological integrity of water." This statement does not define pollution as the chemical contamination of water. Rather, it is a more comprehensive definition that includes any human action (or the result of a human action) that degrades a water resource. Thus, humans may degrade or pollute water resources by chemical contamination or by destruction of aquatic habitats. They may pollute by withdrawing water for irrigation, by overharvesting fish populations, or by introducing exotic species that alter the structure of resident aquatic communities. These kinds of detrimental actions influence the water resource by altering one or more of five major classes of variables (Figure 16.1). Historically, water pollution control programs have concentrated on only one of those sets of variables—water quality. If the biological integrity of water resources is to be protected, the broader approach implicit in Figure 16.1 must be widely adopted.

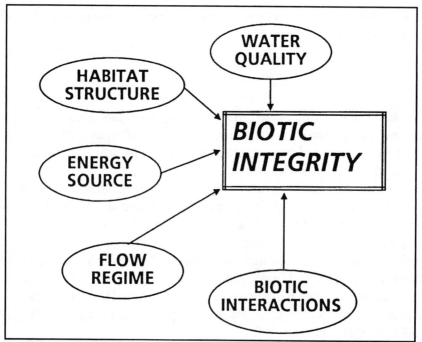

**Figure 16.1.** Five primary sets of variables that influence the biological integrity of water resources.

**Why use biological monitoring in assessing water resources?** First, evidence that monitoring approaches currently in use do not protect the nation's water resources is both substantial and widespread. Fully two-thirds of the 135 fish species known from the Illinois River watershed and 44% of 98 species known from the Maumee River have declined substantially or disappeared since 1850 (Karr et al. 1985). A recent survey of fishes of a California river indicates that more than 60% of the fauna has been harmed by human actions (Moyle and Williams 1990). Nationally, six fish species have become extinct, and nearly 150 others are threatened or endangered (Williams et al. 1989). Mussel faunas are threatened, in particular, as are populations of other aquatic invertebrates.

Second, biological monitoring has repeatedly proved to be sensitive and reliable as an assessment tool. Early this century, biological evaluations clearly showed the negative impact of the 1908 sewage diversion that shunted Chicago's wastewater from Lake Michigan to the Illinois River (Forbes and Richardson 1919, Richardson 1928). By 1909-1911, the river was polluted for 170 km, and by 1920, the river was anoxic for 235 km and heavily polluted for 360 km. The pollution slug moved downstream at a rate of 13 to 26 km per year. Forbes and Richardson (1919), using a variety of approaches and taxa, demonstrated three important facts: (1) substantial negative effects of the sewage diversion were obvious; (2) biological assessments can be both direct and accurate in assessing environmental conditions; and (3) numerical abundances are not as useful as relative abundances in assessing biological cond:tions. Overall occurrences of specific taxa were especially useful. More recently, Ohio EPA (1988, 1990) demonstrated that use of biological monitoring improved their ability to detect stream impacts. Biological monitoring provided evidence that at 36% of sites "instream chemical data implied no impairment while instream biological communities showed impairment" (Ohio EPA 1990).

Furthermore, Ohio is the first state to incorporate an ecologically robust biological approach to water resource protection in their water quality regulations. Ohio EPA routinely applies biological criteria to such diverse activities as maintaining water quality standards under existing use designation; Section 319 Clean Water Act (CWA) nonpoint source program activities; Section 305(B) CWA water quality inventory report; and NPDES discharge permits (Dudley 1991).

**Will biological monitoring serve as a magic bullet or panacea to solve water resource problems?** The answer is an unequivocal no. No such panacea exists. Water resource problems are too complex for such a simplistic solution. But ambient biological monitoring is an essential supplement to existing approaches (physical-chemistry parameters, toxicity testing).

**Will biological monitoring and biocriteria improve our track record?** Yes. Although no perfect solution is likely, sufficient information about the biology of water resources is available to ensure that proper use of biological data can better protect the resource.

## IMPEDIMENTS TO USE OF BIOLOGICAL MONITORING

Despite substantial evidence of the value and sensitivity of biological monitoring (illustrated by the data of Forbes and Richardson (1919), Kolkwitz and Marsson (1908), and many others more recently), little success in using biological monitoring was realized before the 1980s. The reasons for this lack of success are complex (Karr 1991). They include the dominance of reductionist viewpoints. Engineering approaches typically deny or ignore common signs of biological impairment, and ecologists are reluctant to develop and implement criteria and standards or to challenge the conceptual underpinnings of nonbiological approaches. Legal and regulatory programs are so conceptually constrained that resource degradation seems inevitable. Examples include dependence on technology-based controls of pollution, dominance of a narrow chemical-contaminant definition of pollution, and the doctrine of burden of proof under which the government must demonstrate ecosystem degradation and the causal links responsible for it. Although the Clean Water Act passed in 1972 (PL 92-500) clearly stated a goal of biological integrity, that goal remained ill defined. Even worse was the lack of an integrative, broadly applicable approach to the measurement of biological integrity. Ever-expanding human influences on water resources resulted in a diverse and changing array of problems. Early biological monitoring was inadequate because it tended to focus narrowly. A closely related factor was the lack of a region-based quantitative definition of ecological health. Criteria developed for many chemical contaminants were (now obviously incorrectly) applied uniformly, and the lack of uniformity in biological expectations was seen as a sign of weakness in the techniques. That lack of uniformity is now widely recognized as reflective of the true water resource situation. In retrospect, the idea that the same criteria should apply to all waters is ludicrous. In the absence of a general approach to assessment of ecological health, lack of standardized field methods were an inevitable result. Standardization of field methods, including rigorous quality control of both the sampling methods and data handling and analysis, is essential, as is the linking of those field measurements and analyses to enforceable management options. Finally, the belief persists that biological monitoring is too expensive. All of these impediments persist in some quarters, but the fallacy of their underpinnings is more widely recognized (Karr 1991).

## KEY TO EFFECTIVE BIOLOGICAL MONITORING

The key to effective biological monitoring was the development of a more comprehensive approach to assessing the effects of human activities on aquatic ecosystems. The index of biological integrity (IBI), developed more than a decade ago (Karr 1981), provided an ecologically robust and broadly conceived approach to evaluating the full range of human impacts on water resource systems. The success of that index in evaluating water resources in a wide diversity of areas in North America and in selected other regions is based on tests of hypotheses about the influence of human activities on the stream biota (Fausch et al. 1990). In numerous tests, those hypotheses seem to account for pattern in nature (Table 16.1). Each hypothesis is grounded in observations of the behavior of the biota and in the understanding of ecological processes (Karr et al. 1986, Miller et al. 1988). Because those principles seem to be general, the use of IBI is not limited to the region in which it was developed. Rather, by considering the ecological foundation of an IBI metric developed for the midwestern United States, an appropriate replacement metric can be developed. Examples include the use of sculpins (Cottidae) as a replacement metric in some regions without darters (Karr et al. 1986), the use of suckers in place of insectivorous minnows in large rivers (Ohio EPA 1988), and the use of *Rutilus rutilus* as a tolerant omnivore instead of green sunfish (Oberdorff and Hughes, 1992).

Similar efforts to develop ecologically equivalent metrics in estuarine environments have met with considerable success (Thompson and Fitzhugh 1986) as have efforts to apply the IBI approach to reservoir and lake systems (Karr and Dionne 1991, Dionne and Karr 1992). Finally, a number of efforts to adopt IBI approaches for use with invertebrate communities have also made considerable progress (e.g., Ohio EPA 1988, Plafkin et al. 1989, Kerans and Karr in press).

Table 16.1. Underlying assumptions about the influence of human actions on the aquatic biota. (Modified from Fausch et al. 1990.)

| | |
|---|---|
| 1) | Number of native species (total) and number of taxa in selected fish families or habitat guilds declines with human disturbance. |
| 2) | Number of sensitive or intolerant species declines with human disturbance. |
| 3) | Proportion of individuals that are members of tolerant species increases with human disturbance. |
| 4) | Proportion of trophic specialists, such as insectivores or top carnivores, decreases with human disturbance. |
| 5) | Proportion of trophic generalists (especially omnivores) increases with human disturbance. |
| 6) | Total fish abundance declines with human disturbance. |
| 7) | Proportion of individuals in reproductive guilds requiring silt-free spawning substrate declines with human disturbance. |
| 8) | Incidence of hybrids increases with human disturbance. |
| 9) | Incidence of externally evident disease, parasites, and morphological anomalies increases with human disturbance. |
| 10) | Proportion of individuals that are members of introduced species increases with human disturbance. |

# PRINCIPLES OF BIOLOGICAL MONITORING

1) **Biological monitoring should have a firm conceptual foundation that uses the most up-to-date ecological principles.** Existing knowledge is sufficient to improve programs to protect water resources. Substantial information is available in historical data bases of state agencies, universities, and natural history surveys. The professional staff and collections of such institutions are storehouses of information about the distribution of organisms and their natural histories.

The conceptual foundations of ecology provide additional information essential to effective biological monitoring. The strength and vitality of aquatic ecology research is clear (Cummins 1994). Three major advances in understanding aquatic ecosystems contribute to the importance of biological monitoring: (1) the stream continuum hypothesis; (2) path dynamics, or landscape ecology; and (3) the importance of variation in the physical environment, especially annual shifts in distribution and amount of rainfall and runoff.

Finally, biological monitoring approaches based on ecological principles should never be so rigorously implemented that knowledge of special circumstances at a site cannot be applied to provide a more informed biological evaluation. That is, biological monitoring and analysis should be flexible enough to produce sound water resource decisions, not too rigorous to prevent quality results.

2) **The best biological monitoring does not depend on a one-dimensional approach to system evaluation.** An assessment should be based on a number of metrics, which serve as a vector of environmental conditions. By including a variety of issues, the quality of the monitoring effort can depend on a "weight of evidence" approach. Similar conclusions from different taxa or different ecological perspectives lend more weight to the overall assessment. Each taxon (e.g., fish, invertebrates, algae, diatoms, or birds) has special strengths and weaknesses as a biological monitoring tool (Karr 1991). Unfortunately, some continue to argue for narrow approaches. For example, some claim that organism mobility disqualifies a taxon for use in biological monitoring; others suggest that only organisms of narrow size ranges or short life cycles are appropriate for biological monitoring. Fortunately, most individuals have moved beyond these shallow arguments. The strength of a biological monitoring approach is the ecological foundation behind the use of a taxon, not the taxon itself. Further, it is essential to include metrics that encompass the conventional, if somewhat simplistic dichotomy between structure and function. Some researchers argue that studies of processes are not as good as studies of structure (Schindler 1987); others claim that an ecosystem approach is sufficient for informed biological assessment (Ulanowicz 1990). The only appropriate strategy to protect biological resources is to incorporate both in biological monitoring.

3) **Biological monitoring must, incorporate both the elements and processes that all biological systems depend on for their persistence.** Nowhere is this principle more important than in the

protection of biological diversity (Karr 1990b) (Table 16.2). To protect elements (or items) without processes or vice versa is to establish a monitoring program that is likely to fail before it begins (Karr 1990b).

**Table 16.2.** Components of biological integrity.

*Elements*

    Genes within a population
    Populations within species[a]
    Species within communities/ecosystems
    Communities/ecosystems within landscapes
    Landscapes with the biosphere

*Processes*
    Nutrient cycling
    Water cycling
    Atmospheric cycle
    Photosynthesis
    Decomposition
    Competition
    Predation
    Speciation/Evolution

[a]Biodiversity is a commonly used word that often refers to this level (Reid and Miller 1989), although some define it more broadly.

Appropriately, biological scientists tend to seek general principles. However, biologists also tend to seek exceptions to another person's rules. How many of us have heard biologists argue against a generality with the statement, "But I remember a day at such-and-such a place when that pattern did not hold"? Yet many general rules are accurate under most circumstances. We should be accumulating insights and cataloging them. A robust biological assessment can be accomplished if we can show, as noted before, that the preponderance of evidence suggests that a biological system is degraded. Where would science be if we discarded the idea of thermometers after finding that a thermometer accurate at temperatures from 0° to 30°C did not work in the Antarctic winter. Rather, we should use a principle when it accurately represents the real world and find other tools when it does not. Because of the breadth of environmental conditions and organismal adaptations, I do not expect any useful generality to always be true. Simultaneous use of many metrics, each with 80% reliability, provides a robust approach.

4) **Quality biological monitoring must also include quality— control provisions at all stages.** Only highly trained and skilled professionals should collect and interpret biological data for monitoring and assessment purposes. Professional biologists must ensure that relevant field data are collected, analyzed, and interpreted appropriately. The environmental impact statement process of the last 20 years is replete with examples of massive data collection without serious thought to the kind of data to be collected and the kind of analysis and synthesis to be applied. The data must be robust for the problem at hand, and the analysis must be grounded in sound ecological science.

5) **Biological monitoring is not more expensive than chemical monitoring.** The presumed high cost of biological monitoring is a red herring. The conventional wisdom of the past two decades was that chemical approaches were less expensive than biological monitoring. Recent detailed studies by Ohio EPA (cited in Karr 1991) undermine that argument (Table 16.3). Indeed, biological monitoring for survey use is often cheaper than other approaches, and it is more likely to detect degradation than sole use of chemical approaches. Thus, under many circumstances, biological monitoring is more effective than other methods.

**Table 16.3.** Comparative cost analysis for sample collection, processing, and analysis for evaluating of the quality of a water resource.[a] Data from Ohio EPA, provided by C. O. Yoder.

| Approach | Cost per Evaluation[b] |
|---|---|
| Chemical Physical Water Quality | $8,600 to 12,900 |
| Bioassay | $3,500 to 18,300 |
| Ambient Biomonitoring (Fish and Benthic Macroinvertebrates) | $7,800 |

[a]Table is simplified from Karr 1990a.
[b]The cost to evaluate the impact of an entity; this example assumes sampling five stream sites and one effluent discharge.

6) **Because environmental conditions are declining rapidly, we cannot afford the cost of further neglect.** The deferred cost of environmental regulation is often much higher than the cost of a quick-and-easy fix before the problem worsens. No rational person would suggest a sloppy design phase for the construction of a building or bridge. Design of environmental regulations should receive as much thought and rigorous development as engineering of structures. Society can no longer ignore a growing environmental deficit, the ultimate impact of which will likely eclipse current economic deficits.

7) **Complex, multimetric approaches to biological monitoring are both necessary and appropriate for the protection of complex biological systems.** The ecological systems that environmental regulations are designed to protect are very complex. The use of indexes to evaluate the condition of those systems is not inherently bad, as some have suggested. Rather, the skeptic should carefully evaluate the ecological foundations of either simple or complex indexes. If the metrics are firmly grounded in ecology and interpreted robustly, then the resulting index will be a strong representation of ecological conditions. Diversity measurements, for example, seem to have a strong theoretical foundation, but the theory behind them is not robust. Moreover, diversity indexes combine species richness and relative abundances in ways that create index behavior that is difficult to discern. A metric based on number of tolerant species present, however, is valuable because it uses detailed knowledge of the biology of individual species and it reflects a known sensitivity to the influence of human actions.

Although it can be argued that the detailed behavior of many metrics or indexes may not convey complete truth without interpretation and without exception, those problems are more than overcome by the reality that society will continue to make decisions about the use and abuse of natural resources and environmental systems. Ecologists can participate in those decisions or they can allow decisions to be made without ecology. Participation may not create perfect decisions, but the decisions will be more robust with biological input than wihtout.

## PRIORITY RESEARCH AREAS

Although research needs are diverse, a few issues require early attention.

1) **Standarde sampling.** A fundamental underpinning of any quality monitoring program is the establishment of standardized sampling programs. Such programs must include proper seasonality of sampling and gear type and sufficient but not excessive sample effort. The first effort should be to express sampling goals in a clear and concise way. Then guidelines can be provided to attain those goals; highly trained and skilled professional biologists will make the final sampling protocol determinations. Documents prepared by Ohio EPA (1988) are especially strong. They provide general guidelines as well as sufficient detail to ensure that quality control is rigorous. They illustrate that the individuals charged with carrying out a program must have suitable expertise and training. That training allows these individuals to apply appropriate sampling rigor while remaining flexible and insightful enough to modify those protocols so as to assure quality data collection and interpretation at a site. In the end, quality control must be in the hands of the field and laboratory crews, not in the manual.

2) **Document natural variation.** Physical, chemical, and biological attributes of water resources change continuously. Some variation is due to natural events, some to the influences of human society. To protect water resources, individuals and agencies charged with the responsibility of evaluating them must learn to cope with and use information about that variation. Success in accomplishing this goal requires a substantial infusion of resources into integrative research programs designed to document and understand that variation. For more than a decade, biological monitoring has been hampered by fact and fiction about variation. The time is past for considering the mean value of an ecological attribute as "signal" and its variation to be "noise." Several research programs have demonstrated that high variation in an attribute may signal the presence of human influences (Karr et al. 1987, Rankin and Yoder 1990) that have degraded the natural

system's ability to withstand, for example, natural climatic variation.

To be more effective, biological monitoring must seek ecological attributes for which natural variation is low but that are influenced to some significant extent by human actions. Those attributes can be used to identify degradation and the causes of that degradation.

3) **What is being compromised when we undertake "rapid" bioassessment?** Our first goal in biological monitoring is to evaluate the quality of a water resource. Accomplishing that goal requires a high-quality assessment process. Only with such a process can we expect the decision to be well informed. Some in the water resource community suggest that biological monitoring can only be used as a screening tool. It can indeed serve that purpose admirably, but to suggest that it is limited to that role is wrong and to seek rapid bioassessment as an endpoint is likely to relegate biological monitoring to a secondary role. It should be front and center and the only way to ensure that is to not compromise quality in the search for rapid bioassessment.

## CHALLENGES

1) **Degradation of biological systems underlies all environmental problems.** Societal recognition of this fact is necessary before a comprehensive response to environmental problems can be developed. All life on earth depends on complex interactions of physical, chemical, and biological processes. Usually, society recognizes an environmental threat only when a critical biological process is damaged or destroyed. Societal responses to environmental problems too frequently focus on physical rather than biological solutions. As long as humans view these problems as physical problems, the tendency will be to conceive technological solutions; efforts to engineer solutions will dominate biological and preventative approaches. We need to understand and to protect the integrity of biological processes rather than trying to replace those processes with engineered solutions.

2) **Environmental degradation results from how we carry out normal day-to-day activities.** As individuals, we tend to externalize environmental problems and to blame environmental disasters on the behavior of industry, government, or some other entity. In fact, the day-to-day activities of people like ourselves threaten our welfare. The choices we make about suburban living, green lawns, throw-away consumer goods, and land use are central to most environmental problems. Thus, we must integrate environmental goals with social, economic, and political goals and processes. Attention to both environmental and economical deficits is essential to the continuing success of human society.

3) **Fragmentation is at the core of many environmental problems.** Habitat fragmentation reduces biological diversity and biological integrity. Fragmented laws and regulations also do not provide an integrative framework for the protection of natural resources. Biological monitoring can and should add to any effort to circumvent this fragmentation. Fragmentation of goals and approaches at local, regional, national, and international levels constitutes another problem. Different levels of government, even different agencies within the same governmental bureaucracy, can work at cross-purposes. Within state governments, for example, agencies with overlapping mandates rarely function in an integrated fashion. Illinois is a notable exception because natural resource and environmental protection agencies work together to their mutual benefit.

This problem of fragmentation is further illustrated by the piecemeal thinking that often dominates water resource problems. It also permeates our educational systems, land-use planning, and regulatory programs (e.g., single versus multimedia). Neglect of cumulative impacts of many stresses is another consequence of a narrow perspective. Engineers and ecologists are often guilty of one-dimensional thinking when it comes to resolving natural resource problems. Finally, the tendency for the regulatory community to seek simple, legally defensible approaches to regulation has added to the weaknesses of current solutions to water resource problems. Too frequently, a decisive approach, or an approach with inadequate ecological underpinnings, limits the success of environmental monitoring and protection programs.

4) **Most modern ecological research is neutral to the solution of environmental problems.** As a discipline, ecology has not been especially responsive to societal needs in recent decades. Many basic or theoretical ecologists have concentrated on the dynamics of ecological systems and processes in nearly pristine environments. Their goal, to document ecological pattern and process, has often been constrained by focusing too closely on narrow ecological phenomena (e.g., the role of interspecific competition). Conversely, other groups concerned about environmental problems have not been effective at using existing ecological information. The relative isolation of these two groups must be changed.

5) **The complicated nature of biological systems makes traditionally defined risk assessment (i.e., a direct quantitative relationship between a hazard and its effect) inappropriate.** Many hazards cannot be measured "at the pipe" and the impacts of multiple hazards require direct biological assessment under field conditions. Further, conventional risk assessment rarely incorporates risk aversion theory, which is applied in many biological situations (Gillespie 1977, Frank and Slatkin 1990, Seger and Brockmann 1987). These models of risk aversion, or bet hedging, involve a strategy to reduce losses by avoiding risk (Boyce 1988) and involve use of geometric rather than arithmetic means in assessing risk.

6) **Biological monitoring goes well beyond collection of field data.** Biologists have a tendency to amass large quantities of unorganized field data. An inability or reluctance to distill the biological meaning from large quantities of data with rigorous, accurate, yet easily understood analyses has diminished the role of biology in environmental protection. Appropriate data must be collected, analyzed, and interpreted in ways that will detect relevant variation in biological phenomena. Those interpretations must then be conveyed clearly and unequivocally to society. Some recent efforts to use ecological indicators based on biological monitoring emphasize sampling design and field methods while neglecting the later stages of the process. The development of truly integrative biological monitoring programs is perhaps the greatest challenge to biological and ecological sciences.

# REFERENCES

Boyce, M. 1988. Bet hedging in avian life histories. *Proceedings International Ornithological Congress* 19:2131-2139.

Cummins, K. W. 1994. Bioassessment and Analysis of Functional Organization of Running Water Ecosystems. *In*: S. L. Loeb and A. Spacie, eds., *Biological Monitoring of Aquatic Systems*. Lewis Publishers, Boca Raton, FL.

Dionne, M. and J. R. Karr. 1992. Ecological monitoring of fish assemblages in Tennessee River reservoirs. pp. 259-281. *In:* D. H. McKenzie, D. E. Hyatt, and V. J. McDonald (eds.). *Ecological Indicators*. Vol. I, Elsevier Applied Science, London.

Dudley, D. R. 1991. A state perspective on biological criteria in regulation. pp. 15-18. *In*: *Biological Criteria: Research and Regulation*. U.S. Environmental Protection Agency, Washington, D.C. EPA-440/5-91-005. 810 p.

Fausch, K. D., J. Lyons, J. R. Karr, and P. L. Angermeier. 1990. Fish communities as indicators of environmental degradation. *American Fisheries Society Symposium* 8:123-144.

Forbes, S. A. and R. E. Richardson. 1919. Some recent changes in Illinois River biology. *Bulletin Illinois Natural History Survey* 13:139-156.

Frank, S. A. and M. Slatkin. 1990. Evolution in a variable environment. *American Naturalist* 136:244-260.

Gillespie, J. 1977. Natural selection for variances in offspring numbers: a new evolutionary principle. *American Naturalist* 111:1010-1014.

Karr, J. R. 1981. Assessment of biotic integrity using fish communities. *Fisheries* 6(6):21-27.

Karr, J. R. 1990a. Bioassessment and non-point source pollution: An overview. Pp. 4-1 to 4-18. *In*: *Second National Symposium on Water Quality Assessment. Meeting Summary*. U.S. Environmental Protection Agency, Office of Water, Washington, D.C.

Karr, J. R. 1990b. Biological integrity and the goal of environmental legislation: lessons for conservation biology. *Conservation Biology* 4:244-250.

Karr, J. R. 1991. Biotic integrity: A long-neglected aspect of water resource management. *Ecological Applications* 1:66-85.

Karr, J. R. and M. Dionne. 1991. Designing surveys to assess biological integrity in lakes and reservoirs. pp. 62-74 *In: Biological Criteria: Research and Regulation*. U.S. Environmental Protection Agency, Washington, D.C. EPA-440/5-91-005. 171 p.

Karr, J. R., L. A. Toth, and D. R. Dudley. 1985. Fish communities of midwestern rivers: a history of degradation. *BioScience* 35:90-95.

Karr, J. R., K. D. Fausch, P. L. Angermeier, P. R. Yant, and I. J. Schlosser. 1986. Assessing biological integrity in running waters: a method and its rationale. Special Publication 5:1-28. *Illinois Natural History Survey.* Champaign, IL.

Karr, J. R., P. R. Yant, K. D. Fausch, and I. J. Schlosser. 1987. Spatial and temporal variability of the index of biotic integrity in three midwestern streams. *Transactions American Fisheries Society* 116:1-11.

Kerans, B. L. and J. R. Karr. ms. Development and testing of a benthic index of biotic integrity (B-IBI) for rivers of the Tennessee Valley. Ecological Applications. In press.

Kolkwitz, R. and M. Marsson. 1908. Okologie der pflanzlichen Saprobien. *Berichte der Deutschen Botanischen Gessellschaft* 26:505-519.

Miller, D. L., P. M. Leonard, R. M. Hughes, J. R. Karr, P. B. Moyle, L. H. Schrader, B. A. Thompson, R. A. Daniels, K. D. Fausch, G. A. Fitshugh, J. R. Gammon, D. B. Haliwell, P. L. Angermeier, and D. J. Orth. 1988. Regional applications of an index of biotic integrity for use in water resource management. *Fisheries* 13:12-20.

Moyle, P. B. and J. E. Williams. 1990. Biodiversity loss in the temperate zone: decline of the native fish fauna of California. *Conservation Biology* 4:275-284.

Oberdorff, T. and R. M. Hughes. In press. Modification of an index of biotic integrity based on fish assemblages to characterize rivers of the Seine-Normadie Basin, France. *Hydrobiologie.*

Ohio Environmental Protection Agency. 1988. *Biological Criteria for the Protection of Aquatic Life. Volume I-III.* Ohio Environmental Protection Agency, Division of Water Quality Monitoring and Assessment, Surface Water Section, Columbus, OH.

Ohio Environmental Protection Agency. 1990. *The Use of Biocriteria in the Ohio EPA Surface Water Monitoring and Assessment Program.* Ecological Assessment Section, Division of Water Quality, Planning and Assessment, Columbus, Ohio. 52 p.

Plafkin, J. L., M. T. Barbour, K. D. Porter, S. K. Gross, and R. M. Hughes. 1989. *Rapid Bioassessment Protocols for Use in Streams and Rivers: Benthic Macroinvertebrates and Fish.* EPA/444/4-89-001. U.S. Environmental Protection Agency, Washington, D.C. 168 p.

Rankin, E. T. and C. O. Yoder. 1990. The nature of sampling variability in the Index on Biotic Integrity (IBI) in Ohio Streams. pp. 9-18. *In: Proceedings of the 1990 Midwest Pollution Control Biologists Meeting, Chicago, Illinois.* EPA 905/9-90-005. United States Environmental Protection Agency. 142 p.

Reid, W. V. and K. R. Miller. 1989. *Keeping Options Alive: The Scientific Basis for Conserving Biodiversity.* Word Resources Institute, Washington, D.C. 128 p.

Richardson, R. E. 1928. The bottom fauna of the Middle Illinois River, 1913-1925: its distribution, abundance, valuation, and index value in the study of stream pollution. *Bulletin Illinois Natural History Survey* 17:387-472.

Schindler, D. W. 1987. Detecting ecosystem responses to anthropogenic stress. *Canadian Journal Fisheries Aquatic Sciences* 44:6-25.

Seger, J. and H. J. Brockmann. 1987. What is bet-hedging? *Oxford Survival Evolution Biology* 4:182-211.

Thompson, B. A. and G. R. Fitzhugh. 1986. *A Use Attainability Study: An Evaluation of Fish and Macroinvertebrate Assemblages of the Lower Calcasieu River, Louisiana.* Office of Water Resources, Louisiana Department of Environmental Quality.

Ulanowicz, R. E. 1990. Ecosystem integrity and network theory. pp. 69-77. *In:* C. J. Edwards and H. A. Regier, eds., *An Ecosystem Approach to the Integrity of the Great Lakes in Turbulent Times.* Proceedings of a 1988 Workshop Supported by the Great Lakes Fishery Commission and the Science Advisory Board of the International Joint Commission. Great Lakes Fishery Commission, Special Publication 90-4.

U.S. Environmental Protection Agency. 1990a. *Biological Criteria: National Program Guidance for Surface Waters.* EPA-440/5-90-004. U.S. Environmental Protection Agency, Office of Water Regulations and Standards, Washington, D.C. 57 p.

Williams, J. E., J. E. Johnson, D. A. Henrickson, S. Contreras-Balderas, J. D. Williams, M. Navarro-Mendoza, D. A. McAllister, and J. E. Deacon. 1989. Fisheries of North America endangered, threatened, or of special concern: 1989. *Fisheries* 14(6):2-20.

# INDEX